Monographs on oceanographic methodology 6

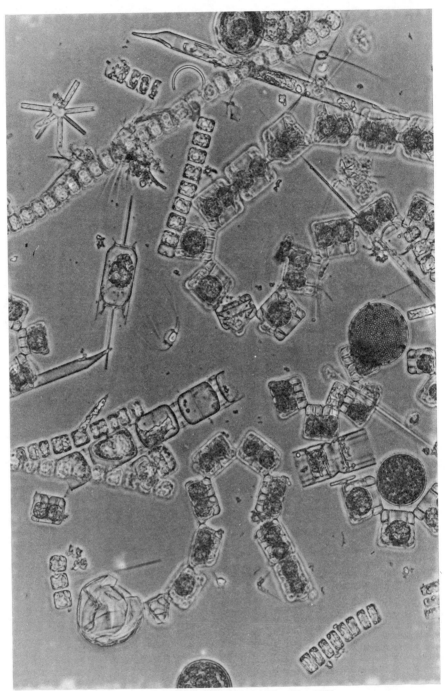

Net sample of phytoplankton from Rhode Island Sound, November 1976.
Phase-contrast photomicrograph ($\times 225$) by Paul W. Johnson and John McN. Sieburth

Phytoplankton manual

Edited by A. Sournia

Muséum National d'Histoire Naturelle, Paris

In this series:

1. *Determination of photosynthetic pigments in sea-water*
2. *Zooplankton sampling*
3. *A guide to the measurement of marine primary production under some special conditions*
4. *Zooplankton fixation and preservation*
5. *Coral reefs: research methods*
6. *Phytoplankton manual*

The designations employed and the presentation of the material in this publication do not imply the expression of any opinion whatsoever on the part of the publishers concerning the legal status of any country or territory, or of its authorities, or concerning the frontiers of any country or territory.

Published in 1978 by the United Nations Educational, Scientific
and Cultural Organization,
7 Place de Fontenoy, 75700 Paris
Printed by Page Brothers (Norwich) Ltd

ISBN 92-3-101572-9

Preface

Over the past twelve years Unesco, in collaboration with the Scientific Committee on Oceanic Research (SCOR) has synthesized the available field and laboratory research techniques that are needed to examine some of the most relevant marine scientific problems through this series of 'Monographs on Oceanographic Methodology'. Such syntheses are a vital component of the modern process of technology transfer.

The five preceding titles in the series have described techniques related to zooplankton sampling, zooplankton fixation and preservation, primary production, photosynthetic pigments, and coral reef research.

As the major marine advisory body to Unesco, the Scientific Committee on Oceanic Research (SCOR) prepares the manuscripts for the series. Most certainly, without this collaboration, dissemination of current research information to the scientific community would be impaired.

Following recommendations from SCOR that a given topic is of particular relevance to current research trends, international experts are appointed to a joint working group to decide which approach is best suited to the subject in hand and to prepare the detailed methodological descriptions. Thus, by the time a manuscript is submitted to Unesco for publication, the SCOR specialists have spent long hours of research, comparison and review as their personal contribution to international marine science.

Unesco is highly appreciative of the efforts of the scientists who prepared the present volume and wishes to express its particular thanks to Dr Alain Sournia, the editor, for his devotion to this project.

The scientific opinions expressed in this work are those of the authors and are not necessarily those of Unesco. Equipment and materials have been cited as examples of those most currently used by the authors, and their inclusion does not imply that they should be considered as preferable to others available at that time or developed since.

Contents

A*

Introduction

At its tenth general meeting, held in Mexico, January 1969, the Scientific Committee on Oceanic Research (SCOR) established Working Group 33 on Phytoplankton Methods. K. Banse was named chairman of the group, which was composed of M. Bernhard, R. W. Eppley, G. R. Hasle, R. Marumo, G. A. Robinson, H. J. Semina and T. J. Smayda.

Working Group 33 was assigned the following terms of reference, namely (1) to review the current, non-chemical methods for the quantitative study of phytoplankton, (2) to select the most satisfactory of them, (3) to recommend detailed procedures for the latter, and (4) to prepare a report that might serve as a basis for a manual.

Although the group held only one meeting (University of Rhode Island, United States, December 1970), an active correspondence was kept up among members and in addition some individual contacts were made. An attempt to perform intercalibration tests for concentrating and counting phytoplankton proved to be rather disappointing in spite of the fact that they had been intentionally oversimplified. Eventually the first and fourth terms of reference were fulfilled through an interim (1972) and final (1973) report that was soon after published by Unesco (see Anon., 1974). Working Group 33 was ultimately disbanded by SCOR at its Texel meeting (May 1973) with the recognition that more specialized activities were needed (see *SCOR Proceedings,* Vol. 9, No. 1, 1973, and previous issues for further details). The group provided as usual a few recommendations, most of which have not yet been followed: to reprint a selection of basic papers; to test fixatives and preservatives; and to establish another group in order to fulfil the second and third objectives of Working Group 33. A recommendation to hold a phytoplankton course, however, did result in a course being given at the University of Oslo in the summer of 1976, and the idea of producing a manual survived in the following way.

In March 1973 K. Banse, acting on behalf of Working Group 33, approached a potential editor for the manual (as to why my name was chosen, I must leave the explanation to Professor Banse or to Working Group 33, as I feel it was both an honour and a cruelty). Then a number of letters were exchanged between K. Banse, the secretary of SCOR and the editor, and several preliminary tables of contents were successively elaborated. Another step forward was taken when some people met in Oslo under the auspices of SCOR (November 1974). There, a pleiad of phytoplanktologists such as T. Braarud, K. R. Gaarder, G. R. Hasle, R. Margalef, E. Paasche, K. Tangen, J. Throndsen, E. L. Venrick and a few others came to the editor's help and, thanks to them, an improved contents, along with names of potential authors, was drawn up. After some further correspondence, final approval of the project was given by SCOR and the International Association for Biological Oceanography (IABO). In October 1975, authors were invited to contribute and, in the case of their acceptance, asked to send a first draft contribution by May 1976. Most of the drafts were received around September, although a few did not arrive until September of the next year. Starting from September 1976, these provisional contributions were duplicated and sent to some authors of the manual as well as to other outside scientists for criticism and review; thus each contribution has been read by an average number of 5·3 colleagues, the editor included. Comments were then forwarded to the respective authors (ten manuscripts, however, were refused or fully rewritten, with the whole procedure starting again for them). Final manuscripts were requested by April 1977 and reached the editor between that date and February 1978. A number of minor editorial improvements or retouchings were still made at this stage, authors being consulted again when this was judged necessary.

SCOPE OF THE MANUAL

As can be seen from the above account, selecting the most satisfactory methods for the quantitative study of phytoplankton and recommending a detailed procedure for them is such a long and exacting task that it took an international group of experts about three years just to circumscribe it (Anon., 1974). Whatever the length of time that might be needed effectively to develop a manual of methodology, and whoever might accept such a task, it was felt that, in the meantime, a detailed evaluation of the existing methods was called for.

The background data are scattered in an unlimited number of original publications. Individual techniques or groups of them are presented or summarized in manuals which were aimed at either different or wider purposes, such as Anon. (1968b), Anon. (1969), Vollenweider et al. (1969) and Steedman (1976). It is hoped that the present manual will provide a description and an evaluation of all current methods, together with some information or recommendations which are seldom, if ever, provided in the textbooks.

An attempt has been made here to cover the quantitative study of phytoplankton at all stages of research, starting even before the act of collecting a sample (i.e. planning the study), and ending somewhat after the act of enumerating the organisms (i.e. interpreting the results). Within this continuum, the following stages are treated: sampling at sea, preserving and storing the samples, concentrating the phytoplankters, identifying them and counting them. As can be seen, only phytoplankton and phytoplankters will be considered, excluding any biochemical or physiological measurements such as pigment concentration, chemical content, photosynthesis, growth rate or nutrient uptake. For these latter aspects the reader may refer to the following choice of handbooks: Strickland, 1960, *Measuring the Production of Marine Phytoplankton*; Anon., 1966, *Determination of Photosynthetic Pigments in Seawater*; Anon., 1969, *Recommended Procedures for Measuring the Productivity of Plankton Standing Stock and Related Oceanic Properties*; Vollenweider *et al.,* 1969, *A Manual on Methods for Measuring Primary Production in Aquatic Environments, Including a Chapter on Bacteria*; Strickland and Parsons, 1972, *A Practical Handbook of Seawater Analysis*; Stein, 1973, *Handbook of Phycological Methods. Culture Methods and Growth Measurements.*

Although this manual was written largely by oceanographers or marine biologists, and in spite of its appearing in an oceanographic series under the sponsorship of two oceanographic organizations, in no place has it been assumed that marine and freshwater planktology are two distinct or inimical sciences. Rather, reference is made to limnological techniques and experience as often as necessary. Conversely, it is expected that the manual may offer some help to limnologists as well.

ACKNOWLEDGEMENTS

The continuity of action from the late Working Group 33 to the editor was ensured through the kindness of Professor K. Banse well after his task as chairman of the working group had been completed.

Contributors deserve more thanks than usual, for two reasons. First, they also acted as reviewers. Then, they had to cope with a particularly finicky editor. Should it be mentioned here that the correspondence to and from authors for the period 1975–77 amounts to more than 300 letters, circular letters omitted?

Twelve external reviewers kindly agreed to spend their time on the provisional manuscripts: K. Banse, M.-J. Dinet-Chrétiennot, S. Frontier, M. Gillbricht, J. T. Hardy, E. Paasche, T. R. Parsons, T. Platt, M. Roux, P. Tett, L. Tinnberg and M. Travers.

ALAIN SOURNIA

How to use the manual[1]

First, the scope of marine planktology is briefly outlined and some reasons for studying phytoplankton are set forth here for the sake of those who have as yet no interest in this matter (Chapter 1).

The prerequisites of a proper sampling design are considered in Chapter 2.1, together with the definition of the objectives and an assessment of alternative strategies. Chapter 2.2 illustrates the practical application of sampling design to some current problems in oceanography.

Routine sampling at sea can be either discrete (water-bottles) or continuous (pumps). The two possibilities are extensively discussed in Chapters 3.1 and 3.2 respectively. In addition, three other sampling techniques deserved consideration for one reason or another. Although nets (Chapter 3.3) are no longer expected to give reliable samples with respect to either composition or abundance of the phytoplankton, they still may be used for certain qualitative purposes. The Continuous Plankton Recorder, obviously, is less recommendable for phytoplankton than for zooplankton studies, but it stands as such an original and inventive device that no planktologist should ignore it (Chapter 3.4). Finally, the surface film of the sea has inspired so much research in many fields during the last decade that phytoplanktologists now need to know how to collect it (Chapter 3.5).

Fixation of the samples is a crucial and still unsolved problem and may well remain so for a long time—if, indeed, a satisfactory method for fixing all the organisms in any kind of sample is ever developed. Facing such a situation, all that can be done here (Chapter 4) is to discuss the respective advantages and disadvantages of the current fixatives and provide appropriate recommendations for their use. Some indications on staining are also relevant to this section.

The statistical implications of subsampling are generally neglected, if not ignored. Their importance justifies a detailed examination as found in Chapter 5.1. Then, the choice of a procedure for concentrating phytoplankton is open: should this be made by settling, or centrifuging, or filtering? As far as preserved samples are concerned, most phytoplanktologists of the world seem to agree

---

1. NB. This can be used by bibliographers as an abstract or summary.

on the first possibility, provided that an inverted microscope and appropriate chambers are available; thus several advantages are combined (Chapter **5.2.1**). If an inverted microscope is not available, then a multi-step settling procedure may take place, with decreased practicability and increased risk of loss (Chapter **5.2.2**). Centrifuging has other advantages and other shortcomings (Chapter **5.3**) and may be used for preserved samples or live samples as well. The reverse-filtration technique (Chapter **5.4.1**) is recommended when a concentrate of living and undamaged cells is needed for some qualitative or physiological study. The technique of membrane filters, in spite of its specific possibilities (Chapter **5.4.2**) can hardly be followed, as it now stands, for the quantitative assessment of populations in routine studies. Finally, Chapter **5.5** treats the different ways of obtaining a permanent record of a phytoplankton sample, either quantitative or qualitative.

The subdivisions made under 'Identification Problems' are self-explanatory. (The beginner should not neglect the general recommendations in Chapter **6.1**, on account of the somewhat trivial heading: this is information that one can get from experience but can hardly ever find in a book.) It sometimes happens that a given fraction or even an individual cell must be separated from the rest of the population for subsequent study; then Chapters **6.2.1** and **6.2.2** may be consulted. Being different kinds of creatures, the different groups of phytoplankters which are likely to appear in a sample will pose problems of their own and this will lead to the choice of specific techniques. Thus the next four chapters are meant as an introduction to the microscopical study of diatoms (**6.3.1**), dinoflagellates (**6.3.2**), coccolithophorids (**6.3.3**) and other flagellates (**6.3.4**), light microscopy and electron microscopy being treated separately in each case. A guide to the basic literature follows (Chapter **6.4**). Not only students, but also senior scientists seem to be unaware of the rules for naming organisms—or even to ignore the existence of such rules! As a matter of fact, the codes of nomenclature are so terrifyingly esoteric and tortuous to the unfamiliar reader that, compared to them, the present account of nomenclature (Chapter **6.5**) should be quite refreshing.

Again, there is a choice of techniques for estimating cell numbers but, whichever is used, some general principles should be followed as to the categories of objects to be recognized (Chapter **7.1.1**) and the numbers of them to be enumerated (Chapter **7.1.2**). Then:

If a settling method has been adopted at the concentrating step, go on to Chapter **7.3** if you have an inverted microscope, and to Chapters **7.2.1** or **7.2.2**, if you do not;

If membrane filters were used, turn to Chapter **7.2.3**;

If phytoplankton has been concentrated by centrifugation, or if it was so abundant that no concentration was needed, then a counting slide will be used (Chapter **7.2.2**).

Fluorescence microscopy provides a means of distinguishing between chlorophyll-bearing organisms and non-pigmented cells or debris (Chapter

7.4); although it was originally combined with centrifugation, its potential application is not confined to any given concentrating procedure. Cell numbers may also be estimated by non-microscopical methods. Two possibilities are discussed in this respect: one consists of a series of automated, electronic counters which are already used on a routine basis (Chapter **7.5.1**) or can be considered as promising (Chapters **7.5.2** and **7.5.3**). The other is the biological method of serial dilution, a rather arduous and selective method (Chapter **7.6**) which helps for those fragile plankters that cannot be preserved or concentrated by any other means.

As for interpreting the results of a phytoplankton study, it is suggested that attention should first be drawn to the geographical, ecological or physiological information that can be derived from the mere presence or absence of species with respect to the world supply of species (Chapter **8.1.1**); from the morphological habitus of a given species with respect to the spectrum of specific variability (Chapter **8.1.2**); and from the average cell size of the population (Chapter **8.1.3**). Estimating any feature in a quantitative manner implies that statistical caution was observed and that the choice of one or another type of statistical treatment is made appropriately (Chapter **8.2**). Diversity of the species in a population is both a biological paradox and a useful index; Chapter **8.3** tells how to calculate and how to interpret it. After the significance of plankton associations has been briefly discussed, the different approaches for characterizing them numerically are described and evaluated (Chapter **8.4**). The vertiginous range of size of the phytoplankton cells has led some investigators to transform data on cell abundance into an estimate of the actual biomass; there are more or less sophisticated procedures to serve this purpose (Chapter **8.5**). Finally, those who make use of computers or plan to do it may welcome some advice on data storage and retrieval (Chapter **8.6**).

The authors of the last part of the manual wrote their contributions after being provided with a (partially complete) set of the previous parts. This was meant to broaden the views expressed on marine phytoplankton with comments on three closely related fields. Just as it is true that bacterioplankton (Chapter **9.1**), microzooplankton (Chapter **9.2**) and phytoplankton overlap one another in the food web as regards sizes as well as functions, it is also true that freshwater micro-algae (Chapter **9.3**) and those from marine waters share the same pains and joys. Inclusion of this ancillary material will be more than justified if the reader picks up some unexpected information therein.

Addresses of the manufacturers cited in the manual are found on page 329.

1

Why study phytoplankton?

Bernt Zeitzschel

The suspended particulate matter in the sea consists of living organisms called plankton and dead particles commonly referred to as detritus. Plankton was defined by Hensen in 1887 and can be summarized as a comprehensive term which includes all organisms, plants and animals that are passively 'drifting' along with water movements. The plant component of plankton—the phytoplankton—is made up of unicellular (exceptionally: multicellular) algae which are either solitary or colonial. The main components of phytoplankton in the sea are diatoms, dinoflagellates, coccolithophorids and some other flagellates. The blue-green and green algae are very abundant in freshwater but are of lesser significance in the sea.

Phytoplankton organisms are autotrophs, i.e. they fix solar energy by photosynthesis, using carbon dioxide, nutrients and trace metals. All these autotrophs contain photosynthetic pigments such as chlorophylls and carotenoids. Some phytoplankton organisms, mainly species of the dinoflagellates, can be temporarily heterotrophic, i.e. they build up organic particulate matter from dissolved organic substances (osmotrophy) or even particulate organic matter (phagotrophy).

Plankton may be arbitrarily classified by size in nanoplankton (or nannoplankton as originally coined by Lohmann), cells < 20 µm; microplankton, organisms between 20 and 200 µm; then mesoplankton, macroplankton and megaplankton (Dussart, 1965, 1966; Lenz, 1968; Sournia, 1968). Phytoplankters belong mainly to the nano- and microplankton fractions. The larger phytoplankton species may be concentrated by plankton nets. Caution has to be applied, however, if net hauls are used for quantitative work. Large, bulky species like *Ceratium tripos* may be caught quantitatively, whereas other species of the same genus with different morphology will slip through even small meshes. In general, the use of plankton nets gives considerable underestimates of the total standing stock of phytoplankton. Malone (1971) compared the nanoplankton and net-plankton primary productivity and standing stock in neritic and oceanic waters. He came to the conclusion that nanoplankters were the most important producers in all the environments studied, but net-plankton productivity was significantly higher in neritic than in oceanic regions. Never-

1

theless, phytoplankters captured in fine-mesh nets can be extremely useful for morphological and taxonomic studies, because net hauls provide large numbers of certain types of phytoplankton which are generally sampled infrequently by water-bottles. For most quantitative investigations, however, phytoplankton must be concentrated by other methods, as described in this volume.

The fantastic diversity in the shape of plankton organisms has attracted naturalists for over one hundred years. It has been traditionally held that the diverse surfaces characterizing diatoms and dinoflagellates are factors directly related to their suspension. Smayda (1970) reviewed the literature on suspension and sinking of phytoplankton in the sea. He states that three principal categories of mechanisms can be recognized: morphological, physiological and physical. Smayda formulates a phytoplankton suspension hypothesis that 'the various morphological adaptations of the diatoms in particular are to be taken not as aids to suspension (flotation) per se, as commonly held, but as mechanisms to permit twisting and vertical movements within the water. The problem for phytoplankton is not to float, but to sink or rise and rotate.' Smayda concludes that biological suspension mechanisms are of an ambiguous nature and that physical mechanisms appear to provide a satisfactory explanation for phyto-plankton suspension.

Most phytoplankton organisms have a density greater than that of water. The higher density is in part caused by skeletons which consist of silica, calcium carbonate and cellulose. Water turbulence combined with other factors such as shape or physiological state reduce the sinking rate of non-motile organisms such as diatoms. There is recent evidence that the settling of phytoplankton to the bottom, at least in neritic waters, is not uniform but occurs at irregular intervals. Motile phytoplankters, like most dinoflagellates, may actively swim to compensate for sinking.

Autotrophic algae are most abundant in the euphotic zone. The latter is defined as a zone reaching from the surface of the sea to a depth where the energy intensity is such that production of organic matter by photosynthesis in respect to an individual phytoplankton cell balances destruction by respiration. This depth is called the compensation depth, where light intensity generally ranges between 0·1 and 1 per cent of the incident radiation reaching the surface of the water or very approximately 1 to 10 ly day^{-1}. The zone beneath the compensation depth is called the dysphotic zone and reaches down to approxi-mately 200 m or more. The aphotic zone is defined as a region further down where no daylight penetrates. Gran and Braarud (1935) introduced the concept of critical depth. According to Sverdrup (1953) the critical depth is defined as the depth to which phytoplankters may be mixed and at which the total pro-duction for the water column is equal to the respiration of primary producers for a period of 24 hours. It follows that a net increase in production can take place only if the critical depth is greater than the depth of mixing.

Phytoplankton is not distributed evenly in the oceans. It is believed to occur in three-dimensional patches of various size, caused by biological and

2

physical phenomena. Steele (1976) suggests that although variability of spatial heterogeneity occurs at all scales, there may be patches with, typically, dimensions of 10 to 100 km. According to Steele, many of these features can be explained by a combination of accumulation due to phytoplankton growth and dispersion due to turbulent diffusion. However, combined phytoplankton and zooplankton patches are less easy to explain. Riley (1976) proposes a model of plankton patchiness. He states that 'the interaction of diel migration with tides and residual drift can lead to cyclical variations in the rate of drift of zooplankton. This in turn varies the length of time that a so-called "mesoscale" patch of zooplankton will be associated with any given parcel of surface phytoplankton, and resulting differences in grazing pressure can produce phytoplankton patchiness. The latter may then stimulate the development of patchiness in zooplankton, producing the well known inverse relation between the two populations.'

Species in biology are comparable to elements in chemistry. To understand the structure and functioning of an ecosystem it is essential to know the different elements of which it is composed, i.e. the distribution of organisms in space and time. Phytoplankton normally consists of a heterogeneous collection of algae and the problems posed by the distribution and seasonal succession of the species present are not only of interest in themselves. Such qualitative differences may have effects on the higher components of the food chain and may thus also be of economic importance.

If quantitative data of the standing stock of phytoplankton are required three approaches are applicable:

1. Bulk measurements like particulate carbon, nitrogen or phosphorus, chlorophyll, ATP, etc. (Strickland and Parsons, 1972). These more-or-less standardized methods may differentiate roughly between phytoplankton, zooplankton and detritus. They do not give any indication, however, of species composition and abundance of species.
2. Particle counters can provide the size spectrum of particles in suspension (Parsons, 1969; Parsons and Seki, 1969). A distinction between dead and living matter and the identification of species is generally impossible.
3. Microscopic methods are up to now the only means to identify and count phytoplankton at the species level. These counts may be used to define phytoplankton communities and patterns of distribution in space and time. The counts can also be used to convert phytoplankton numbers to biomass or energy, e.g. in terms of organic carbon or calories respectively. Reliable data on phytoplankton carbon are essential for trophodynamic models.

Phytoplankters are of great ecological significance because they comprise the major portion of primary producers in the sea. They are, like the plants on land, the basic food in the sea for all consumers such as zooplankton and fish. It is a well-accepted fact that primary production by phytoplankton contributes to the energy required by benthic animals in shallow waters. There is a controversy, however, as to whether or not there is a direct input of organic matter produced in the euphotic zone to the bottom of the deep sea.

According to Koblentz-Mishke *et al.* (1970) the total annual net primary production by phytoplankton of the world ocean amounts to 15 to 18×10^9 t of carbon. The variation of plankton production in the pelagic environment is considerable: primary production in the open ocean (corresponding to about 90 per cent of the whole sea area) amounts on the average to $50 \, g \, C \, m^{-2} \, y^{-1}$; production in the continental-shelf regions, which make up about 10 per cent of the whole ocean, and where the water is up to 200 m deep, is of the order of 100 to $150 \, g \, C \, m^{-2} \, y^{-1}$. The highest production values are measured in up-welling areas where meteorological and hydrographical processes cause nutrient-rich water from several hundred-metre depths to rise to the surface. These values amount on the average to 300 to $500 \, g \, C \, m^{-2} \, y^{-1}$, but the areas concerned make up only a fraction of 1 per cent of the whole surface of the ocean (Ryther, 1969).

The main difference between primary production in the sea and on land is that phytoplankton in the open ocean is eaten almost entirely by zooplankton, whereas on land only about 10 per cent of plant material is eaten by herbivores. On land we find a large long-lived plant population, in the form of grasses, bushes and trees, while in the sea the population of primary producers has a short generation time of about one day (Sournia, 1974).

Phytoplankton organisms may be used to identify 'natural regions' of the oceans. These regions can be characterized by typical species or species groups. Up to now most work in biogeography in the sea has been carried out with zooplankton (McGowan, 1971; Zeitzschel, 1978). It is very likely, and there are some indications in the literature, that phytoplankton organisms are also good indicators of 'natural regions' as defined by latitude (e.g. boreal, subtropical, tropical) and by oceanographic features like the oceanic gyres (Braarud *et al.*, 1953; Smayda, 1958). Such biogeographical investigations are carried out on the species or even infraspecific level.

Phytoplankton may also be used to trace climatic changes in different geological periods. In palaeontology, the environmental factors, e.g. temperature of recent species of skeleton-bearing algae like diatoms or coccolithophores, are used to identify changes in the environment of former periods. Fractions of phytoplankton skeletons like coccoliths of coccolithophores are often sufficient for micropalaeontologists to identify the species (Funnell and Riedel, 1971).

In recent years, applied aspects of phytoplankton research have become more and more important. On the one hand, experiments on phytoplankton cultures are conducted to obtain basic background on maximal phytoplankton growth under optimal environmental conditions (Ryther *et al.*, 1972; Goldman *et al.*, 1975). The results of these experiments give guidelines for aquaculture work. On the other hand, experiments on low-level perturbations are carried out in the laboratory or in *in situ* to obtain data on the effect of pollutants in the sea (Parsons, 1974). There is clear evidence from experiments in large

4

plastic bags in Saanich Inlet, British Columbia, that pollution stress is indicated mainly by the population structure and the succession of phytoplankton species rather than by changes of standing stocks in terms of chlorophyll or rate measurements of phytoplankton (Menzel, 1977). Specialists are looking for specific phytoplankters which might be useful as test organisms to identify the degree of pollution by harmful substances.

The question why study phytoplankton is obviously dependent on the particular scientific question which has to be answered. There is one important fact, however, which holds true for all phytoplankton work: the need for simultaneous sampling and analysis of collateral parameters in order to acquire an understanding of phytoplankton ecology.

2
Sampling design

2.1 Sampling strategies

Elizabeth L. Venrick

A *sample* is an estimate of a larger body of information, the *population*. In the present discussion, the statistical concept of a population will often refer to a collection of organisms, i.e. a biological population; but the two concepts are distinct. In the statistical context a population may refer to all possible temperature measurements in a body of water, the lengths of all bluefin tuna in the Pacific Ocean, or any other complete body of facts about which knowledge is desired. The term *statistic* refers to a characteristic of the sample; *parameter* refers to a characteristic of the population from which the sample is drawn and which is to be estimated by the sample statistic. Thus, the mean of several samples is a statistic which estimates the population mean, the parameter.

The purpose of sampling design is to increase the amount of information obtained about the population for the effort expended to collect and analyse a series of samples. Ideally, the information content of the samples should be measured by the *accuracy* with which the samples reflect the parent population, i.e. by the deviation of the sample statistic from the corresponding population parameter. In the absence of information on the value of the parameter under consideration, accuracy is difficult or impossible to determine. Thus, in reality, information is measured by *precision*, the variability of repeated sample statistics. The concepts of accuracy and precision are distinct and are not necessarily synonymous. Net tows may give precise estimates of abundance but have poor accuracy because they underestimate the true abundance of the smaller organisms which are only partially retained by the mesh. Such consistent deviation between a parameter and the corresponding statistic is termed *bias*.

Much of the theory of sampling design is dependent upon knowledge of the underlying frequency distribution of the population which, in the case of phytoplankton, is seldom known. Studies of microdistributions of populations of zooplankton and phytoplankton have been reviewed by Cassie (1962; 1963); in the following discussion, no effort has been made to distinguish the two types of organisms. Theoretical expressions used to describe field distributions have

7

included the normal (e.g. Vollenweider, 1956; Cassie, 1959c), the Poisson (Student, 1907; Lund et al., 1958), the negative binomial (e.g. Bliss and Fisher, 1953; Holmes and Widrig, 1956; Kutkuhn, 1958), the log-normal (e.g. Winsor and Clarke, 1940), the double Poisson (Neyman, 1939; Thomas, 1949; Barnes and Marshall, 1951), the Poisson-log-normal (Cassie, 1962), and the log^p-normal with p varying according to the scale of observations (Frontier, 1973).

Other work (reviewed by Cassie, 1963) has been focused on the environmental influences on the microdistribution of plankton, including general hydrodynamic and advective processes (Rassoulzadegan and Gostan, 1976), horizontal gradients (Hasle, 1954; Cassie, 1959a, 1963), upwelling (Beers et al., 1971; Steele, 1973), and physical convective processes such as turbulence (Lund et al., 1958; Platt, 1972), wind-induced convection cells (Bary, 1953), and internal waves (Venrick, 1972b; Kamykowski, 1974). Ibanez (1976) has found heterogeneity varying according to the time of day, while Margalef (1958) has related it to differential succession. A few studies have attempted to model the dynamics of a plankton patch (Kierstead and Slobodkin, 1953; Platt and Denman, 1975; Wroblewski et al., 1975; Wroblewski and O'Brien, 1976).

The primary generalization to be made from the numerous studies of microdistribution is that plankton organisms are rarely if ever distributed according to Poisson, or random, expectation (Cassie, 1962; Margalef, 1969a; Frontier, 1972), although rare organisms will often appear random because of insufficient sample size (Cassie, 1959b). The critical abundance for non-randomness may be a function of the size of the organism (Rassoulzadegan and Gostan, 1976). In the pelagic environment, deviation from random is nearly always towards overdispersion (aggregation). Patchiness has been detected on scales from a few centimetres (e.g. Cassie, 1959b) to hundreds of miles (Bainbridge, 1957), but there may be discrete scales of aggregation superimposed on a continuum (Platt et al., 1970). The degree of heterogeneity has often been correlated with environmental variability (Venrick, 1969) and with phytoplankton abundance (e.g. Cassie, 1971; Rassoulzadegan and Gostan, 1976); the latter may reflect heterogeneity of growth rate or production efficiency (Platt and Filion, 1973). In general, one may expect greater phytoplankton heterogeneity in regions of rapid growth and high standing crop as well as in regions of environmental heterogeneity.

STEPS IN A SAMPLING PROGRAMME

The goals of a sampling programme may be purely descriptive, such as a list of species present, or a map of abundance, but usually they include statistical tests of hypotheses about causal or predictive relationships. The requirements of this latter type of programme are far more stringent than the requirements of a descriptive programme for which the following procedures may be relaxed somewhat.

The important steps in a quantitative programme have been detailed by Cochran (1963, p. 5–8). They may be summarized as follows:

1. *A rigorous statement of objectives*, including: (a) the ultimate goal—the relationships to be examined, the hypotheses to be tested, the predictions to be made, etc.; (b) the analytical methods to be employed; (c) the precision desired.

Consideration of the first two aspects ensures that the necessary data are collected and that they are appropriate to the study. There is a deplorable tendency among biological oceanographers to collect as much information as possible in the hope that 'something interesting' will emerge. Such non-selective acquisition of data may be justifiable as a preliminary survey or an accessory study, but it is grossly inefficient in the case of a primary programme.

A statement of acceptable precision is necessary to determine the number of samples to be taken. The relationships of precision to number of samples and population variance are discussed in Chapter **7.1.2**. By substituting some measure of the variance of the population in the field, the formulae of Chapter **7.1.2** may be used to determine the minimum number of samples needed to obtain a specified precision. Methods of measuring or estimating this variance are discussed at the end of the present section.

Most formulae for precision are based upon the normal distribution. This is unlikely to be a rigorous representation of natural populations. However, means of samples tend to approach normality regardless of the distribution of the underlying population, so that moderate deviations of the population from normality may be unimportant in the case of formulae for sample number (and many other statistical formulae) which are based on sample means. For instance, Serfling (1949) has shown that the means of replicate samples from a Poisson distribution are approximately normal when the mean is greater than 5. Deviations of means from a normal distribution will usually result in an underestimation of the necessary number of samples. Alternatively, the data may be transformed to a normal distribution (cf. Chapter **8.2**). Sample number determined from transformed data, however, will give the expected precision only if the subsequent analysis is performed on similarly transformed data.

2. *Definition of the target population*: In a statistical sense, the target population is that body of facts about which information is desired and to which conclusions will apply. To assure that a given study samples a representative fraction of the target population it is necessary to: (a) define the target population; (b) specify the sampling unit; (c) set up the sampling frame; (d) select random samples.

The sampling unit is usually dictated by available gear; it may be the organisms in a surface bucket sample, a water-bottle sample, a net tow, etc. The sampling frame is the assemblage of all possible sampling units which comprise the target population. Thus, if one defines the target population to be the phytoplankton in a particular bay, and the sampling unit is the phytoplankton in a Nansen-bottle sample, the sampling frame is all possible Nansen-bottle samples which potentially could be collected from the bay. The target population is, in reality, defined by the dimensions of the sampling frame. If the Nansen-bottle samples are to be collected only from a depth of 10 metres, the target population is correspondingly reduced to the phytoplankton at 10 metres depth.

Statistical theory for determination of precision demands some element of randomization in the selection of samples from the sampling frame. In practice, this requirement is frequently violated in plankton work. The logistics of random sampling in the ocean are difficult to overcome and, therefore, systematic sampling is substituted. The consequences of this are discussed below. Ibanez (1973a, 1973b, 1976) has investigated the problem of random sampling at sea and has suggested practical alternatives.

Improper definition of the target population is a frequent source of inefficiency in field studies. For instance, if the intention is to investigate estuarine phytoplankton, it is not acceptable to collect samples from the nearest estuary, or even from a 'typical' estuary, and thence extrapolate to all estuaries of the region. The one or more estuary to be sampled must be selected at random from all possible estuaries in the target population. If the nearest estuary is selected it must be the result of chance, not of convenience. Any estuaries which have no probability of selection are not members of the target population and any conclusions cannot legitimately be applied to them.

At the same time, the target population may be unnecessarily large, causing inefficient expenditure of effort. It is often possible to investigate a concisely formulated hypothesis from carefully selected stations represented by one or usually a few replicate samples. Thus, two estuaries may be compared by means of one or more analogous locations within each, rather than by describing each in its entirety before the comparison. The success of this approach depends upon the skill with which the locations are selected. Conclusions about differences and similarities refer only to the paired sites, not to the entire estuaries, but such a restricted definition of the target population may eliminate the need to consider the complex spatial heterogeneity within a broader population.

SOME ALTERNATIVE STRATEGIES

There are many sampling strategies available which provide the required randomization and which offer various advantages in terms of precision or effort. The theoretical aspects of each are discussed by Cochran (1963), and Ibanez (1976) has investigated several strategies empirically.

Simple random sampling

Each sampling unit in the population has an equal and independent probability of being selected. The most direct method of accomplishing this is to number each element of the frame and select the desired number of elements by means of a random numbers table. This is the strategy used in most subsampling designs, but it is one of the least effective approaches for field-work.

Stratified random sampling

The population is divided into strata (non-overlapping subpopulations whose sum is the target population) and one or more samples are collected at random from each stratum. For maximum efficiency, strata should be internally homogeneous with maximum variability between strata. This requires narrower strata in regions of rapid change. Furthermore, if the cost per sample is constant, it is most efficient to take more samples from those strata which are more heterogeneous, allotting the number of samples in proportion to the square root of the variance. The advantage of stratified random sampling is that it ensures complete coverage of the population without sacrificing the element of randomness.

In the absence of knowledge about the population, environmental gradients are often used to delineate strata. Distance from shore or gradients of temperature and salinity may be useful for horizontal stratification, while seasonal cycles of biomass or productivity may be used to stratify populations in time. Strata may be located along the vertical axis according to the strength of any of the numerous vertical gradients. The isothermal mixed layer is one obvious stratum. A well-developed subsurface maximum layer of chlorophyll or biomass may warrant several narrow strata, or a broad stratum with several samples. If it is necessary to sample a particular depth, this may be considered as one very narrow stratum, but care must be taken to ensure that all depths are assigned to one stratum or another.

Cluster sampling

The population is subdivided into units, each of which is a miniature of the parent population, i.e. there is minimum variability between clusters and maximum variability within. One or more clusters are selected at random. These may be completely analysed or may in turn be subsampled. The major advantage of cluster sampling is the saving in time and effort. In the case of an estuarine investigation, the target population includes all estuaries, while each individual estuary represents a cluster. So long as the estuaries to be sampled are selected at random, future work may be focused on the selected populations without invalidating the general applicability of the results.

Systematic sampling

The samples are collected from the population at regular intervals, and thus there is no element of randomization inherent in the sampling. The success of this approach depends upon the distribution of the population being studied. When systematic sampling is imposed on a random distribution, the results may be equivalent to simple random sampling. With other distributions, the results are likely to be biased unless the frequency distribution is sufficiently

11

known to allow special adjustments of the sampling regime. This is rarely the case in phytoplankton work. The primary advantages of systematic sampling lie in the ease of locating sampling sites and the ultimate simplicity of data presentation—tabulation or mapping.

The use of systematic sampling should be carefully evaluated. It is possible that in some instances the complexities of the natural distributions do provide a satisfactory approximation to randomness (Milne, 1959). To the extent that this is not true, the probability levels associated with the resultant statistics will be incorrect. The problem is confounded by the fact that systematic sampling may give more precise results (Cochran, 1963, p. 221–4); however, the presence of a linear trend, or periodic fluctuations, may introduce undetected biases into the data. The potential errors of systematic sampling are well illustrated by Uehlinger (1964).

Ratio and regression sampling

These two techniques offer considerable gain in efficiency, especially when sampling is for total biomass rather than species composition. Both techniques depend upon the use of auxiliary data which must be highly correlated with the phytoplankton data and which are easier to obtain, or which have already been obtained. Such data may include related environmental parameters or more extensive phytoplankton data from the same location during a comparable season. Ratio sampling assumes a constant ratio between the two sets of data (i.e. a linear relationship which passes through the origin) while regression sampling assumes only a linear relationship. Cochran (1963, p. 154–205) discusses both procedures, and Cassie (1968, 1971) discusses regression sampling in terms of plankton studies. In either method, the basic phytoplankton data are adjusted according to information contained in the more extensive set of auxiliary data. If a hundred measurements of surface chlorophyll are available from a given region, and if ten of these are accompanied by phytoplankton samples, we may obtain a more precise estimate of the true phytoplankton abundance by adjusting the mean of the ten samples according to whether the mean of the ten corresponding chlorophyll samples falls above or below the mean of the hundred samples. Obviously, the success of these methods depends upon the validity and stability of the underlying relationships.

Iterative procedures

The most useful procedure often involves successive application of different sampling designs as, for instance, when a broad systematic sampling programme forms the basis for subsequent stratified or cluster sampling. In this way the information gained from a general sampling programme allows one to improve the efficiency of subsequent programmes.

12

Simulation studies

Access to a computer facilitates an empirical approach to the design of sampling strategies. In a simulation study, an artificial population with a specified distribution pattern is created and sampled by the computer. Wiebe and Holland (1968) investigated the efficiency of various net tow strategies in sampling zooplankton distributed in patches of various sizes. Ibanez (1973b) studied the spatial–temporal interactions produced by different sequences of the same sixteen stations. Colebrook (1975b) examined the efficiency of the present strategy used in the North Sea–North Atlantic plankton recorder surveys, constructing his computer population from frequency distributions observed during the previous twenty years of sampling. The validity of the results of such simulation models depends upon the accuracy with which the underlying population has been reproduced. The advantages lie in the ease with which one may examine interactions between a large number of alternate sampling strategies and a large number of population structures.

GENERAL CONSIDERATIONS

The principles of sample allocation for stratified random sampling may be generalized to all strategies: the most precise estimate of most parameters is obtained by allotting greater numbers of samples to the most variable components of the population. Thus, sampling should be intensified in regions of environmental transition. Furthermore, frequent correlation between heterogeneity and abundance leads to the general rule that more abundant populations should be more intensively sampled. Thus, other factors being equal, for comparable precision more samples should be collected from the neritic zone than from the oceanic, more samples from subarctic populations than from subtropical, and more samples during spring and autumn blooms than during the more stable summer and winter populations.

A different allocation of samples may be appropriate to study some attributes. Accurate determination of rank order of abundance, for instance, will necessitate a larger sample from a more diverse association (Margalef, 1958) which will often have a lower mean abundance (Hulburt, 1963).

Samples must also be allocated in the vertical direction. In some instances a satisfactory test of a hypothesis may be based on samples from a single depth, and this will greatly reduce the number of samples to be collected. If information about the vertical structure of phytoplankton is desired, the number of samples needed will depend upon the amount of vertical structure. In the presence of a deep mixed layer, wind-driven turbulence may be sufficient to maintain a random or nearly random distribution of cells (Lund *et al.*, 1958) and a few samples may give adequate precision. The presence of stratification usually results in increased vertical structure within the phytoplankton (Hulburt, 1962, 1966) often characterized by a pronounced subsurface maximum (e.g. Karsten,

1907; Allen, 1940; Venrick, 1969), necessitating an increase in the number of samples.

Because cells are continually sinking from the euphotic zone, there is no absolute lower boundary for phytoplankton populations. The depth of 1 per cent of the surface light is often used to determine the depth of the deepest sample. This represents the approximate lower limit of production in many environments. In oligotrophic areas, however, significant populations may occur at greater depths (Venrick et al., 1973), and the sampling must be correspondingly deepened. Another criterion is the maximum depth of the winter mixed layer, which is the maximum depth from which cells are likely to be returned to the euphotic layer.

In situations where estimates of mean values are of primary interest, savings in time and effort can be realized by physically integrating the samples before counting, rather than arithmetically averaging the individual counts. Samples must be mixed in proportion to the volume of water (or horizontal area, or vertical extent) they represent. Möller and Bernhard (1974) use an objective criterion of population homogeneity as a basis for combining samples. When the parameters of interest represent the entire water column (such as the number of cells per m^2), the samples from a single vertical cast may be combined into one sample (Riley, 1957) or into several composite samples representing broad strata (Beers et al., 1975). In all cases, the savings in time must be weighed against the potential loss of information. However, a portion of each sample may be retained for future analysis in greater detail.

DETERMINATION OF SAMPLING VARIABILITY

Most statistical procedures are based upon some measure of internal variability against which to measure differences between samples from different populations. This internal variability, or error, reflects the heterogeneity within the target populations, but the scale on which this is measured is often somewhat arbitrary. It is not satisfactory to use only the analytical variability of subsamples within a single sample (unless the target population is no larger than that sample); some number of replicate field samples must be taken.

The appropriate measure of variability is the mean square between replicate samples. If the samples are enumerated entirely, this mean square will be the variance of the population in the field. If the data are obtained from subsamples of the original sample, the mean square will include subsampling error as well (Chapter 5.1).

In some cases it may be possible to collect replicate samples on a routine basis (e.g. Barnes and Hasle, 1957). From statistical considerations this is the preferred approach, as it gives a direct measure of the sampling error and a balanced design which facilitates most statistical analyses. Unfortunately, the large number of samples required may be prohibitive.

14

One alternative is to establish a satisfactory theoretical relationship between the variance and the mean so that the former can be estimated from abundance data. Several formulae have been tried. In the simplest case $\sigma^2 = \mu^2$, which gives the familiar coefficient of variation, $CV = (s/\bar{x}) \, 100\%$, i.e. the standard deviation as a percentage of the mean (Winsor and Clarke, 1940). A more general equation suggested by Taylor (1961) is of the form $\sigma^2 = a\mu^b$ (Frontier, 1973). This reduces to the Poisson when $a = b = 1$ and to the log-normal when $b = 2$ (Cassie, 1963). Cassie (1963, 1971) recommended the model $\sigma^2 = \mu + c\mu^2$, which is related to the negative binomial and the Poisson-log-normal and which approximates the Poisson for small values of μ and the form of the coefficient of variation when μ is larger. The value of c falls between 0·1 and 1·0 for a wide variety of plankton species (Cassie, 1959b). Applicability of these or other models must be determined for the population under consideration.

In the absence of a satisfactory model, the relationship between the variance and the mean may be determined directly by means of a pilot study. A series of replicate samples will produce a frequency distribution from which selected percentiles can be determined. Their relationship to the mean (or median) may be used to estimate confidence bands around single samples (Venrick, 1972b). One advantage to this empirical approach is that the data will yield precision estimates for numerous parameters not amenable to theoretical treatment (such as median abundance, diversity, coefficients of correlation and autocorrelation, etc.).

It is sometimes desirable to minimize sampling error. This can be accomplished by collecting more samples from the target population as demonstrated by Verduin (1951). An alternative, discussed above, is to integrate samples over some area, physically removing the variability within the area integrated (e.g. Bernhard and Rampi, 1966; Hrbaček, 1966). The use of a pump facilitates this, but numerous discrete samples may be pooled before enumeration to achieve approximately the same results. The loss of information occasioned by this pooling must be weighed against the gain in precision by the reduction of an error component.

INFORMATION AND STATISTICAL RIGOUR

In conclusion, although the preceding discussion has emphasized the requirements for rigorous statistical analysis, in reality, the conditions demanded by classical statistical methods are rarely, if ever, realized in the planktonic environment (Ibanez, 1976). For one thing, the distributional complexities of the plankton, confounded by their three- or four-dimensional nature, are far removed from the theoretical distributions underlying parametric statistical procedures. In addition, the difficulties of imposing a specific sampling strategy on an invisible, mobile population are enormous; rarely, if ever, can one be assured of

15

sampling the same population (or of not sampling the same population) on subsequent attempts. Our inability to sample the same population repeatedly (in the case of temporal analyses) or to sample locations simultaneously (in the case of spatial studies) imposes an element of spatial–temporal interaction on the data which may be difficult to extract (Ibanez, 1973a, 1973b, 1976).

Many excellent studies have attempted to describe the complexities of planktonic distributions and to adapt them to statistical methods. At the present time, there is no evidence that the specific results from one study can be extended to different organisms (i.e. from macroplankton to phytoplankton) or to different environments (i.e. from neritic to oceanic) or to different scales of sampling (i.e. from net tows to bottle samples). Previous studies should serve primarily as models for future studies to be directed at the particular population of interest.

In the interest of gaining biological information it may be necessary to relax the statistical requirements to a greater or lesser degree and to use statistical procedures as qualitative rather than probabilistic tools. But statistical requirements must be understood before they can be ignored; only if the researcher has a firm understanding of the principles of classical statistics can he violate them without risk of fooling himself or his reader.

16

2.2   Some examples

Ramón Margalef

The first step in sampling design is to state the problem in a clear and logical manner, for the nature of the problem will define the scales of time and space that should be contemplated. A sampling programme has to result in an acceptable allocation of the effort that goes into counting from 100 to 1,000 samples, a typical range. The universe to be sampled cannot be considered statistically uniform; neither can the samples be taken at random. This poses difficulties in the statistical treatment of the data and the need for simplification, change and compromise is always present. Limitations of equipment, time and manpower also make it difficult to project or to adhere strictly to a programme involving an ideal distribution of sampling points. In fact, plankton is an organization, and the sampling programme has to be flexible enough to follow the organization as and if it is revealed.

Usually there is available some information about the same or similar areas from which one can estimate the expected variability of results. Samples collected as part of routine surveys may provide some background for taxonomic studies, as well as for the anticipated variability. Although it may be wise to start with a location of prospective sampling points more or less uniform and at random, later on sampling is usually made more dense in the directions, areas or seasons where more change is expected, so that the variance becomes more uniform. Distribution of sampling points can be arranged to cover and discriminate better the effect of suspected factors of variation. It is a good rule to use samples for counting that are well documented by simultaneous study of many physical and chemical parameters.

For a total of 100 to 1,000 samples it is possible to recommend average distances between samples in relation to the different scales of the phenomena under study. The limits and nomenclature are arbitrary and merely indicative:

	Approximate size	Separation between samples		
		Horizontal	Vertical	Time
Upwelling	100 to 1,000 km	10 to 100 km	10 to 50 m	100 days
Coastal areas	10 to 100 km	1 to 10 km	1 to 10 m	10 days
Red tides, pollution	1 to 10 km	100 m to 1 km	0·1 to 1 m	0·1 to 1 day
'Microdistribution'	100 m to 1 km	10 to 100 m	0·01 to 0·1 m	0·01 to 0·1 day

Studies in microdistribution provide the basis for understanding medium-scale distributions in coastal waters, and these distributions help to explain large-scale structures. It is good practice to supplement each particular research

B

with a more dense sampling over a selected small area, to give a convenient connection with a lower range. But any strict advice may happen to be unsound in a given instance. Usually, after the research programme has been carried out, one feels that one could plan a similar investigation better, but this is probably wishful thinking.

SURVEYING AN UPWELLING AREA

The four upwelling areas off California, Peru, north-west Africa and south-west Africa have several characteristics in common, all of them, for example, being associated with eastern boundary currents at the border of major oceanic gyres. It is helpful to have four replicas of almost the same phenomenon, and a comparative method can be introduced. Its usefulness is enhanced by the differences between the upwelling areas; for instance, the South American system seems to be more productive than the others. In their initial phase, studies on the phytoplankton of the different regions attempt to characterize the communities and to describe the differences between the upwelled water and the peripheral areas. Priority research subjects are the response of plankton to upwelling, mixing and enrichment of water, recycling of nutrients and the extent in space and time of populations of high density.

Marine areas directly influenced by upwelling extend over hundreds of kilometres, and any survey made on a small scale would require an impossible number of working hours for the study of samples. Obviously one has to sacrifice detail to obtain a synoptic overview. The data which will be commented upon here, with the help of Figures 1 and 2, involve two cruises. In the first cruise 31 stations with 13 depths were sampled; in the second cruise 27 stations with 8 levels. This made a total of 557 samples (a few others being lost for various reasons) to be sedimented and counted in a reasonable time, in order that the results could be used in the preparation for the next phase of the research programme. Actually, 1,000 working hours by a relatively experienced observer could not produce extremely accurate descriptions of the communities, but only partial lists apt for making comparisons between medium-sized areas. The total number of taxa exceeds 300, although usually more than one half of the cells present in each sample could not be identified after preservation under the optical microscope. A more precise taxonomic study would require more time, special resources (scanning electron microscope) and the cooperation of several people. In our case, the fact that the counts were made by the same person may have resulted in a better comparability of the data, but not in their accuracy. In fact, posterior checks and revisions revealed important inaccuracies.

Plankton populations are controlled by the physical environment, and interaction among species plays a secondary role. This is especially obvious in upwelling systems, where nutrient input and turbulence are paramount. Stations placed 60 miles (one degree of latitude) apart were considered adequate for the

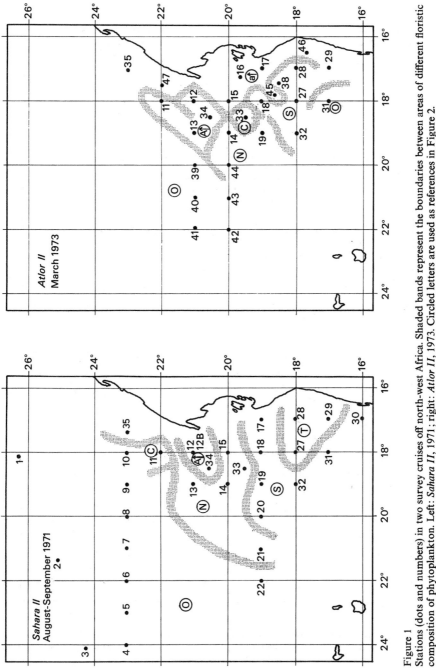

Figure 1
Stations (dots and numbers) in two survey cruises off north-west Africa. Shaded bands represent the boundaries between areas of different floristic composition of phytoplankton. Left: *Sahara II*, 1971; right: *Atlor II*, 1973. Circled letters are used as references in Figure 2.

19

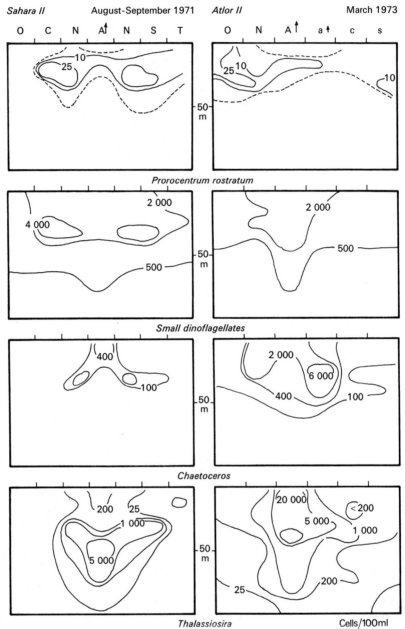

Sahara II   August-September 1971   Atlor II   March 1973

Prorocentrum rostratum

Small dinoflagellates

Chaetoceros

Thalassiosira   Cells/100ml

Figure 2
Distribution (cells/100 ml) of one dinoflagellate (*Prorocentrum rostratum*), miscellaneous small
dinoflagellates, and two genera of diatoms (*Chaetoceros* spp., *Thalassiosira* spp.), as a function of
depth (vertical dimension) and of 'area' (as defined in Figure 1). The symbols for the areas are
reported on the top of the figures and ordered along the abscissas. Only a few lines of equal
concentration, freely drawn, have been retained.

20

purpose of providing a large-scale image of the phytoplankton distribution in relation to the physical oceanography of the area. But this density of sampling stations is too low to provide an adequate image of the changing conditions and communities closer to the major foci of upwelling. There are also other highly dynamic boundaries or fronts that require a greater density of observations around them. Any accepted net of sampling sites has to be scaled and validated by the examination of a denser net in a lower order of dimension, over some small areas.

As is usually the case, some compromise had to be reached with the requirements of physical and chemical oceanographers. As a consequence, the sampling depths were those adopted in the general oceanographical programme that provided sufficient amounts of water (100 ml) from the depths of 0, 10, 20, 30, 40, 50, 75, 100 and 150 metres. Samples of water from deeper levels were also available, but they were not routinely studied. Sampling the populations at the same points from which physical and chemical data were secured provides an appropriate basis for the study of statistical correlations between properties of the environment and presence and abundance of species, or composition of communities. But, of course, some aspects of the knowledge of phytoplankton would have benefited from additional or different sampling depths.

In most surveys, the distributions and phenomena that are clearly recognized belong to a scale larger than the one which was embodied in the sampling design. The present study was no exception and could serve only to recognize differences in plankton composition on a very large scale. The results can be presented on a geographic basis, defining large areas, and comparing the characteristics of plankton inside each area. This simplified picture not only has a descriptive value, but may be also of some help in designing sampling programmes in the same or in other areas.

The compression of the information provided by the original lists has been done in the following way: the plankton communities of each station were compared with the composition of plankton in the neighbouring stations, making use of rank correlation. As density of phytoplankton is higher in the photic zone, affinities or differences shown by the upper layers have been made to overrule affinities and differences observed in deeper levels. The latter anyway are less reliable, because of the larger errors associated with the small number of cells counted in deep samples. This procedure led to drawing tentative boundaries that divide the whole region into a number of smaller areas (Fig. 1). Most of the major discontinuities observed in the distribution of phytoplankton are backed up by hydrographic structures. More detailed transects based on surface samples collected along the tracks between stations are helpful in defining the position of the major discontinuities.

The average vertical distribution of each species within each area can be combined in idealized patterns (Fig. 2) that may help to understand the distribution of species and life forms of plankton in relation to environmental factors. Our example can be compared with the result of surveys that have led to the

21

recognition of plankton types associated with types of water, sometimes with practical implications, for instance when organisms are used as indicators of different situations, as has been done in the North Sea (Braarud *et al.*, 1953).

Each area emerges with particular features. Where upwelling is strong, the number of species is rather low and diatoms are predominant; visible extracellular secretions are abundant, a rather uncommon situation for marine phytoplankton, although usual in freshwater phytoplankton. The comparison of the patterns of distribution over different seasons shows a sort of expansion and contraction of the entire pattern. This may be useful in the understanding of the correspondence between the composition of plankton and the upwelling events, but is not sufficient to explain the fine details of the pattern and how it is generated.

A number of surveys, about four per year, and sampling stations 50 km apart can provide only a rough picture of the size and main fluctuations of the upwelling systems.

STUDYING AN ANNUAL CYCLE

Major expeditions are not geared to this sort of work. Studies of annual cycles are more appropriate for coastal laboratories, which can use small craft in the neighbouring areas. In coastal waters fertilizing processes and mixing are rapid, bottom topography and coastal line introduce many irregularities in water movements, and from the start a rapid change in the populations and a considerable local heterogeneity have to be taken for granted. It is necessary to cover adequately both time and space, and the appropriate design includes stations placed approximately 10 km apart and studied at weekly or fortnightly intervals. Where tides are important, a more detailed study at least over one tidal period is necessary, and care should be taken that interference between sampling periodicity and tidal rhythms is not interpreted as some peculiar periodicity. Intervals of one month between samples can miss important events. As depth is not great in coastal waters, the number of sampled levels can be small. In areas where the sea is really rough, series based on samples obtained from small boats show frequent interruptions, often, and unfortunately, at the most interesting times.

With limited manpower, an efficient allocation of effort may call for mixed strategies. During the seasons when water is mixed, or displaced rapidly, it is convenient to sample more often, but there is less spatial heterogeneity, and the number of stations can be reduced. When stratification develops, changes in time become less important, but assemblages become more different from place to place, and it is worth studying more stations at the expense of reducing the frequency in sampling. Such a strategy tends to counterbalance the differences between samples, over space and time, and may be helpful if the sampling programme continues over many years. It may make statistical evaluation of the results difficult, but often is the only way to allocate sensibly the manpower involved in identifying and counting cells.

Studies of the annual cycle of phytoplankton are so numerous that only a few of them are cited here as examples. As a matter of fact, almost all marine biological laboratories of the world have produced one or several studies of this kind; investigations in the open sea, however, are more scarce. The following choice of references will at least give an idea of the geographical scattering of data: Braarud, 1945 and Braarud *et al.*, 1958 (Norwegian fjords); Holmes, 1956 (Labrador Sea); Riley, 1957 (Sargasso Sea); Pratt, 1959 (Narragansett Bay, United States); V. Cassie, 1960 (New Zealand coast); Kollmer, 1962 (Walvis Bay, Namibia); Robinson, 1965, 1970 and Reid, 1975 (North Sea and North Atlantic), Kawarada *et al.*, 1966, 1968 (Sea of Japan); and Avaria, 1971 (Valparaiso Bay, Chile). Most of the references available from the Mediterranean Sea may be found in Sournia's review (1973) of the productivity of that area, and a number of data on the tropical seas were summarized in another review by the same author (1969).

Such programmes, as a result of concentrating much effort on the study of a small area, produce characteristically long lists of taxa. In the area of Barcelona and Castellón, a checklist with more than 350 species is used, and about the same number of species is identified in north-east Venezuela. Travers and Travers (1975) list 600 species from the area of Marseille, including Tintinnoinea, but excluding the taxa inferior to species, that amount to one hundred. A list of Adriatic plankton algae, prepared by Keržan (1976), includes more than 850 names, with subspecific taxa and some benthic species that occasionally appear in the plankton samples. It seems that the number of species is roughly proportional to the time spent in identification and to the number of examined cells. Such quantities of information are difficult to compress. Broad communities can be characterized and recognized using a rather short list of species, but the distribution of less common species may be highly significant in relation to definite environmental factors.

Continuation of the sampling programmes over many years always reveals considerable differences between successive yearly cycles. Comparison between years is valuable in revealing the actual mechanisms of fertilization, and of the selection operating on the different groups of species (Margalef, 1957*a*; Reid, 1975). Primary production in coastal waters of the north-west Mediterranean is determined by the sum or the combination of different fertilizing events, among them the breakdown of thermocline in the autumn; moderate inflow of deep, rich water in winter; and eventual extension of Atlantic water in spring. According to the relative importance of these events, the average composition of the communities, on the whole, changes from year to year. Although sampling design can be changed as experience advises, it is helpful to keep some station fixed over the years as a reference.

It is dangerous to accept multi-annual cycles or periodicities until very long series are available, and even in this case one can only speak of stochastic, quasi-periodic fluctuations. Long-term studies should be encouraged as they can show sustained trends due to climatic change and, in coastal

23

waters, to eutrophication or other cultural effects. Unfortunately, it is impossible from a practical point of view to continue an intensive study over many years: after a period of time the number of sampled stations is inevitably reduced, and the visits are spaced at greater intervals. The experience gathered during the years of intensive study can be used in making the best of whatever sampling possibilities are available.

MONITORING A POLLUTED AREA

In surveys of perturbed or stressed areas, the planktologist is usually asked to provide complementary information, rather than to study plankton distribution for its own sake. In so far as the marine environment is concerned, it is more correct to speak of fertilization or eutrophication than of pollution. Eutrophication is the consequence of an increased input of nutrients, leading to increased primary production, depletion of oxygen in deep water and eventual denitrification. Pollution means a direct input of organic matter, leading to more oxygen consumption and the development of heterotrophs, and also the introduction of substances of negative and relatively high activity, as organochlorides or heavy metals. Additions of heat and of sources of ionizing radiation are considered as special kinds of pollution.

Eutrophication can be observed over large areas, as in the Baltic Sea, but usually is much more local, affecting a bay, a harbour or a fjord (Braarud, 1945). Under eutrophic conditions, the species that were common in the local plankton are usually recognized in small numbers, dispersed among great populations of 'weeds', organisms that belong to a small set of selected genera (*Prorocentrum, Scrippsiella, Olisthodiscus, Eutreptiella*, etc.).

The eutrophicated systems can be considered or modelled as reactors or as chemostats. Although high populations are maintained in a quasi-steady state by the supply of nutrients, the 'plumes' expanding offshore are subjected to frequent and rapid modification.

The purpose of the research defines the work of the planktologist. An idea of the species present is helpful in the evaluation of the effects of pollution. If the practical orientation of the study demands a large amount of data, plus rapid availability and less emphasis on correct identification of the kinds of organisms, the adoption of automatic devices like fluorometers and Coulter counters is preferable to the counting of plankton samples.

'RED TIDE' STUDIES

The combination of high nutrient concentration and low turbulence is frequently noticed in inland waters, but less commonly at sea, where it is characterized by the development of a particular plankton community, with a large

24

proportion of swimming organisms of a rather rounded shape, rich in chlorophyll, often reddish in colour, and sometimes toxic. Per unit of surface their biomass is usually lower than in most diatom plankton populations, but as they congregate in the top layers they are visible, and often noxious. The populations may be subjected to rapid changes, as the conditions that lead to their accumulations are transitory. These 'red tide' outbreaks usually show a high horizontal heterogeneity. Although some observations of red water have been made offshore (Bainbridge, 1957; Hart, 1966), red tides are essentially coastal phenomena, in part because the supply of freshwater enhances stratification or reduces turbulence. Toxicity of the planktonic species can be transferred to suspension feeders, and shellfish growing in areas subjected to red tides are suspect of being dangerous for human consumption. The toxicity of organisms associated with red tides may be felt also in the effects of sea spray on man, and in fish mortality rates.

Red tides are endemic in certain areas, such as the coasts of Florida (Gunter et·al., 1948) or the bays of Galicia in north-west Spain (Margalef, 1956b), and occur more sporadically in many places. They have also been reported around areas of upwelling (California, north-west Africa, Peru). Their popular name implies that they are detected visually, but it is in fact impossible to set limits, in terms of a minimum density, to qualify as red water, in so far as periods of dispersal and aggregation may alternate as an essential aspect of the phenomenon. Where red tides are unexpected, no sampling programme can be anticipated and they should be treated and studied as exceptional and, scientifically, highly rewarding events. Only in the areas where red tides are common can the events be studied in the frame of a coastal survey, adding fine-grained observations.

The practical consequences of red tides have stimulated an interest perhaps out of proportion to their extent (LoCicero, 1975). In a given area, the dominant organisms are generally few and more or less recurrent, for example, *Gymnodinium breve* in Florida, *Gonyaulax polyedra* in southern California. Actually, the distribution of phytoplankton in red tides is not simple. A good approach would be to draw a minimal net of stations separated only from 100 to 1,000 m, and to take as many samples as possible in between, especially if patches are visible or detected through fluorometry, or photography from the air. Patches are usually only a few metres thick, some amount of vertical migration of the organisms is to be expected, and the evolution of the whole pattern is rapid, so that sampling must be repeated frequently, in hours or days. One can give a cursory examination to the majority of samples, but one has to keep an eye open for unexpected complications in the composition of communities.

In the Ría de Vigo, in north-west Spain (Margalef, 1956b), red tides are frequent during periods of stagnation, usually at the end of summer, in the form of superficial patches of changing shapes, associated with Langmuir cells and perhaps with other kinds of hydrographic structures. *Gonyaulax* is the dominant genus, although it sometimes is associated with *Ceratium furca* and other dino-

25

B*

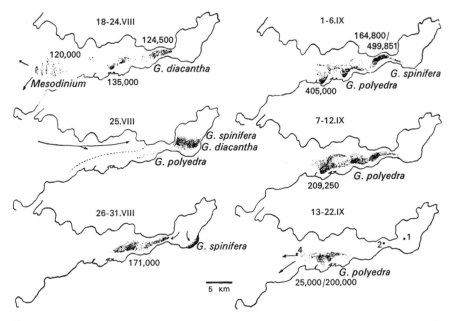

Figure 3
Distribution of 'red water' in the Ría de Vigo, north-west Spain in August and September 1955.
The dominant species was *Gonyaulax polyedra*, but other species were also present with local
and ecological segregation. The figures express the number of Harvey units of pigment per cubic
metre. (Margalef, 1956*b*.)

flagellates. Different species of *Gonyaulax* may be present, with a certain amount
of local and ecological segregation (Fig. 3): *G. diacantha* is common at the start
of the red tide and in the most interior part of the bay, giving a rather olive
colour to the patches in which it dominates. *G. polyedra* is more reddish and
forms the bulk of the 'tide'. *G. spinifera* is more common where temperature
fluctuates the most. Other species of the genus are present in smaller numbers.
Mesodinium rubrum is a ciliate containing algal cells or organelles as symbionts
and is a producer of red tides, sometimes in association with *Gonyaulax*
(Fig. 3).

SAMPLING AT A FIXED STATION:
DIEL STUDIES, VERTICAL MIGRATION

For the study of the effects of horizontal transport of plankton with water and
of the vertical movements of water and/or active migration of organisms,
collections are repeated with great frequency at the same geographic station and
at different depths. The results of advection (movement of water) and of migration cannot be distinguished if there is only one station of observation. The situa-

26

tion is worse in inshore waters, where horizontal differences in the composition and abundance of phytoplankton are important (Hasle, 1950; Sournia, 1968, 1974). Offshore, horizontal gradients are usually less important and the sequence of observations reflects with less distortion the vertical migration of organisms and the reorganization of phytoplankton.

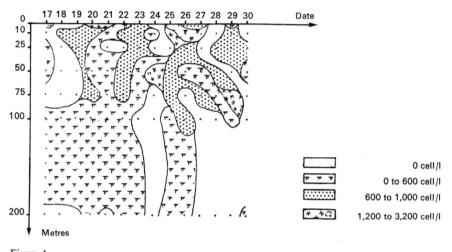

Figure 4
Density of populations of *Exuviaella baltica*, as observed from the 'Bouée-Laboratoire', in July 1964 (Léger, 1971b).

The COMEXO 'Bouée-Laboratoire', a moored vertical cylinder of steel, providing direct access to a thick layer of the photic zone, was used by Léger (1971a, 1971b) for studying the vertical distribution of phytoplankton in the north-west Mediterranean Sea. Samples were taken at daily intervals during 7 to 14 days at different periods of the year (Fig. 4). A number of intake taps were provided, but became useless because of contamination with periphyton growth. This is not the only example of the observer changing the observables, because even when sampling was done with clean conventional water-bottles, there was some suspicion that the composition of the phytoplankton around the buoy was influenced by the very presence of it; at least the pattern of turbulence was perturbed (this criticism does not apply to the samples collected below the buoy).

The laboratory buoy is fixed in space, while the water drifts by it, so that the pattern resulting from repeated sampling from the buoy reflects a combination of vertical movements of plankton with horizontal variability in the cross-section of water sampled from the column. With the observed current speeds, ten days of sampling to 300 m depth may be equivalent to a non-synchronous study of a section or transect of water 300 m deep and 86 km long.

27

But the situation is not exactly so, since speed and direction of current change with depth.

Intensification of sampling during selected periods could have permitted a more detailed analysis, and an effort could have been made to detect vertical migration of circadian periodicity, but this would have required sedimentation and counting of many more samples.

MICRODISTRIBUTION STUDIES

Red tides provide a visual impression of heterogeneity over a rather small scale (10 to 1,000 m) and this fact, as well as the observation of swarms of pelagic animals, has helped to introduce the expressions 'patches of plankton', and 'patchiness', when referring to the lack of uniformity in distributions. Of course, non-uniformity does not imply patches, but the wording may be retained, because it conveys the notion that the whole pattern of distribution, on analysis, does not break down in equivalent pieces like a tessellate mosaic, but rather has the form of peaks of high density, forming more or less rounded patches that are dispersed over a background of lesser abundance. As such distribution is re-peated for the different species and in different ways, the whole pattern is the sum of a number of patterns not necessarily of the same scale, although re-taining up to a certain point the same general or topological properties.

Along any transect, the plot of plankton densities in successive points has the form of an irregular mountain-range pattern, with peaks of different orders. The plot becomes more symmetrical if the values are plotted as the logs of densities (Fig. 5). Such distributions are found along transects running in all directions and this is consistent with the existence of a pattern, simple or com-posite, in which discontinuous regions of high density are dispersed over a more uniform phase of low density, netlike or in the form of a honeycomb. This pattern, repeated on many scales, is far from the hypothesis of a random dis-tribution, sometimes used in elementary statistical approaches. In consequence, the sets of plankton counts appear always as distributed in a contagious or clumped way, and the ratio variance/mean is consistently higher than one. In fact, it is appropriate to accept a linear relationship between log of the mean and log of the variance.

The study of distribution of plankton on a small scale provides empirical information for the development of conceptual models of plankton organization (Platt, 1975; Powell et al., 1975), but it is difficult to introduce as part of regular surveys, since it involves a tremendous amount of effort.

Two approaches have been used: (a) a number of point samples are obtained simultaneously with collecting bottles fixed along a rod or any suitable tri-dimensional frame and trapped simultaneously; (b) samples are collected sequentially along a transect, using discrete samples (although the use of wire and bottle can hardly provide the required speed of operation) or by pumping in

28

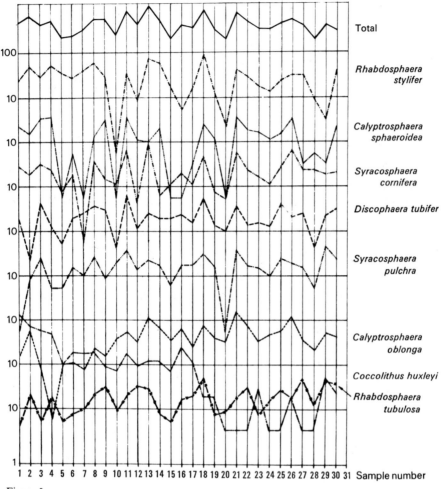

Figure 5
Horizontal distribution of the predominant species of Coccolithophoridae in Ligurian plankton, at 40 m depth, over a transect of 100 m, June 1964 (Bernhard and Rampi, 1966).

water continuously and drawing discrete samples from the flow. This allows one to apply automatic sensors like fluorometers and Coulter counters to the flow of sampled water. Such monitoring fills in substantial information when samples for plankton identification and counts are taken less frequently.

It is difficult to make specific recommendations about the distance between samples in space and time. Most of the results of the research on microdistribution are appropriate for the computation of variances for different groups of samples. Such variances can be compared with either the respective mean densities of the different groups or with the space sampled. The plot of the variance (absolute or relative to the respective means) against the sampled space is particularly useful when related to physical parameters such as turbulence. One should always come as close as possible to the lowest dimension. The perturbation introduced by the equipment is important. For instance, the mixing of water in the pipes acts as an integrating filter, and removes variations at the fine end of the scale.

As the sampling points are not at random, difficulties can be foreseen in the statistical interpretation of data, as well as those arising from the combination of variation in space with changes in time.

It is difficult to sample in two dimensions. In practice, continuous sampling is restricted to one level, or the problem is dealt with by arranging for a sinu-soidal path, on a vertical plane, of the water intake. This, however, gives rise to attendant problems of fixing the position of the inlet and allowing for the unavoidable mixing. As the spectrum of the distribution of plankton shows some coherence with the spectrum of turbulence, if sampling is done along two dimensions, x, z, it is advisable to keep distances between samples proportional to the respective coefficients of eddy diffusion (A_x, A_z). This means simply that samples have to be spaced more along the horizontal axis than along the vertical, as every planktologist knows and practices.

The surface film and the areas close to the surface film are apt, in consequence, to show the greatest heterogeneity, thus sampling on a minimal scale is necessary in the study of neuston.

Studies on the distribution of plankton on a small scale are understandably not common. The following may be cited as examples: Hasle, 1950; Cassie, 1963; Bernhard et al., 1969; Margalef, 1969a, 1971; McAlice, 1970; Sakshaugh, 1970; Venrick, 1972b; Kamykowski, 1974. Outlooks and results were different,

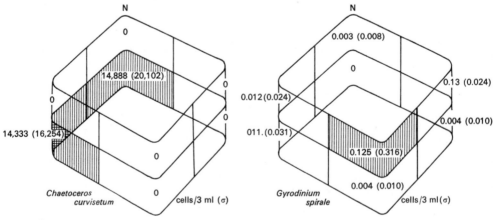

Figure 6
Distribution of the diatom *Chaetoceros curvisetum*, and the dinoflagellate *Gyrodinium spirale*, over the sides of a prism 2 miles square and 40 m in depth, in the western Mediterranean, July 1968. Average number of cells per 3 ml and (in brackets) standard deviation of untransformed counts (Margalef, 1971).

but always worthwhile. Nowhere is there uniformity nor complete confusion, and everywhere interesting and ecologically meaningful distributions may be revealed.

As a specific example, the following can be considered. The sides of a rectangular prism 2 × 2 miles in the western Mediterranean (Fig. 6) were explored down to 40 m depth, by collecting water through a submerged pump and hose, manoeuvred in a sinusoidal path. The area was circumscribed three times during 8 hours (Margalef, 1969a, 1971). A general pattern emerged as constant, although as time passed small distortions were recognized. The important features were: (a) population rich in diatoms in the western inferior corner, in correspondence with local mixing due to internal waves breaking over the nearby sloping bottom; (b) less dense population with a high proportion of dinoflagellates in the more stratified and poor water of the upper eastern corner. A detailed analysis of about 150 samples could serve only to hint at structures at the scale of 1,000 m, one order of magnitude above the distances between the sampled points, and left one wishing for a more detailed analysis. In all, and almost everywhere, a large number of species was present, as a potential seed for rapid and substantial changes in the composition of populations.

3

Sampling techniques

3.1   Water-bottles

Elizabeth L. Venrick

The first instrument 'for fetching up water from the depths of the sea' was designed by Robert Hook in 1666 (Fig. 7a). Since then, water-bottles of various designs have been the principal means of sampling water from below the sea surface, although pumping systems have superseded discrete samplers in some recent programmes. A comparison of the advantages and disadvantages of pumps and water-bottles for phytoplankton work is presented in Chapter **3.2** and will not be repeated here. The statistical implications of discrete and continuous (pump) sampling have been investigated empirically by Ibanez (1976).

BASIC DESIGNS

The requirement for a useful water sampler is that it collects a representative sample of suitable volume from a specified depth and retains it free from contamination during retrieval (Riley, 1965; Martin, 1968). There are numerous designs available which fulfil these requirements to a greater or lesser degree. These have been reviewed by several authors (ZoBell, 1946; Lund and Talling, 1957; Herdman, 1963; Riley, 1965; Schwoerbel, 1966; Saraceni and Ruggiu, 1969; Golterman and Clymo, 1969).

The most frequent design is a plastic or metal cylinder which is lowered to depth with both ends open. Closure, usually triggered mechanically by a messenger, is effected in one of two ways. In reversing-bottles such as the Nansen bottle (Helland-Hansen and Nansen, 1925; Wüst, 1932; Anon., 1968a) the messenger strikes and releases the upper clamp allowing the bottle to swing 180° about the lower clamp (Fig. 7b). This reversal closes valves in the top and bottom of the bottle. Other reversing bottles include the Ekman and Knudsen bottles (Ekman, 1905; Knudsen, 1929). In non-reversing bottles such as the Van Dorn bottle (Van Dorn, 1957; Finucane and May, 1961; Stephens, 1962),

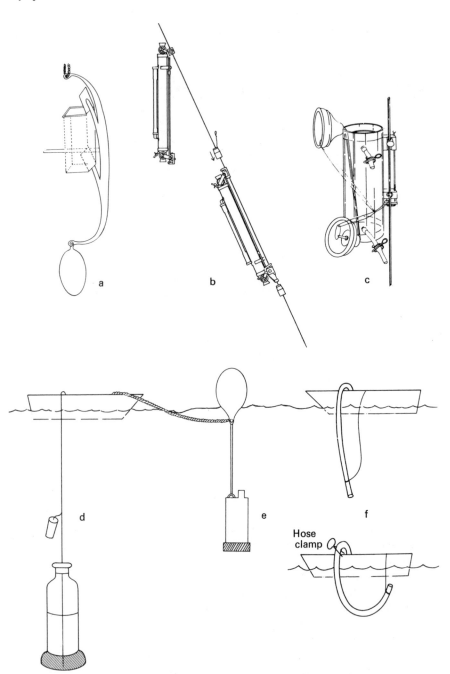

Figure 7
Water-bottle designs: (a) historical sampler (Hook, 1666); (b) reversing bottle (Nansen bottle); (c) non-reversing bottle (Van Dorn bottle); (d) 'snatch' sampler; (e) surface sampler; (f) integrating sampler.

the Fjarlie bottle (Fjarlie, 1953), the NIO sampler (Institute of Oceanographic Sciences,[1] United Kingdom) and the Niskin bottle (General Oceanics Inc., United States), the top and bottom caps are held open by a clamp against the tension of a spring or rubber connecting them through the bottle. The action of the messenger releases the clamp, and the caps are pulled into position, closing off the top and bottom of the bottle (Fig. 7c).

A second group of water-bottles includes those which are lowered in a closed or collapsed position and are opened at depth. Many of these samplers have been designed for special purposes such as sterile sampling (ZoBell, 1941; Aabye Jensen and Steemann Nielsen, 1953; Niskin, 1962; Sieburth, in press), avoidance of contamination from the surface layer (Niskin 'Go-Flow' bottle, S. Niskin, pers. comm.), non-toxic sampling (Throndsen, 1970b) or large-volume sampling (Schink and Anderson, 1969). They are generally adequate for phytoplankton sampling, although use of the rigid samplers is restricted to shallow depths.

Horizontal samplers, designed to sample a narrow stratum (Joeris, 1964; Ottmann, 1965; Duursma, 1967; Howmiller and Sloey, 1969), are especially useful in very shallow water, or for epibenthic samples. Samplers have been designed for use from a moving ship (Ekman, 1905; Lumby, 1927; Spilhaus and Miller, 1948; Doty and Oguri, 1958) and from a helicopter (Pinon and Pijck, 1975).

For studies of vertical and horizontal microdistribution, instruments have been designed to take simultaneously a series of closely spaced samples (Cassie, 1959b; Kjensmo, 1967; Broenkow, 1969; Sholkovitz, 1970).

Insulated samplers (Barnes, 1959), originally designed for determination of subsurface temperatures, are now used mainly for physiological studies.

The potential for electronic or automated sampling systems has not yet been fully explored. Systems in which water samples are collected in conjunction with an STD (electronic salinity-temperature-depth recorder) have been developed (Gerard and Amos, 1968; Niskin, 1968). Several water-bottles are mounted on a rosette frame which is clamped on the wire just above the STD. Bottles are triggered one at a time by electronic signals from the ship. Since the return signals from the STD can be monitored, water samplers may be accurately positioned with respect to either temperature, salinity or depth.

Automated samplers have been designed to take a series of samples at a single location at pre-set time intervals (Jensen and Sakshaug, 1970). Each of several narrow-necked containers (such as glass bottles) is evacuated and connected to the intake manifold by a flexible tube which passes through a pinch valve. A time clock releases the pinch valve at the desired times, causing a sample to be pulled into the container. For phytoplankton work, preservative may be added to each bottle ahead of time.

---

1. Addresses of manufacturers are found at the end of the volume, beginning on page 329.

SELECTION CRITERIA

Collection of water for preservation of phytoplankton places few restrictions on the sampler used. Most materials are suitable for construction so long as the sampler is free of particulate contaminants such as rust or paint flakes. Samplers which are lowered to depth with both ends open may be adequately flushed during the lowering (Fjarlie, 1953; Weiss, 1971); in addition, bottles may be repeatedly raised and lowered prior to closure to further ensure flushing. If the samplers are to be used for purposes other than phytoplankton preservation, additional requirements may be necessary. Most commercial bottles are now constructed from, or coated with, an inert material such as PVC or teflon to eliminate chemical contamination, but, if the samples are to be used for physiological work, additional restrictions may be put on the composition of the sampler. Rubber and plastic, for instance, may be toxic to some phytoplankton species (Throndsen, 1970b). In order to remove organisms such as fungi (Willingham and Buck, 1965) and bacteria (Bogoyavlenskii, 1962) which can grow on sampler surfaces, the sampler may be scrubbed mechanically with an acid such as N/10 HCl or 70 per cent ethanol just before use. While this is not generally required for phytoplankton work, it is recommended to give valid bacteriological samples (Sorokin, 1971a; Sieburth et al., 1976).

The volume of the sampler is a second consideration. The minimum volume of sample from which cells must be concentrated to give reasonable data (Chapter **7.1.2**) varies from a few millilitres or less in rich coastal or estuarine environments to a litre or more in oligotrophic regions. Also, it is usually desirable to preserve additional material for systematic work or for duplicate counts, and additional water may be required for concomitant chemical or biological work. Water samplers are available in sizes from less than one to thousands of litres. Most laboratories have found 1- to 3-litre bottles adequate, although 5- and 30-litre bottles are becoming standard on United States research vessels. The larger samplers become difficult to handle, especially when filled, and may require special facilities. The statistical implications of different initial sample volumes are discussed in Chapters **2.1** and **5.1**.

One desirable feature absent from standard samplers is a mechanism for mixing the samples before the removal of a subsample in order to reduce heterogeneity of particulate material, increasing the probability of a representative subsample. Small samplers may be agitated by hand (although not if there are attached thermometers), but larger samplers present a real problem.

A lower drain valve and an upper air vent are provided with most designs. These are convenient for drawing samples, but the valves tend to be susceptible to damage, and leaking valves are a frequent complaint with many models.

Many water samplers are designed to be used serially on the wire, providing simultaneous samples at several depths. This can represent considerable savings in time and effort over a single bottle lowered repeatedly to different depths. However, a multiple-bottle cast usually necessitates use of a winch and racking facilities for the bottles, which are standard on oceanographic vessels but may not be available on a small skiff.

Most samplers are closed mechanically by means of a brass messenger slid down the wire. When bottles are used in a series, the closure of one bottle releases a messenger below it which travels down the wire and trips the next bottle. Electronic signals (Niskin, 1968) and pressure-sensitive devices (Spilhaus and Miller, 1948) are also used, while a few bottles are closed on contact with the bottom. Bottles which are tripped mechanically must have a minimum separation distance between adjacent bottles (usually 3 to 5 m) in order for the messenger to develop sufficient momentum to trip the succeeding bottle. This may be a limitation on the investigation of fine-scale vertical stratification with standard bottles.

Direct determination of the actual depth of sampling is highly desirable. This allows immediate detection of malfunctions and aids the interpretation of seemingly aberrant samples. This is most frequently accomplished by using a set of protected and unprotected reversing thermometers (Sverdrup *et al.*, 1942) which provide information on the sample depth as well as accessory data on the *in situ* temperature. Non-reversing water samplers may be equipped with reversing thermometer racks (Niskin, 1964). Although initial cost and subsequent maintenance of a complete set of high-performance thermometers may be outside the capabilities of a small laboratory, use of thermometers on at least the lowest sampler is recommended as a check on the success of the cast.

Closure of the samplers may also be monitored electronically, acoustically or, in many cases, directly by feeling the vibration transmitted through the wire.

In the absence of direct determination, sample depth must be estimated from the amount of wire and the wire angle, using the relationship

$$\text{true depth} = (W)(\cos \theta),$$

where W is the amount of wire out, below the sea surface, and θ is the angle of the wire with the vertical, measured at the time the bottles are tripped (messenger time). The relation between true depth and wire out for selected values of θ is presented in Figure 8. The relationship for other values is easily calculated with the cosine scale on a slide rule. The author has compared depths determined thermometrically and estimated from wire angles (unpublished data) for 130 samples between 50 and 200 metres depth and with wire angles up to $26°$. The discrepancy averaged 1·5 per cent of the thermometric depth, with the wire-angle estimate tending to overestimate the thermometric depth. This is probably sufficient precision for most field studies.

COMMERCIAL AND HOME-MADE SOURCES

Most, if not all, of the numerous commercial water samplers, available from most oceanic and limnological supply firms, are satisfactory for the collection

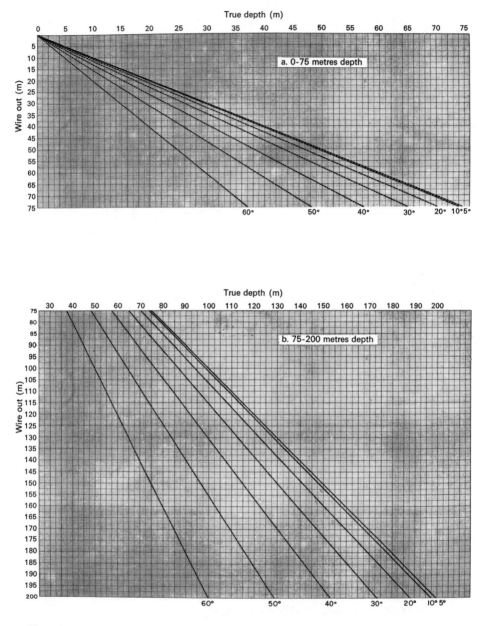

Figure 8
The relationship between metres of wire out from the surface to the sample and the true depth
of the sample as determined from the relationship: true depth = wire out (cos θ), where θ is the
wire angle.

of phytoplankton samples. A list of some of the better known firms is given in Table 1. It must be kept in mind that this list is both incomplete and unstable. Advertisements in technical journals may provide more current information.

At the time of writing (1977), most commercial samplers range in price from U.S.$100 to several thousand dollars per sampler. If a well-equipped machine shop is available, suitable samplers may be built for less. Designs such as the ZoBell J-Z bacteriological sampler (ZoBell, 1941; Sieburth, 1963; Sieburth et al., 1963) and the Van Dorn bottle (Van Dorn, 1957) are particularly amenable to construction and modification.

TABLE 1. Some commercial sources[1] of water samplers

Country	Firm[1]	Sampler designs[2] offered
Denmark	Laboratoire Océanographique	1, 10
France	Mécabolier	1
	Nereïdes	1, 5, 6
Federal Republic of Germany	Hydro-Bios Apparatebau	1, 2, 3, 5, 6, 10
Japan	Fuyo Sangyo	11
	Rigosha	11
	Tsurumi-Seiki	1, 3, 5, 6, 7
Norway	Bergen Nautik	1
Sweden	Machinator	3, 9, 10
United Kingdom	Institute of Oceanographic Sciences	2
United States	General Oceanics	2, 4, 5, 8, 10
	Hydro Products	2
	InterOcean Systems	1, 2
	Kahl Scientific Instrument	1, 2, 4, 5, 10
	Wildlife Supply	1, 2, 3, 5, 6, 10

1. Addresses of manufacturers are listed at the end of the manual.
2. Sampler designs: (1) reversing; (2) non-reversing, used serially; (3) non-reversing, used singly; (4) sterile samplers; (5) large-volume (>25 litres); (6) horizontal samplers; (7) under-way samplers; (8) electronic samplers; (9) automatic samplers; (10) other; (11) complete catalogue not examined.

There are useful samplers which may be easily constructed from materials at hand. The Meyer sampler, or 'snatch bottle' (Aabye Jensen and Steemann Nielsen, 1953; Doty and Oguri, 1958) consists of a stoppered glass bottle in a harness or frame which is weighted at the bottom to keep the bottle upright (Fig. 7d). The empty bottle is lowered to depth by means of a line attached to the stopper. A quick jerk removes the stopper, allowing the bottle to fill, while a safety line to the bottle prevents it from sinking and allows retrieval. Use of this sampler is restricted to depths of 15 m or less; at greater depths, hydrostatic pressure prevents removal of the stopper. Since the bottle is retrieved 'open', a narrow neck is important to minimize contamination.

A variant of this design facilitates collection of surface samples at some distance away from the ship or pier, minimizing contamination from these sources. A weighted, unstoppered bottle is attached to a surface float, in turn fastened to a retrieval line (Fig. 7e). The float and bottle are tossed free of the vessel and the bottle allowed to fill before recovery (Sheldon et al., 1973).

A sample integrated through the upper 5 m or less is easily obtained with a length of garden hose or flexible tubing which is weighted on one end (Fig. 7f) (Lund, 1949). The weighted end is lowered to the desired depth, 'coring' the water. The surface end is then pinched off or plugged and the lower end raised to the surface by means of a line attached near that end. The volume of sample can be regulated to some extent by means of the tube diameter. A 2·5-cm tube is a convenient diameter (F. J. Taylor, pers. comm.), and this gives a 5-m integrated sample in excess of 2 litres.

Additional water-sampler designs of simple construction are undoubtedly awaiting invention.

3.2 Pump sampling

John R. Beers[1]

The use of pumping systems for phytoplankton sampling can be traced back to Hensen (1887) in the earliest days of quantitative plankton study. It should be noted, however, that Hensen relied mainly on net sampling for his study materials and that it was a decade or so later that the use of pumps was advocated to overcome the inadequacy of towed nets to sample the complete spectrum of phytoplankton. In the intervening years, references in the literature to the application of pumps for collecting plankton samples are found regularly, albeit perhaps not commonly. Aron (1958, 1962) provided a synopsis of the use of pumps, including those in zooplankton study, through the early 1960s. Interesting accounts of the early use of pumps are given in Kofoid (1897) and Dakin (1908). Lisitsyn (1962) provides a thoughtful discussion of the use of pumping systems for obtaining large-volume seawater samples for various purposes, including the study of plankton. In addition, considerable detailed technical information is given on some types of pumps and accounts of first-hand experience with several systems are provided.

In recent years, pumping as a means of continuous sampling for studying biological (e.g. chlorophyll *a* by *in vivo* fluorometry; Lorenzen, 1966) and chemical (e.g. inorganic nutrients, Strickland *et al.*, 1970; Strickland and Parsons, 1972) parameters of the pelagic environment has become widespread. It is not unusual for samples for the study of phytoplankton population abundance and taxonomic composition to be taken from pumped water (e.g. Reid *et al.*, 1970; Beers *et al.*, 1971; Mulford, 1972; and Kiefer, 1973).

CHARACTERISTICS OF PUMP SAMPLING

Phytoplankton samples are collected using water-bottles (see Chapter **3.1**) and nets (see Chapter **3.3**) as well as pumps. As with many aspects of plankton study, there are both advantages and drawbacks to pumping relative to the other means of sampling and these must be considered when deciding upon the sampling procedures to be used for any particular purpose.

Advantages

1. The entire size spectrum of the phytoplankton can be sampled from the same source with a pump. The water, as it is being pumped, can be divided between

1. This work was supported by the Biological Oceanography Program of the Oceanography Section, National Science Foundation (United States), Grant OCE71–00306 A02.

A

Figure 9
Two examples of concentrators that can be used for sampling the larger phytoplankton (A, original; B, from Beers et al., 1967).
(A) A pair of 35-μm-mesh nets of an inverted-pyramid configuration used in sampling the large plastic 'controlled experimental ecosystem' enclosures (1,300 m^3) of the Controlled Ecosystem Pollution Experiment (CEPEX) conducted in Saanich Inlet, British Columbia (ref. Beers et al., 1977). All pipe is PVC. (1) Net of Nitex nylon with PVC rods used to support the corners of the pyramid-shaped design; (2) water meter (plastic-lined); (3) discharge pipe with numerous exit ports for gentle, uniform flow of pumped water on to the nets; (4) ball valve at the net cod end, which allows for quick release of the concentrated sample; (5) receptacle, into which a preserving bottle fits; the sample material is washed quantitatively into the bottle with filtered seawater (under pressure) directed at the inside of the net; (6) valve to direct flow of pumped water to right or left net; (7) valve to allow removal of a part of the pumped water before it enters the concentrator; and (8) valve to direct flow to bypass completely the concentrating unit.

(B) The concentrator described by Beers et al. (1967) and used by Beers and Stewart (e.g. 1969, 1971) for sampling various size classes of the microzooplankton and phytoplankton. (1) Circular acrylic plastic unit; (2) Nitex nylon filter cloth; (3) metal base section; (4) hose from 10·2 cm outflow valve to ship's scuppers; (5) water meter; and (6) concentrator bypass valve. Filter cloths commonly used were, in descending order of increasing fineness, of 363, 202, 102 and 35 μm mesh. A small valved exit-port (not shown) in the base of the unit is used to sample the water passing the finest-mesh cloth. The pumped water enters the concentrator at the top and flows by gravity through the series of filter cloths. The cloths provide barriers which slow the water but still permit effectual flow, providing proper sorting of the organisms into size classes. The filter cloths, mounted on flexible 1·6-mm acrylic rings with 3·2-mm neoprene gaskets on either side, are removed from the unit to clear the sample by releasing the several 'past-centre' clamps that have provided a watertight seal during pumping. The filter cloths are placed in large-volume receptacles where the concentrated material may be washed off using a stream of filtered seawater directed on to the face of the cloth at an oblique angle. Volumes of less than 1 litre were generally adequate to clear the sample from even the finest-mesh filter.

42

B

unconcentrated and concentrated fractions in whatever proportions are appropriate. Samples for organisms at the lower end of the spectrum are generally not concentrated and are treated similarly to those taken from water-bottles. The larger, usually less abundant forms can be concentrated on netting of any desired mesh. Various forms of deck-mounted concentrators can be designed and, if desired, may incorporate several nets of different mesh dimensions (Fig. 9). With the nets arranged sequentially in order of increasing fineness it is possible to obtain samples of various size classes of organisms. Organisms passing the finest mesh net can be sampled by collecting a constant fraction of the filtrate throughout the sampling interval.

Gentle filtration of the pumped water reduces the possibility of damage to organisms and can be effected by submerging the concentrator net in a tank of water. However, this may not be necessary with fine-mesh nets since, depending

43

upon the pumping rate, etc., a layer of pumped water may build up on the net and provide a cushion for further flow.

2. Since concentration of the pumped material using fine-mesh netting is done on deck, clogging problems can be seen and measures taken to relieve them. This is a particularly important advantage of pumping over towed nets when sampling the phytoplankton in waters where there is an equal or even greater abundance of non-living seston material in the same size range.

3. Pump systems can be good quantitative samplers. Pumped-water volumes can be measured accurately with a water meter or in a calibrated container. It is important, however, that each pumping system be tested relative to water-bottle and/or net samples in order to determine its particular quantitative sampling characteristics.

4. The deployment of pump systems is flexible and allows one to obtain samples from point sources or to integrate over spatial intervals (either horizontal, vertical, or oblique). Hence, pump samples can be relatively discrete or can average across the levels of small-scale patchiness of the phytoplankters. The discreteness of 'point' samples will depend upon the ship's movement (i.e. motion due to sea state, surface currents, wind, etc.) during the time of sampling.

With both 'integrated' and 'point' sampling, the actual collection of the samples on deck must be accurately coordinated with the time required for the water to travel through the hose from the point of intake to discharge. While this 'delay' time can be calculated from a knowledge of the hose diameter and length, the pump rate, etc., it is best determined empirically by measuring, with a stop-watch, the time required for dyestuff introduced into the intake of the system to pass to the point of discharge.

Various size samples can be collected with pumping systems dependent upon the capacity of the pump used and the time of pumping. Volumes up to hundreds of litres per minute are practical to pump and process. Hence, pump systems can be used to collect large amounts of materials such as may be needed for chemical measurements, etc. Large-volume water-bottles are heavy and cumbersome and can be difficult to remove from the hydrowire when full.

Disadvantages

1. Some physical damage and/or adverse physiological effects can be anticipated, especially in the case of delicate organisms (e.g. athecate dinoflagellates), and chains of cells may be broken. However, by the proper choice of gear this can be kept minimal to the point that some pump systems are currently used for obtaining samples for primary productivity measurements and other purposes which require that the physiological condition of the organisms is not affected by the collecting procedure (e.g. Controlled Ecosystem Pollution Experiment; Thomas *et al.*, submitted for publication).

2. Avoidance of the intake by some of the more highly mobile forms may be possible, but this presumably is a less serious problem with phytoplankters than with zooplankters which, in general, have better locomotor capabilities.

3. Because transport of water through the tubing may not be uniform, owing to factors such as frictional resistance, there may be some 'smearing' of organisms taken at various depths when collecting from different strata as in a vertical or horizontal series. When determining the 'delay' time (see above), a comparison of the time of discharge of the dye relative to the time of uptake provides a rough indication of the smearing effect along the length of hose.

Another source of possible contamination of samples includes the growth of organisms inside the tubing between periods of use, since it is often difficult to clear the tubing of all water. With regard to the photosynthetic forms, this possibility can be reduced by using opaque tubing; for marine sampling, filling the tubing with freshwater when not in use should also reduce contamination.

4. With any given pump system the amount of hose that can be used and, hence, the depth of sampling possible, is restricted to the length for which the pump capacity is sufficient to overcome the frictional resistance of the hosing and to provide the necessary head (i.e. the lift from the water surface to the highest point of the system on deck).

5. In many cases, a pump system may be more expensive than nets or water-bottles.

PUMPING-SYSTEM COMPONENTS

Pump systems, in addition to the pump itself, include tubing (i.e. hose, flexible pipe) for conveying the water to and from the pump. The tubing on the intake side of the pump is the suction line while that on the other is the discharge line. Hardware, such as valves for regulating the water flow and meters for determining the volumes sampled, may be incorporated into the system. Many components of a typical system are currently available in plastic which has the dual advantage of being non-corrosive in seawater and generally less, if at all, toxic to the organisms than many other materials. Other components can be plastic-lined or coated so that the samples never come in contact with metal. It cannot be emphasized too strongly that each pumping system developed should be thoroughly tested for toxicity if it is to be used for live-organism studies. If any harmful effects are noted the system can be examined part by part for the toxic component(s).

The pumping system will also include a sample collecting unit. This can be custom-designed to meet the demands of its specific use. In the case of phytoplankton sampling, the collector may be relatively simple, especially when

45

concentration of the sample materials is not necessary, and may consist of nothing more than a receptacle (e.g. bottle, carboy) for the pumped water. However, when sampling the total size spectrum of the phytoplankton it is often desirable to concentrate, by net filtering, the larger, less abundant forms from a considerably greater volume of seawater than is required for study of the smaller, more abundant taxa (Fig. 9). A wide variety of plastics that can be machined into various configurations as desired are available for fabricating concentrators. Fine-mesh nylon netting (e.g. Nitex; Tobler, Ernst and Traber Inc.) ranging upward in mesh from approximately 10 μm (side dimension) can be secured to the plastic with a variety of commercially available bonding agents to provide a filtering surface. The use of clear plastics allows for visual monitoring of the filter during sample concentration.

Pumps

Various types of pumps have been used in plankton work (Aron, 1958, 1962). In contrast to zooplankton studies, phytoplankton work generally requires only a small volume of water to sample adequately the populations and, hence, pumps of relatively low capacity usually are sufficient. Centrifugal pumps have been among the most commonly employed (e.g. Cassie, 1958; O'Connell and Leong, 1963; Beers *et al.*, 1967; Kuwahara *et al.*, 1973). In these, the seawater enters the pump near the axis of an impeller and is thrust radially outward into the casing. The energy imparted is dependent upon such factors as the design of the impeller and casing, the rotational speed of the impeller, and the number of impeller stages employed. Most other types of pumps used in plankton study are classed as positive displacement units, either moving the seawater through the action of gears, lobes, vanes, screws or flexible impellers (i.e. rotary pumps) or displacing it by changing the volume of the pump such as in piston and diaphragm units (i.e. reciprocating pumps). Other 'pumping' principles for moving liquids that have been applied to sampling plankton include the 'air-lift' pump in which compressed air is used to force the movement of the seawater (Bernhard and Rampi, 1966). Recently Lenz (1972) developed a pump system for use at relatively shallow depths which simply draws the water by use of a vacuum, without its having to pass through any moving parts.

Because of the manner in which the water is propelled, centrifugal pumps are potentially highly damaging to phytoplankters. Diaphragm pumps have been found to result in minimal damage to the plankton. Furthermore, diaphragm pumps as well as most other positive displacement pumps, in contrast with centrifugal units, do not require priming, i.e. air in the suction line or pump does not have to be displaced before pumping can commence.

The position of the pump in plankton systems may be near the intake, as in the case of submersible centrifugal pumps, which thus 'push' the water through the length of tubing up to the surface, or on deck with water being 'pulled' from the desired depth. Energy to operate most conventional pumps may be electrical or derived from a petrol-driven motor. A drawback to the use of submerged pumps is that the need to get power to them generally requires underwater electrical lines which are subject to breakage and can be dangerous. On-deck, hand-operated pumps can be used in cases where other power is not available and when only relatively small volumes of water are needed.

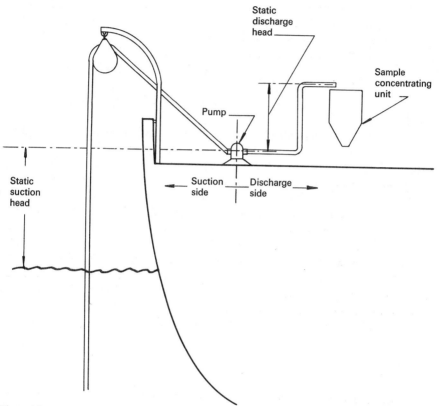

Figure 10
A simple on-deck pumping system for plankton sampling showing some of the components of the total dynamic head.

Efficient use of a pumping system requires an understanding of the various static and dynamic forces operating in the system. Figure 10 shows the basic orientation of a pumping system. The pump must have sufficient suction lift and be able to develop the 'total dynamic head' required for the flow rate desired in the system. The 'total dynamic head' is the sum of the 'dynamic suction head' (i.e. 'static suction head' and frictional head loss in the suction line), 'dynamic discharge head' (i.e. 'static discharge head' and frictional head loss in the discharge line), and the 'velocity head'.

Frictional head loss in both the suction and discharge lines results from restriction of flow and is a function of the size, length, inner smoothness and configuration of the tubing, the types and numbers of fittings including such components as water meters and filters, and the velocity of the seawater being moved.

The 'static suction head' (Fig. 10), is the vertical distance from the free level of the water in the tubing on the intake side and the centre of the pump. In the case of marine plankton systems with an on-deck pump, the free level would be approximately at the ocean's surface and a negative head (i.e. a lift) would exist. With a pump which does not require priming, the initial pull must be sufficient to lift the water to the highest point on the suction side of the pump. This may be above the centre of the pump. Once the prime has been obtained, however, the vertical distance over which lift must be effective is only that between the free-water level and the centre line of the pump.

The 'static discharge head' (Fig. 10) is the vertical distance from the centre of the pump and the point of discharge from the system. If the actual position of discharge of the water is below the highest point on the discharge side of the pump, the initial push must be sufficient to propel the water to its maximum height.

'Velocity head' is a function of the discharge velocity and is usually negligible in plankton pumping systems.

In operating a pump, care should be taken to avoid cavitation since this not only causes rapid wear in the pump but also increases the risk of damage and injury to the pumped organisms. Cavitation is the vibration and noise occurring when gas-filled spaces in the water collapse when carried into higher pressure areas of the pump. It can occur when the pump is not completely filled with water as the result of air leaks into the system, e.g. through the pump seals.

Valves

While there is a wide variety of types of valves (Holland and Chapman, 1966) that can be used in systems pumping liquids, those which have straight-through flow such as ball and gate valves are recommended. When completely opened they present little, if any, restriction to the passage of water and, hence, would increase frictional loss only minimally.

Tubing

A wide range of plastic tubing is available and is preferred over rubber and other natural products for its smooth, non-porous surface which gives relatively little frictional resistance to water flow. In addition, most synthetic tubing is relatively lightweight, withstands wear and ages well under seawater use. In general, it is preferable to use the largest diameter tubing practicable to deliver the desired volume of seawater per unit time since a slower flow rate through a larger tube will have less total frictional loss than a higher flow rate through a smaller tube.

In deploying the tubing care must be taken to avoid its constriction. A particular point of potential trouble is where it goes over a fairlead or through a sheave. The sheave must be of an appropriate diameter relative to the rigidity of the tube wall to avoid any blockage of water flow by compression. Tubing can be reinforced, for example with nylon, which impedes it from collapsing when suction is applied or when curved (i.e. bent).

A convenient means of deploying tubing from vessels equipped with a hydrographic winch is by attachment to the hydrowire. By this means the depth of the intake can be known accurately from the meter wheel readings and the hydroweight can be used to maintain the tubing at depth. Weights attached directly to the tubing sufficient to maintain it at depth may cause stretching with a resulting undesirable decrease in diameter. When using the hydrowire to deploy tubing, fasteners or clamps can be used at appropriate intervals along the length of the tubing in order to prevent 'bowing' due to subsurface currents. Where continuous profiling is desired, the fasteners should be of a quick connect/disconnect type.

If a long length of tubing is needed it is desirable to have it stored on a winch, the simplest form of which is perhaps a hand-operated windlass. In addition to being a deterrent to accidental kinking or compression of the tubing, a winch also allows for a very uniform deployment rate. If pumping is to continue during deployment from a winch, the tubing outlet must be equipped with a rotary water joint. Also, a slip-ring assembly is required if electrical leads are being deployed from a winch. The rotary water joint and slip-ring assembly circumvent, respectively, the problems of twisting of the water discharge tubing and electrical input leads as the winch rotates around its axis.

3.3 Nets

Karl Tangen

Plankton nets have been widely used as sampling devices in phytoplankton investigations. The advantage of nets is the ease with which large volumes of water can be filtered and organisms concentrated. The main disadvantage is the distorted species composition shown by net samples. Only a few of the cells entering the net are caught by the gauze; one can assume that only about 10 per cent of all cells are retained by nets with a mesh-size of 40 µm (Margalef, 1969c). Nets with very fine meshes (e.g. 5 or 10 µm) catch small cells more effectively than coarse nets. However, a quantitatively important component of the phytoplankton may also pass through modern nets with fine meshes (McCarthy et al., 1974; Durbin et al., 1975).

The filtering properties of a net are also determined by the species composition of the plankton. When chain-forming species (e.g. *Nitzschia delicatissima*, *Skeletonema costatum*) or species with spines or setae (e.g. *Chaetoceros* spp.) are abundant, the plankton itself may form a fine network inside the gauze. Small, solitary cells, which in other cases would have passed through the net, might then be retained.

Because of their selective and non-predictable filtering properties, nets should not be employed in quantitative phytoplankton sampling. Methods for the evaluation of the volume of water filtered through a plankton net (e.g. use of flow-meters), which have been developed primarily for quantitative zooplankton sampling, are of little or no value in phytoplankton investigations and will not be discussed here. Quantitative zooplankton nets have been thoroughly discussed elsewhere (see Anon., 1968b).

Sampling with nets may, however, be useful in providing material for qualitative purposes which may be combined with quantitative methods. Gauzes of the same mesh size as those used in plankton nets may also be employed as screens or filters in phytoplankton size-fractionation procedures (see McCarthy et al. (1974), Durbin et al. (1975) and Chapter **6.2.1** herein).

CONSTRUCTION AND PROPERTIES OF NETS

Net types

A great variety of plankton nets have been designed, but only a few have achieved general use in phytoplankton sampling. The most common, which might be called the *standard net* (Fig. 11a), consists of a cone-shaped gauze bag equipped with a metal or plastic ring at the wider end and closed at the narrow end by a plankton-collecting vessel ('bucket'). The open end of the net (the net mouth) is

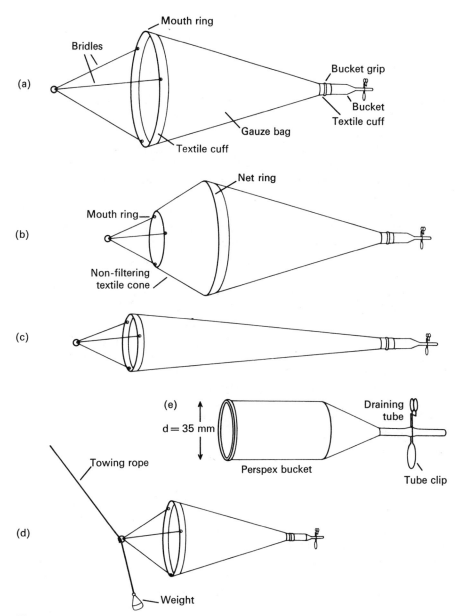

Figure 11
Nets with accessories: (a) Standard net. The length of standard nets is normally 2 to 3 times the mouth diameter. (b) Fine-mesh net with reduced mouth diameter. A tapering non-filtering textile sleeve is inserted between the large net ring and the smaller mouth ring. (c) Extra long, fine-mesh standard net. (d) Standard net attached to the towing rope, with the weight in front of the mouth. (e) Plankton collecting bucket made of clear perspex material. Diameter of the bucket is 30 to 100 mm (here 35 mm); length of the cylindrical part is 50 to 200 mm (here 65 mm). The bucket is attached to the net tail by textile tape or a specially made metal grip.

51

attached to the towing line by a number, usually three, of rope bridles from the mouth ring. The front and tail part of the net are reinforced with non-porous textile cuffs.

The bucket is either a glass jar or a specially designed vessel which is tied to the net by a band or fastened by use of a special grip (e.g. a hose clamp). The bucket should be detachable in order to facilitate the handling of samples and washing of the net after use. A perspex bucket with a rubber or silicon draining tube (Fig. 11e) is practical for use with fine-mesh nets, particularly in areas where clogging often occurs. When the haul is finished, the sample is tapped directly into a bottle through the draining tube. Permanently attached buckets may hinder effective washing, and they often cause harmful abrasion during the washing procedure.

In fine-mesh nets with low filtering efficiency the standard design is modi-fied; the inflow of water is reduced by reducing the diameter of the mouth ring and inserting a tapered non-porous textile cone in the front of the net (Fig. 11b). By this modification the ratio between the filtering area and the mouth area is increased. This can also be achieved by making the filtering cone (gauze bag) particularly long, as compared with the mouth diameter (Fig. 11c). One or both of these modifications are recommended when using gauze with meshes of less than 25 μm. Clogging is reduced by increasing the filtering area without increasing the mouth area; for moderate clogging a doubling of the gauze area will increase the time of effective sampling by a factor of approximately six (Tranter and Smith, 1968).

Nets especially constructed for vertical hauls may have a weight attach-ment device behind the tail. The shape of the net may change if the weight is attached to the filtering cone, and the attachment device (e.g. a metal ring or hook) is therefore connected with the net mouth ring or the bridles by a number (usually three) of steel or nylon wires.

Gauze types

The gauzes used in plankton nets have been developed and produced mainly for industrial purposes (sifting or screening of various products). Bolting silk has been widely used, but this may rot and shrink when exposed to seawater (Wiborg, 1948). Modern gauzes made of synthetic filaments are more resistant to chemi-cals, are stronger, and the meshes are more stable; they should therefore replace silk gauzes in plankton nets. Various synthetic gauzes made of polyamids (e.g. nylon and perlon) or polyester are satisfactory for use in plankton nets (Table 2). Monofilament nylon and polyester gauzes are manufactured in a wide range of sizes, from coarse meshes with apertures larger than 5 mm to ultra-fine meshes down to 0·5 μm. Apart from their low flexibility, metal gauzes also appear to meet all requirements; however, some of them are toxic and thus not suitable for collecting live material.

Sampling with small (mouth diameter 15 cm; length 110 cm) fine-mesh (5 or 10 µm) monofilament nylon nets in the Oslo Fjord and other inshore waters has provided phytoplankton material of high quality in which the amount of small species (e.g. *Emiliania huxleyi* and small flagellates) is markedly larger than in samples collected with coarser nets (35 or 60 µm). Comparison with results from quantitative water samples collected simultaneously (counted on an inverted microscope) has, however, made it quite clear that a large number of small cells also escape from these fine-mesh nets. Ultra-fine gauzes with mesh-size of 1 µm or 0·5 µm are unsuitable for even qualitative plankton sampling because of their very poor filtering properties.

TABLE 2. Some manufacturers[1] of synthetic gauzes which are suitable for plankton nets. Only the lower size of mesh opening is indicated

Country	Firm[1]	Gauzes offered
France	Tripette & Renaud	Nylon Blutex down to 25 µm
		Polyester Monocron down to 28 µm
		Also fine-mesh steel screens
	Union Générale des Gazes à Bluter	Polyamide Nytrel-Ti down to 1 µm
Federal Republic of Germany	Vereinigte Seidenwebereien	Nylon Monodur down to 15 µm
		Polyester Monodur down to 3 µm
		Also fine-mesh steel screens
Japan	Nippon-Nakano Bolting Cloth	Nylon Fuji
		Polyethylene Pylene
Switzerland	Schweiz. Seidengazefabrik	Nylon Nytal and Estal down to 5 µm
	Züricher Beuteltuchfabrik	Nylon Scrynel down to 22 µm
		Polyester Scrynel down to 1 µm
		Monyl?
United Kingdom	Henry Simon	Nylon Simon (St Martins) down to 5 µm
	Plastok	Polyamide Nytrel-Ti down to 1 µm
		Polyester Mono-E down to 28 µm
		Nylon Blunyl?
		Also fine-mesh metal screens
United States	Newark Wire Cloth Company	Polyester down to 21 µm
		Also fine-mesh steel screens
	Tobler, Ernst & Traber	Nylon Nitex

1. Full addresses are listed at the end of the manual.

The filtering efficiency of a gauze is a function of the porosity (free sifting surface), which is the ratio between the open area and the total area. Manufacturers specify the porosity as a percentage. The porosities of some Nytal monofilament nylon gauzes are shown in Table 3. Fine-mesh gauzes have very low porosities, e.g. only 1 per cent in a 5 µm gauze. Monodur polyester gauzes with apertures of less than 20 µm have approximately the same porosities as the corresponding Nytal gauzes.

Gauzes which are to be used in plankton nets should have square meshes of constant size. Some types of weave give meshes of varying shape and size (Fig. 12), and relatively large deviations from the specifications given by the manufacturers may occur, especially in fine-mesh gauzes. Gauzes should there-

53

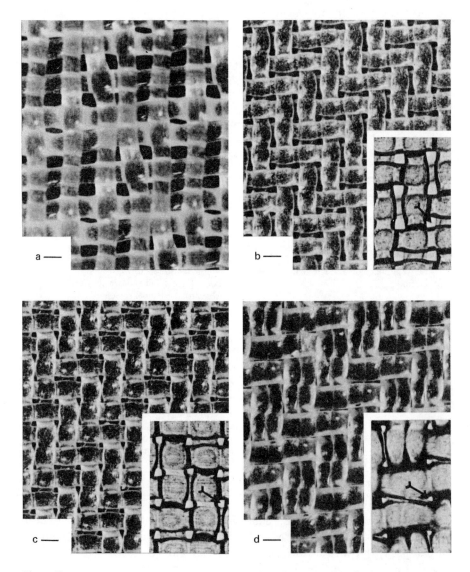

Figure 12
Micrographs of some fine-mesh Nytal monofilament nylon gauzes: (a), 35 μm; (b) 20 μm;
(c) 10 μm; (d) 5 μm. Note the variation in size and shape of the apertures (meshes) in the 35-μm
gauze. Insertions of slightly higher magnification show the apertures (arrows) of 20-μm, 10-μm
and 5-μm gauzes. Scales: — = 50 μm.

TABLE 3. Porosities of some Nytal monofilament nylon gauzes (specifications from the manufacturer, Schweiz. Seidengazefabrik, Switzerland)

Mesh size (μm)	Gauze number	Gauze type	Porosity (free sifting surface) (%)
100	13P	431	49
80	17P	390	42
60	HC	405	44
47	P	395	37
41	P	319	33
30	P	414	25
25	P	397	21
20	P	416	16
20	HD	397 kal.	14
15	HD	397 kal.	8
10	HD	397 kal.	4
5	HD	416 kal.	1

fore be examined under a microscope in order to determine the size and uniformity of the meshes; this should also be repeated periodically throughout use of the net.

Properties of synthetic gauzes

The material in gauze filaments should resist all kinds of exposure during sampling, cleaning and storage.

Depending on other conditions the synthetic materials nylon and polyester are unaffected by temperatures up to $100°C$; the filaments also retain their strength at temperatures below $0°C$. Both materials are non-toxic and resistant to many chemicals including chlorinated hydrocarbons and ketones (e.g. acetone). Polyester is more resistant than nylon to acids, whereas nylon is more resistant to alkalis. Both materials are dissolved by phenol and some other organic compounds. Sunlight and strong bleaching as well as oxidizing agents degrade nylon filaments and make the net stiffer, while polyester gauze is almost unharmed by these agents. Plankton nets made of nylon gauze should therefore be stored in a dark place. Heron (1968) has summarized some of the chemical properties of materials used in plankton gauzes.

SAMPLING

The simplest sampling arrangement is a standard net with a weight attached to the end of the towing cable (Fig. 11d). With this arrangement net hauls may be taken near the surface or at greater depths. The collecting bucket receives most of the plankton as the net is towed, but some cells remain on the gauze and must be washed into the bucket after the haul is ended.

A plankton sample is obtained from a *particular depth layer* by towing the net horizontally while the weight holds the net at the selected depth; a depth recorder may be used to monitor the depth of the net while towing. The Clarke–Bumpus net (Clarke and Bumpus, 1940; Paquette and Frolander, 1957) and other zooplankton nets (see McGowan and Brown, 1966) have an opening and closing mechanism especially constructed for sampling at selected depths. Such nets might have some applications in phytoplankton investigations if they were equipped with fine-mesh gauze. Phytoplankton may also be concentrated from a selected depth by pumping water from the depth (see Chapter **3.2**) and filtering the water through a net on board ship. The net should then be hanging with the filtering cone submerged in a water tank while the straining procedure is going on; this precaution reduces the pressure on the net, and organisms which should be retained are not forced through the net or broken.

In a vertical haul the entire *water column,* or part of it, is sampled. The net is lowered from the anchored or drifting ship to the desired depth and slowly hauled up again. If the weight is attached to the towing cable in front of the net (Fig. 11d), the net is also filtering water when lowered; nets with weight attachment behind the tail collect plankton only when they are hauled.

In waters poor in phytoplankton the material collected in a single vertical haul may be insufficient. A larger catch from the same water layer is obtained from an *oblique haul.* In this case the net is lowered when the ship is moving and then hauled again while being towed, so that the net traverses an oblique course downward and upward. The advantage of oblique hauls over simple vertical hauls is that more water is filtered from the same water layer.

The *towing speed* should not exceed $1 \text{ m} \cdot \text{s}^{-1}$ (2 knots). If the towing speed is too high, the pressure stress on the net may in extreme cases be so great that the gauze breaks. When using nets with fine meshes (less than 20 µm) even lower speeds, preferably below $0.3 \text{ m} \cdot \text{s}^{-1}$ (0.5 knot), are recommended in order to reduce clogging to a minimum. Nets which are equipped with a non-filtering cone in the front (Fig. 11b) or long nets (Fig. 11c) may be towed at somewhat higher speeds than the standard net of the same porosity; the low net-mouth/gauze-area ratio in these nets results in reduced pressure on the gauze.

After the tow, phytoplankton material adhering to the net is washed down to the tail and collected in the bucket by spraying the outer or inner net surface with seawater while the net is hanging with the mouth upwards. Filtered water should be used after a deep-water tow (e.g. when using an opening–closing net) to eliminate contamination with surface organisms from the spray water. Unfiltered surface water is sufficient when collecting the plankton material after a surface tow. Freshwater may destroy the organisms and should therefore not be used.

In productive areas, especially when gelatinous species are abundant, *clogging* often hinders effective sampling with nets. In nets with meshes finer than 10 µm clogging regularly occurs near the net tail. Excess water inside the net may then be filtered off by lifting the net tail and bucket, so that water and phytoplankton material flow to the unclogged gauze near the net mouth; the organisms may then be collected in the bucket in the usual manner.

Some phytoplankton material (e.g. diatom frustules, dinoflagellate thecal plates or whole thecae) might still be left in the gauze, the cloth cuffs or the seams after cleaning. Some of this may then contaminate the next or later samples collected with the same net. In order to reduce uncertainties about geographical distribution of species, the use of one particular net should be restricted to a limited geographical area.

WASHING THE NET

The net should be washed carefully as soon as possible after the tow to remove organisms and salts from the gauze and textiles. In order to reduce the mechanical damage to the gauze during the cleaning procedure, the plankton collecting bucket and preferably the mouth ring are detached prior to washing. Nets made of synthetic materials are washed according to the recommendations of the manufacturer. At least three steps are necessary:

1. Pre-washing or rinsing with freshwater to remove the salts.

2. Washing with a cleaning agent dissolved in freshwater. For cleaning nylon, one manufacturer (Henry Simon Ltd) recommends the use of either (a) warm soap or detergent solution, or (b) alkaline solutions up to 15 per cent strength. Polyester should be washed in diluted acids instead of alkaline solutions. In most cases soap will adequately clean the net, but, after extensive clogging or contamination with organic compounds (e.g. resistant oils) other agents (e.g. benzene or acetone) may be necessary.

3. Rinsing with fresh water to remove the cleaning agent.

After washing, the net should be dried in the air and stored in a dark, cool place. Care should be taken to prevent nylon nets from exposure to sunlight when drying.

CARE OF PLANKTON NETS

The use of synthetic gauzes has increased the lifetime of a plankton net, but even synthetic materials are subject to abrasion, so care should be taken to prevent nets from coming into contact with rough surfaces and sharp objects. In case of mechanical damage (e.g. small holes or slits in the gauze), the net may be mended by use of a water-resistant epoxide resin; various types of Araldite are satisfactory. Although synthetic materials used in plankton gauzes are resistant to rot, the nets should be cleaned and dried as soon as possible after use. If the net is allowed to dry prior to the routine washing procedure progressive accumulation of residual plankton and reduced filtering efficiency may result. Synthetic gauzes should not be exposed to excessive heat, since the melting point of the commonly used material (polyester and the polyamids nylon and perlon) is relatively low (c. 250°C); e.g. carelessly discarded cigarette ashes may damage nets made of these materials.

3.4   The Continuous Plankton Recorder survey

G. A. Robinson and A. R. Hiby

The Continuous Plankton Recorder (CPR) (Figs. 13 and 14) was described by Hardy (1939) and no significant alterations have been made to the sampler since then. Since 1930 the CPR has been deployed from ships of opportunity in the North Sea and later in the north Atlantic to provide a long-term synoptic survey.

The primary objective of the survey is to describe and analyse the variability of the plankton of the north Atlantic and North Sea and to interpret this variability in relation to events in fisheries and the marine and atmospheric climate. The data are also used to provide information on the biology of some important species of both phyto- and zooplankton.

SAMPLING

The CPR is towed at a depth of 10 m from merchant ships and weather ships at monthly intervals on a set of standard routes in the north Atlantic and North Sea (see Anon., Edinburgh Oceanographic Laboratory, 1973). Water enters at the front through a small aperture (1·27 cm square) and the plankton is filtered on to a moving band of bolting silk, 15·25 cm wide, with 24 meshes to the centimetre and a mesh aperture of 270 µm (the silk is specified as 'quadruple extra heavy quality grit-gauge No. 60 XXXX'). The plankton is fixed and preserved *in situ* with a weak solution of formaldehyde, which is stored inside the recorder (Fig. 14). The silk band moves through the recorder at a rate of 10·16 cm per 10 miles of tow, irrespective of the speed of the ship.

The combination of the rate of silk advance and mesh size was chosen in order to allow continuous sampling of phyto- and zooplankton over a distance of 500 miles. The selected mesh size results in low retention of phytoplankton. Measurement of chlorophyll in samples taken by the CPR and water samples taken at the same time (by pumping seawater up a tube fitted to the nose of the CPR) showed that chlorophyll retained by the CPR was, on average, only 0·8 per cent of that in the water samples. In common with all filtering systems, the species composition is distorted in favour of the larger species, and it is not possible to compare the abundance of one species with another. However, all the abundant species (including small diatoms, such as *Skeletonema costatum*) were found in both the water and CPR samples. We consider that the survey provides an index of the abundance of species which is sensitive to changes on a monthly time scale and provides the only available information by a uniform

Figure 13
The Continuous Plankton Recorder.

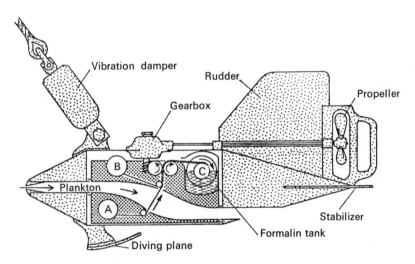

Figure 14
A diagrammatic longitudinal section of the Continuous Plankton Recorder (from Glover, 1967):
(A) filtering silk; (B) covering silk; (C) take-up spool.

method of sampling over a wide area over a long period of years. It has revealed regular and meaningful patterns of distribution and abundance from year to year (Colebrook and Robinson, 1964, 1965).

For analysis, the silk bands are cut into lengths corresponding to 10 miles of towing, representing the filtration of 3 m³ of water (assuming 100 per cent efficiency). The height of the filtering aperture is 5·08 cm, that is, half the length of each sample. As the silk moves continuously, the trailing edge of any sample clears the sampling aperture 15 miles after the leading edge enters the aperture. Thus the plankton density estimate derived from each sample is a weighted average of the plankton density along 15 miles of tow, i.e.

$$1/10 \int_{x_0}^{x_0+15} f(x)\,g(x)\,dx$$

where $f(x)$ = density at point x along the tow (nos/3 m³)

and the weighting function $g(x) = \dfrac{x - x_0}{5}$ for $x = x_0$ to $x_0 + 5$

$$= 1 \text{ for } x = x_0 + 5 \text{ to } x_0 + 10$$

$$= \frac{x_0 + 15 - x}{5} \text{ for } x = x_0 + 10 \text{ to } x_0 + 15$$

Only alternate samples are analysed, so in the present survey variations in plankton abundance on scales of less than 30 miles are not identified. The rate of silk advance and mesh size can be changed so that variations in plankton abundance on smaller scales may be studied. Derivatives of the CPR mechanisms are used in the Longhurst Hardy Plankton Recorder (Longhurst and Williams, 1976) and the Undulating Oceanographic Recorder (Bruce and Aiken, 1975) for this purpose.

COLOUR ANALYSIS

As soon as the samples are returned to the laboratory a visual assessment is made of the quantity of phytoplankton caught. The samples are assessed according to three standard arbitrary categories of greenness to each of which a numerical value has been given, based on acetone extracts of chlorophyll from a considerable number of samples in each category.

PHYTOPLANKTON ANALYSIS

The method of analysis was originated by Colebrook (1960). The abundance of each species (or group of species) is estimated by viewing a subsample of each

61

length of the silk, consisting of 20 microscope fields of 0·295 mm diameter ranged in two diagonals of ten fields. This gives a subsample covering 0·0001 of the sample. It has been found to be quicker and more reliable to record the presence or absence of the species in each field, rather than to attempt to count the cells. A species is defined as present if a given point on the cell lies within the field.

The statistic $h = -\ln(k/20)$ is used as an estimate of the mean number of cells per field, where k equals the number of empty fields observed. If no empty fields are observed (which is extremely rare) the value 4·5 is used for h. If a species is randomly distributed over the silk with a mean density of m cells per field, the probability that a species is absent from a field is e^{-m}, independently for each field, so that k has a binomial distribution $B(20, e^{-m})$. Then for k not equal to 0, h is the maximum likelihood estimator for m.

Approximate values for the expectation and variance for h, given m, may be found by expanding $h(k)$ in a Taylor series about the expectation of k and finding the expectations of the first 3 terms (see, for example, Mood et al., 1974, p. 181). Thus:

$$E(h(k)) \approx h(E(k)) + h''(E(k)).\mathrm{var}(k)/2$$

From the binomial distribution for k we have

$$E(k) = 20e^{-m} \text{ and var}(k) = 20e^{-m}(1 - e^{-m})$$

which, substituting into the expression for $E(h(k))$, gives

$$E(h|m) \approx m + \frac{e^m - 1}{40}$$

Similarly,

$$\mathrm{var}(h|m) \approx \frac{e^m - 1}{20}$$

Over the range of values of m encountered in the samples the expression for $E(h|m)$ provides a close approximation and shows that h is a consistent estimator of m, biased upwards. Counting the number of cells in each field would give an unbiased estimate of m with variance $m/20$ which is less than $\mathrm{var}(h|m)$, but these advantages must be weighed against the extra time and possible error involved in counting cells against the background of the filtering silk.

In practice, many species show non-random distribution of cells over the silk, observable as gradients of density across the silk and aggregations on a scale similar to that of the microscope field. A density estimate obtained by counting the number of cells in randomly selected fields would still be unbiased, but small-scale aggregation of cells would increase its variance considerably. The problem with the presence/absence estimate h under these conditions is that the formula derived above for its expectation is no longer valid, and it is not possible to determine its expectation without knowing the form of, and parameters for, the spatial distribution of the cells. In general, for non-random distribution of the cells, the probability distribution and hence the expectation of h will depend not only on m but also on one or more other parameters, say θ, i.e. we have $E(h|m, \theta)$. The decision to use the estimate h under these conditions then rests upon the assumption that in practice θ is itself a function of m, so that $E(h|m)$ is uniquely determined by m, and further that $E(h|m)$ is strictly increasing with m. In this case h provides an index to the value of m, rather than an estimate of that value, in the same way as the number of cells retained by the filtering silk in the CPR is assumed to provide an index of the density of a species in the sea, rather than providing an estimate of that density. The justification for these assumptions rests on the results of the CPR survey; for example, the observation of similar variations in abundance in different sea areas.

IDENTIFICATION

The level of identification of the phytoplankton in the CPR survey operated by the Institute for Marine Environmental Research is governed by expediency coupled with a consideration of the existing knowledge of a species. For instance, it is considered important to identify the three varieties of *Rhizosolenia alata* Brightwell (var. *alata*, var. *indica* and var. *inermis*) because they have contrasting distributions (see Anon., Edinburgh Oceanographic Laboratory, 1973). But as it is difficult to recognize the varieties of *R. styliformis* Brightwell, identification is not attempted. For similar reasons, diatoms of the genera *Thalassiosira* and *Chaetoceros* and dinoflagellates (except *Ceratium* spp.) are not identified to species level. Inevitably there will be some loss of information, but a balance must be kept between the objectives of a survey, the time available for analysis of samples and the loss of information that is caused by identification of some organisms to genus or family only.

A list of species identified in the CPR survey is given in Lucas (1940, 1941, 1942), Robinson (1961, 1965, 1970) and Anon. (Edinburgh Oceanographic Laboratory) 1973.

3.5 Neuston sampling

Bruce C. Parker

It has long been recognized that organisms inhabit microlayers often associated with environmental discontinuities (Naumann, 1917). However, only recently have investigators devoted major attention to sampling and investigating the microbiology and chemical composition of the microlayer ecosystem associated with the air–seawater interface (Zaitsev, 1971; Baier *et al.*, 1974; Sieburth, 1976; Sieburth *et al.*, 1976). The microlayer ecosystem includes the pleuston (from the Greek πλέω, meaning to sail), used by some investigators for organisms with special float adaptations, and the neuston (from νέω, meaning to swim) (Cheng, 1975); for simplicity, all biota in microlayers between < 1,000 μm depth and the air–water interface will be called neuston.

The variety of neuston-sampling techniques commonly used are screen samplers (Garrett, 1965; Sieburth, 1965, 1971; Duce *et al.* 1972; Sieburth *et al.*, 1976), a rotating drum with collection trough (Harvey, 1966), and a glass plate (Harvey and Burzell, 1972). No universal agreement has been reached as to the ideal neuston-sampling method. However, the sampling effectiveness, based on neuston enrichment and thickness of microlayer sampled, differs according to method: screen (250 to 100 μm microlayer), which is least efficient, rotating drum (50 to 60 μm microlayer), and the glass plate (*c.* 22 μm microlayer), which seems most efficient (Hatcher and Parker, 1974*a*, 1974*b*; Daumas *et al.*, 1976; Blanchard and Parker, 1977). Details of these and other neuston samplers in addition to modern trends in the investigation of the neustonic ecosystem will be found in Sieburth (1976), Sieburth *et al.* (1976), and Blanchard and Parker (1977).

Two major types of samplers are discussed here. Screen samplers have become the most popular and are inexpensive, lightweight and easily sterilized. A wide variety is available as will be shown, and they are capable of sampling relatively uniform (100 to 150 μm) microlayer depths. The glass-plate sampler also has low cost, light weight, easily sterilized features and samples still thinner microlayers (*c.* 22 μm); sample volume, however, is less than with screen samplers. The rotating-drum sampler (Harvey, 1966) will not be discussed, because of its high cost, bulky nature and difficulty of operation in all but perfectly calm waters.

SCREEN-SAMPLER CONSTRUCTION

1. A stainless-steel screen (1·16 mm mesh) may be framed in 6·3 mm stainless-steel rod (29 to 30 cm diam.) with rings for suspension and a drain opening (see Sieburth, 1965, fig. 1). Or:

2. A polyethylene screen (75 × 75 cm, 1·27-mm mesh) may be mounted on a plexiglas (= acrylic plastic) frame (Duce *et al.*, 1972). Or:
3. A nylon screen (70·8 × 70·8 cm, 1·0-mm mesh) may be mounted on a plexiglas frame (Sieburth, pers. comm.), as shown in Figure 15. Sieburth's screen sampler thus consists of a 'kite-frame' construction. The 70·8 cm square (= 0·5 m²) nylon (or substitute polyethylene) 1·0-mm-mesh screen is heat-treated at the edges to prevent unravelling; this feature eliminates excess water from the frame (e.g. 1, 2). Four 2·5-cm-diameter plexiglas corner grommets hold four polyethylene (or fibreglass) struts which are attached to the cross-shaped plexiglas handle. The three components of Sieburth's screen sampler (screen, struts, handle) disassemble, roll up and can be stored in a small plastic fishing-rod case for easy transport.

For all screen samplers, a nylon brush and alcohol are used to clean the screen before and between pooled samplings.

SCREEN-SAMPLING AND CONCENTRATION ESTIMATES PROCEDURE

Screens are scrubbed with a nylon brush dipped frequently in alcohol before use. Sample either by tapping the screen on the water surface or by submerging vertically and withdrawing horizontally. Both methods apparently permit fairly efficient collection of surface (*c.* 100 μm thick) microlayer with the nylon (polyethylene or steel) mesh spaces. Allow the screen to drain up to 10 seconds in a horizontal position. Now, tilt the screen over a clean (sterile) bottle with funnel (Fig. 15) and allow the screen to drain about 60 seconds. Repeat as necessary, without intermediate alcohol-sterilizing, to collect the desired microlayer volume. When collecting microneuston for culturing, it is recommended that screens be handled with sterile gloves. Also, sampling at the lee side of a ship's bow or otherwise to avoid sampling errors or microlayer interferences may be desirable.

Since the surface microlayers are diluted by subsurface waters in the screen sample and the sampler is only some 70 per cent efficient in recovery (Garrett, 1965), this must be accounted for in estimating effective surface microlayer concentrations. Screen enrichment (Duce *et al.*, 1972) equals screen-sample concentration minus subsurface-sample concentration and is multiplied by 1·43 to account for sampling inefficiency. This value is then multiplied by the dilution factor, which is the thickness of the screen sample (about 150 μm) divided by the estimated thickness of the surface microlayer under consideration (Sieburth *et al.*, 1976).

GLASS-PLATE SAMPLING
Sampler

This neuston-sampling device consists of a square sheet of window glass (e.g. 25 × 25 cm) to which a removal plastic-coated clamp is attached. A soft teflon

Figure 15
Neuston sampling with Sieburth's screen sampler.

or silicone rubber window scraper is used to remove water and microlayer film from the glass.

Procedure

Pre-sterilize the glass plate by scrubbing with alcohol and brush before use. Hold the clamp handle with one hand, lower the glass plate vertically into the water so that the upper edge penetrates below the surface. Withdraw in a few seconds, allow the plate to drip for about 5 seconds, then tilt one corner of the plate over a sterile funnel and collecting bottle, and scrape the liquid from one or both sides of the glass. This procedure is repeated many times until the desired volume is obtained.

Although the principles involved in collecting microlayer differ between the glass plate and screen samplers, apparently they collect with reasonable efficiency and produce qualitatively comparable results (Hatcher and Parker, 1974*a*).

4

Preservation and storage

Jahn Throndsen

The handling of samples after collection is a critical stage in most phytoplankton work. It is important to minimize quantitative and qualitative changes in the phytoplankton composition before further treatment of the samples takes place. This can be achieved either by proceeding rapidly to the next step, e.g. fixation of the sample, or by keeping the samples at a low activity rate until the processing of the sample can take place, e.g. when cultures are to be set up. It is also relevant here to mention staining procedures, though more details on stains—as well as fixatives—can be found under Chapter 6.3 for the different taxonomical groups of phytoplankters.

KEEPING SAMPLES ALIVE

This procedure is limited to some special conditions such as inshore investigations (where the distance to the laboratory is reasonably short) or work on board an oceanographic vessel.

Phytoplankton in water samples will keep their viability for some time provided they are not subjected to rise in temperature or light intensity. In net hauls or samples from blooms, however, the viable period may be rather short since most of the photoautotrophic, phagotrophic and saprotrophic species are highly concentrated. In order to extend the period of viability it is recommended that the level of physiological activity of the species should be lowered.

Water samples should be kept at sea temperature if the latter is below 10° to 15°C, or cooled (2° to 10°C) if temperature is above 15°C, and stored in total darkness. Water samples from bloom situations should be dealt with as net hauls.

Net hauls should not be too dense, and they should be kept in a rather large amount of water (250 to 1,000 ml). Some light may be necessary to avoid oxygen depletion in the sample. Temperature should be low, 2° to 5°C, even in summer, because heterotrophic organisms, including bacteria, will multiply rapidly in such a dense sample. The release of dissolved organic matter and the decay of many cells would provide a good substratum for such growth.

Samples may be stored or transported in glass bottles (pyrex quality) with glass stoppers, enclosed either in boxes made from expanded polystyrene (thus preventing mechanical as well as thermic damage) or in wide-mouthed insulated bottles. In the latter case the glass bottles are wrapped in a thin layer of foam plastic to prevent mechanical damage to the outer bottle.

To keep the samples alive in the laboratory, e.g. for demonstration purposes, a refrigerator with fluorescent lighting inside will be satisfactory in most cases. Dense samples, however, such as those obtained from net hauls, start to decay soon; the most fragile species often dies within an hour from the sampling time. Dilution with fresh seawater may be necessary to extend the viability period of net hauls.

FIXATION

A great number of fixatives and preserving agents have been described, but they have all proved to be more or less selective in their fixing capacity. Up to now, only a few have been used extensively enough to give any experience worth drawing upon, though some of the recent ones, particularly glutaraldehyde, have given promising results on a small scale (see Taylor, 1976b). As other fixing agents, they are selective, and extensive surveys with storage periods over a period of several years are needed to reveal their qualities.

The most widely used fixatives for phytoplankton are formaldehyde and potassium iodide plus iodine, both used on their own or with other compounds added:

Neutral, changing to acid with time
   Formaldehyde.
   Potassium iodide plus iodine (Lugol's solution).
Neutral or slightly alkaline
   Formaldehyde with hexamethylenetetramine, sodium borate or sodium carbonate added.
   Potassium iodide plus iodine, with sodium acetate added.
Acid
   Formaldehyde with acetic acid added.
   Potassium iodide plus iodine, with acetic acid added.

From this choice of combinations, formaldehyde (either neutral or acid) and Lugol (preferably acid) may be selected and will be dealt with in detail.

The formaldehyde method

1. Fixing and preserving agent. The agent is a 20 per cent aqueous solution of formaldehyde neutralized with hexamethylenetetramine. (Alternatively it may be acidified with acetic acid when coccolithophorids are absent. Silicified structures may dissolve more or less in alkaline solution.)

70

Prepare the solution by diluting p.a. (pro analysi) grade formalin ($= 40$ per cent formaldehyde HCHO) to 20 per cent with distilled water, or make a 20 per cent solution from paraformaldehyde. For a neutralized (weakly alkaline) solution add 100 g hexamethylenetetramine to 1 litre of the 20 per cent solution; for acid solution mix equal amounts of p.a. grade formalin (40 per cent HCHO) and concentrated acetic acid.

2. Procedure. For water samples, add 100 ml of sample to 2 ml of the fixing/preserving agent (final concentration of HCHO is thus 0·4 per cent); shake the bottle immediately to facilitate an instantaneous fixation.

For net hauls, add fixing/preserving agent to make up about one-third of the volume if the sample is dense. Mix well. It is important that fixation should take place immediately after the collection of each sample (the water sample should be added directly to the fixing agent) in order to prevent any adverse effects of e.g. light and temperature changes. This is particularly important for net hauls where the high cell density may otherwise cause rapid decay.

3. Transport to laboratory or store. The preserved samples should preferably be transported and stored at low temperature, but frost free. They should be kept away from vibrating machinery as vibrations may favour the formation of undesired aggregates as well as disintegration of e.g. coccolith cell covers.

If samples are to be transported for some time or if a prolonged storage time can be anticipated, then it is wise to seal the stopper of the bottles with wax by dipping the inverted bottles to the neck in melted beeswax. This is especially important when plastic bottles are used. Experience has shown that, for some unknown reason, once a sample bottle has been opened, the sample is not good for prolonged storage.

4. Advantages of the method
(a) The neutral formaldehyde method (when properly carried out) renders coccolithophorids, diatoms and thecate dinoflagellates in identifiable condition, and a large number of naked flagellates can be recognized as such. In addition, cells devoid of flagella can be distinguished.
(b) The ingredients of the fixing/preserving agent are usually easy to obtain, and the solution will keep well during storage. (The 20 per cent formaldehyde solution will not produce the precipitations so common in formalin.)
(c) Samples preserved with formaldehyde may keep for years without further attention, when stored properly.

5. Disadvantages of the method
(a) Formaldehyde fixation is qualitatively selective especially in that it distorts cell shape of naked species and causes flagella to be thrown off in many flagellates.

71

(b) Cell content will bleach, making it very difficult to distinguish between pigmented and non-pigmented cells. Organelles, as e.g. flagella, may need phase-contrast microscopy in order to be revealed.

The iodine method (acid solution)

The most commonly used alternative to the above method is Lugol's fixation, i.e. iodine dissolved in a potassium iodide solution. For marine phytoplankton it is often used in its acid version (Lovegrove, 1960, and others), a technique which was first proposed for freshwater studies (Rodhe et al., 1958). The weakly alkaline Lugol's solution (with acetate: see Utermöhl, 1958) will be better for coccolithophorids, but inferior to the acid one (with acetic acid) with regard to flagellates.

1. Fixing and preserving agent. 100 g KI is dissolved in 1 litre of distilled water; then 50 g iodine (crystalline) is dissolved and 100 ml of glacial acetic acid is added. As the solution is near saturation, any possible precipitate should be removed by decanting the solution before use.

2. Procedure. For water samples add enough fixing agent to give the sample a weak brown colour (0·4 to 0·8 ml to 200 ml sample: Willén, 1976), shake well.

Use this fixative for water samples only; net hauls are usually better preserved with formaldehyde. Use clear glass bottles only, since coloured bottles such as brown glass bottles would make it difficult to ascertain the amount of fixing agent needed. Avoid plastic bottles as they will take up iodine from the solution. As for formaldehyde, it is important that fixation should take place immediately after sampling. It is important that an adequate amount of fixing agent be added, as too much iodine may give a heavy staining.

3. Transport to laboratory or store. The preserved samples should be kept frost free and in darkness, preferably at low temperature. Avoid vibrations (see formaldehyde method).

If samples are to be stored for some time, the concentration of iodine in solution should be checked regularly (check the colour) and more preserving agent added if necessary. The loss of molecular iodine from the sample depends on the storage conditions and is increased by light.

4. Advantages of the method
(a) The main advantage is that a large number of flagellates will retain their flagella. Other organisms of the phytoplankton community will also fix reasonably well. The cells will stain brownish yellow and hence be more easy to observe during the counting procedure, though many organic and inorganic particles will be stained as well. The staining will render phase-contrast microscopy less desirable than for the formaldehyde method.

72

(b) The ingredients are relatively easy to obtain, and the stock solution will keep well for many years.

5. Disadvantages of the method

(a) The acid Lugol's solution will dissolve coccoliths; also silica will dissolve with long storage time (G. T. Boalch, Plymouth, pers. comm.). Lugol's solution without any addition will turn slightly acid relatively soon with a resultant elimination of coccolithophorids.

(b) Samples preserved with iodine need attention during storage as iodine is oxidized with time.

(c) Many organisms are frequently overstained; a large amount of starch will often obscure the anatomy, and also make surface structures difficult to observe, e.g. thecal plates of dinoflagellates. However, cells stained by iodine may be cleared up by adding sodium thiosulphate, which reduces the iodine.

Formaldehyde versus iodine

No single method can be expected to be suitable for the preservation of all types of phytoplankton. The fixing and preserving agent should be chosen with regard to the aim of the investigation. For instance, in a typical offshore area with thecate dinoflagellate and diatom domination, an acid formaldehyde fixation may be applied; in a tropical area with coccolithophorids predominating, neutralized formaldehyde is appropriate; in inshore areas with naked flagellates as an important part of the community, the acid iodine method may give the best result from a combined quantitative and qualitative view. For fixation and preservation of special groups of plankton, see Chapter **6.3** in this volume, and Steedman (1976, part VIII).

STAINING

Two types of staining have been used in phytoplankton research: (a) to distinguish between live and dead cells; and (b) to reveal the presence or quality of cell constituents.

Cell staining

Live-dead staining may be made on fresh material by the addition of Evans blue or Neutral red (Crippen and Perrier, 1974). The results are, however, inconsistent.

Distinction between cells and detritus may be accomplished by the addition of Rose Bengal stain (cf. Zeitzschel, 1970) or Acridine orange, the latter giving fluorescence in living cytoplasm (see Chapter **7.4**).

73

Staining of cellular components

A number of more or less specific stains may be used to reveal the presence of specific materials in phytoplankton cells. The reactions should be carried out on fresh microscope preparations (under coverglass on an object slide) so that the results can be observed in the microscope. The most useful reactions are as follows:

1. Starch is stained blue by iodine. An aqueous solution of iodine will work, but Schimper's solution (8 parts of chloral hydrate, 5 parts of water, add a little tincture of iodine) is better (Jane, 1942). A weak Lugol's solution may be used, but will give a brownish colour.
2. Chrysolaminarin, chrysose (soluble β-1:3-glucan) stains reddish purple with Brilliant Cresyl blue. Note that chrysolaminarin will leak out of dead cells because of its solubility in water.
3. Fat and oils stain black with Sudan black (saturated solution in 70 per cent ethanol) or blue with Sudan blue (saturated solution in 50 per cent ethanol; Jane, 1942). Osmic acid will also blacken fat and oils.
4. Cellulose will stain blue when a Lugol solution is followed by concentrated sulphuric acid. Add Lugol solution to a thin preparation (with as little water as possible) until a pale straw colour is obtained. Leave it for one minute and then add a drop of concentrated sulphuric acid (Jane, 1942).

For more comprehensive information on staining techniques, Pearse (1960), Gabe (1968), Romeis (1968), Clark (1973) or Gerlach (1977) may be consulted. Additional information is also available in Chapter **5.5.2** for permanent slides and in Chapter **6.3** for taxonomical peculiarities.

5

Concentrating phytoplankton

5.1   The implications of subsampling

Elizabeth L. Venrick

Between the collection of the initial field sample and the calculation of the final data point there occur several steps (preservation, subsampling, counting, etc.) each of which is a potential source of bias or variability. Some of these sources can be minimized or eliminated with proper care. However, variability introduced by removing a subsample from a larger sample can never be eliminated since we have no way of assuring that a subsample is a perfect representation of the sample.

An understanding of this additional source of variability is essential for the design of an efficient subsampling programme, as well as for the ultimate analysis of the data. However, the subject should not be isolated from other steps in the programme—the design of sampling strategy and ultimate enumeration of cells—which are closely interrelated; the optimization of all steps depends upon the objectives of the programme.

Theoretical considerations of the implications of subsampling depend upon a knowledge of the frequency distribution of the population under consideration. Once the population is contained in the original sample, we are no longer involved with the complex patterns of field distributions (Chapter **2.1**). None the less, models are weakened by deviations between the real distribution and the theoretical. Many workers have evaluated the precision of their subsampling routine directly by analysis of replicate subsamples (e.g. Allen, 1921; Lund *et al.*, 1958; Uehlinger, 1964; Frolander, 1968; Frontier, 1972; Rassoulzadegan and Gostan, 1976). For any given programme this may provide the most straightforward estimate of analytical error, but, as seen from the disparity of results (reviewed by Serfling, 1949; Woelkerling *et al.*, 1976), the empirical conclusions have little generality. At present, the optimum solution is probably a combination of theoretical and empirical approaches.

The following discussion focuses more heavily on theoretical aspects, both because of their broader applicability and because few theoretical treatments have considered the complete sampling strategy employed by many phytoplankton workers. To avoid repeated redefinition of terms, the more frequently used symbols and concepts have been summarized in Table 4.

The common distributions which underlie the statistical procedures, such as the normal, *chi*-square and binomial distributions, are discussed in standard statistical texts (e.g. Snedecor, 1956; Dixon and Massey, 1957; Sokal and Rohlf, 1969; Dagnelie, 1973–75). Pertinent aspects of this topic have been discussed by Cassie (1968, 1971).

TABLE 4. Definitions of commonly used symbols and concepts

Symbol	Definition	Symbol	Definition
n	Number of samples	v	Sample volume
N	Population size, in terms of total number of individuals or total number of potential samples	V	Population volume
		SS	Sum of squares
x_i	Abundance in sample i	df	Degrees of freedom; a mathematical expression of the number of independent variables involved in a calculation (often $df = n - 1$)
$\sum_{i=1}^{n} x_i$	Sum of abundances in all samples		
$\bar{x}$	Sample mean $= \sum_{i=1}^{n} x_i/n$		
		MS	Mean square $=$ SS/df
μ	Population mean	FPC	Finite population correction
s^2	Sample variance $= \sum_{i=1}^{n} (x_i - \bar{x})^2/(n - 1)$		FPC–1 $= (V - nv)/V$ FPC–2 $= (V - v)/V$
		α	Significance level: probability of rejecting a true hypothesis
$s_{\bar{x}}^2$	Variance of the mean $= s^2/n$		
σ^2	Population variance	z_α	Standard normal variate
$\sigma_{\bar{x}}^2$	Theoretical variance of the mean	$t_{(\alpha)(n-1)}$	Studentized normal variate

POISSON DISTRIBUTION

The distribution most often referred to in phytoplankton work is the Poisson distribution (Student, 1907) which, because it describes the distribution of random, rare events, is the standard against which phytoplankton distributions and their departures from randomness are evaluated. In this context, it will be used extensively in the following discussion. The Poisson distribution results when the number of possible loci in space or time is sufficiently large that the probability of any one being occupied is very small, and is independent of the occupancy of other sites.

Although the Poisson is the distribution of 'random, rare events', it may still apply to distributions with large mean values, in which case it approaches a normal distribution. The criteria of rareness must be assessed relative to the maximum possible number of occurrences; a mean of 1,000 cells/ml may be small relative to the total number of cells which could be contained in a 1 ml volume. The criterion of independence excludes the use of the Poisson for single cells of species that occur in chains or colonies (Gilbert, 1942; Holmes and Widrig, 1956; Uehlinger, 1964); in these cases, the appropriate unit is the colony.

Within a thoroughly mixed sample, organisms appear to be random in some cases (Student, 1907; Ricker, 1937; Gilbert, 1942; Serfling, 1949; Holmes and Widrig, 1956; Kutkuhn, 1958) but not in others (Littleford et al., 1940; Serfling, 1949; Holmes and Widrig, 1956; Kutkuhn, 1958). Non-random distributions are more frequent among the more abundant species (Frontier, 1972; Rassoulzadegan and Gostan, 1976). For small mean values, the Poisson is strongly asymmetrical. The terms of the Poisson have been tabulated for means up to 20 or more (Arkin and Colton, 1950; Selby, 1968).

The variance (σ^2) of a Poisson is equal to its mean (μ) and thus the comparison of sample mean and variance ($\bar{x}$ and s^2) is a useful index of fit. Fisher (1963) has shown that the ratio $(n - 1)s^2/\bar{x}$, the index of dispersion of a sample of size n from a Poisson distribution, follows a χ^2_{n-1} distribution, allowing the calculation of the probability that a sample came from a random distribution. A satisfactory fit of this index of dispersion to the χ^2 is a necessary, but not sufficient indication of randomness. Tests which consider the shape of the frequency distribution may be more sensitive, e.g. the χ^2 goodness of fit (various statistical texts: Cassie, 1962, 1963, 1971; Uehlinger, 1964) or the Kolmogorov–Smirnov test (Tate and Clelland, 1957; Uehlinger, 1964; Conover, 1971); for rare species, the index of dispersion will rarely detect non-randomness (Cassie, 1971) and a run test (Tate and Clelland, 1957; Conover, 1971) for presence and absence may be more useful. Autocorrelations have also been used (Cassie, 1959b; Bernhard and Rampi, 1966).

Examples

1. The following data represent a sequence of observations in space or time (cells/ml): 0, 0, 2, 1, 2, 0, 0, 1, 1, 2, 0, 0, 0, 0, 0, 1, 1, 1, 2, 1, 0, 0, 0, 0, 0, 0, 0, 0, 0, $\bar{x} = 0.54$, $s^2 = 0.55$, $n = 28$. The index of dispersion does not lead to the rejection of the hypothesis of Poisson distribution: $\chi^2_{27} = 27(0.55)/0.54 = 27.5$, $p \approx 0.50$. A run test on presence and absence gives 7 runs out of 28 observations; this is too few runs ($p = 0.05$) under the hypothesis of random occurrence of the species, indicating that they are in fact aggregated.

77

2. The following observed frequency distribution is compared with that expected from a Poisson of the same mean and variance:

Abundance (cells/ml)	Frequency observed	Expected (Poisson)
0	20	37
1	75	37
2	0	18
3	0	6
4	0 ⎤	2 ⎤
5	5 ⎦	0 ⎦
$\bar{x}$ =	1·0	1·0
s^2 =	1·0	1·0
n =	100	100

Index of dispersion

$$\chi^2_{99} = 99(1·0)/1·0$$
$$= 1·00$$
$$p \approx 0·50$$

Goodness of fit

$$\chi^2_3 = \frac{(20-37)^2}{37} + \frac{(75-37)^2}{37}$$
$$+ \frac{(0-18)^2}{18} + \frac{(0-6)^2}{6}$$
$$+ \frac{[(0+5)-(2+0)]^2}{(2+0)}$$
$$= 75·34$$
$$p < 0·01$$

Again, the index of dispersion does not lead to rejection of the hypothesis of randomness. Either a Kolmogorov-Smirnov test on the cumulative frequency distributions or a χ^2_{n-1} goodness of fit test indicates that the probability that the observed distribution was drawn from a Poisson is less than 0·01, leading to rejection of the hypothesis of random distribution.

Other distributions have been derived that attempt to describe non-random distributions. These are usually applied to distributions in the field (cf. Chapter **2.1**). While they may provide more realistic descriptions of species distributions in the sample, their general applicability has not been well established. Many workers prefer to transform their data into a distribution which is approximately normal, thereby justifying the use of parametric procedures (cf. Chapter **8.2**). When a Poisson distribution is sampled repeatedly the distribution of the sample means will be approximately normal when the mean is greater than 5 per sample (Serfling, 1949).

MULTI-STAGE SUBSAMPLING

In routine phytoplankton work, it is generally the case that the initial field sample is successively subsampled two or more times so that the final count is based on a fraction of the initial sample. Thus, the initial field sample may be a 1-litre Nansen bottle of which 500 ml are preserved and returned to the laboratory where 100 ml are removed and settled to be counted. For the more abundant species, only a fraction of this settled material may be counted. At each stage, a component of variability is introduced into the data. The appropriate model for this addition of error is the nested analysis of variance which is described in statistical texts. For instance, the following nested analysis of variance model describes the distribution of variance components at each level of a three-stage subsampling procedure as follows:

Level	Source	Example	Degrees of freedom (df)	MS estimates
1	Between n_1 primary samples	1,000 ml water samples	$n_1 - 1$	$\sigma_3^2 + n_3\sigma_2^2 + n_2n_3\sigma_1^2$
2	Between n_2 secondary sub-samples within each primary subsample	500 ml preserved sample	$n_1(n_2 - 1)$	$\sigma_2 + n_3\sigma_2^2$
3	Between n_3 tertiary sub-samples within each secondary subsample	100 ml counted	$n_1n_2(n_3 - 1)$	σ_3^2

where n_1, n_2 and n_3 are the number of primary samples, the number of secondary subsamples per primary sample and the number of tertiary subsamples per secondary subsample, and $\sigma_1^2, \sigma_2^2, \sigma_3^2$ are their respective theoretical variances. The more complicated (but often unavoidable) case of an unbalanced design, as used by Platt et al. (1970), is treated in most standard statistical texts. Although the analysis of variance is the appropriate model for the decomposition of the variance, the usual test between components is valid only if the data have been normalized.

It should be noted that even though an error component cannot be calculated directly, as is the case when there is only a single subsample at that level, the component is still present at all higher levels and can be estimated from them.

In the context of the nested analysis of variance, there are two features peculiar to plankton work which warrant discussion, variance adjustment and finite population corrections.

Variance adjustment

In plankton work, each level of subsampling is based upon a different volume and the means and variances must all be adjusted to some common volume before they can be summed (Venrick, 1972a). If the subsamples at level 2 have some volume, v_2 (i.e. 500 ml), with a mean $\bar{x}_2$, and a variance between replicate subsamples, s_2^2, and if the data are analysed in terms of some standard volume, v_{st} (i.e. cells per ml or cells per litre) then the mean of the subsamples is adjusted by a factor v_{st}/v_2 while the variance is adjusted by a factor $(v_{st}/v_2)^2$. If, at any one level, the variance is estimated by some function of the mean, the volume correction will alter this functional relationship. In the simplest case of a Poisson distribution s_2^2 is approximated by $\bar{x}_2$ but s_{st}^2 is approximated by $\bar{x}_2 (v_{st}/v_2)^2 = \bar{x}_{st}(v_{st}/v_2)$. Under the assumption of a Poisson distribution at each level, the mean square formulae may be expressed in terms of the standardized mean:

Level	MS estimates
1	$\bar{x}_{st}(v_{st}/v_3) + n_3 \bar{x}_{st}(v_{st}/v_2) + n_2 n_3 \bar{x}_{st}(v_{st}/v_1)$
2	$\bar{x}_{st}(v_{st}/v_3) + n_3 \bar{x}_{st}(v_{st}/v_2)$
3	$\bar{x}_{st}(v_{st}/v_3)$

As a consequence of this volume correction, the expected variance about a sample mean expressed as $\bar{x}_{st}$ estimated with a multi-level subsampling procedure may be considerably greater than the mean, even though the population is random, and the relative contribution to this mean square from the successive subsampling stages is a function of the number and volume of the subsamples. Ignorance of the first consequence may lead to an unjustified rejection of the Poisson distribution.

Finite population corrections (FPC)

When population parameters are estimated from a large fraction of the population, it is appropriate to apply a finite population correction. There are two corrections, applicable under different circumstances. The most frequently encountered (hereafter to be designated as FPC-1) is of the form $(V - nv)/V$ where v and V are the sample and population volumes and n is the number of samples. This FPC-1 is applied to the variance of the mean ($s_x^2 = s^2/n$) when the latter is used to determine the precision of the estimate of the population mean from the formula $|\mu - \bar{x}| < t_{n-1}\sqrt{(s^2/n)}$ (Cochran, 1963). As more and more samples are collected from a population the observed mean approaches the true mean, implying that the expression under the square root sign must approach zero. This is achieved by use of FPC-1, which approaches zero as $nv \to V$, so that the estimate of μ is calculated from $\bar{x} \pm t_{n-1}\sqrt{[(s^2/n)(V - vn/V)]}$ (Cassie, 1971). (The effect of FPC-1 on relative precision is shown in Figure 30 of Chapter 7.1.2.)

A second finite population correction (FPC-2) is less frequently encountered in the literature, although it may be of considerable importance in subsampling theory. This correction, derived from the binomial distribution, is applied to the expected variance between samples when a single sample contains a large proportion of the population. In this case, the expected variance between samples is reduced by the factor $(V - v)/V$, i.e. by the fraction of the population not taken (Ricker, 1937).

Although the two finite population corrections are similar in form they may be quite different in function. FPC-1, which is a function of the number of samples, relates an observed variance to an expected variance of the mean. FPC-2 is independent of the number of samples and is applied to a theoretical variance in order to calculate an expected variance.

Incorporation of the appropriate FPC into the formulae for the expected mean square at each level of subsampling results in the following:

Level	MS estimates
1	$\sigma_3^2(\text{FPC-1}) + n_3\sigma_2^2(\text{FPC-1}) + n_2n_3\sigma_1^2(\text{FPC-2})$
2	$\sigma_3^2(\text{FPC-1}) + n_3\sigma_2^2(\text{FPC–2})$
3	$\sigma_3^2(\text{FPC-2})$

Note that the finite population correction changes according to the level of the analysis.[1] Thus, at level 3, FPC-2 is appropriate because the mean square at that level is an estimate of the variance between the tertiary subsamples. But, at level 2, σ_3^2 serves as a measure of the error around the estimate of the true tertiary subsample mean, derived from the observed variance of the mean, and it must be corrected by FPC-1. Under the assumption of a Poisson distribution at each level, the equations should be modified as follows, where v_1, v_2 and v_3 are the sample volumes at levels 1, 2 and 3:

Level	MS estimates
1	$\bar{x}_{st}(v_{st}/v_3)\left[(v_2 - n_3v_3)/v_2\right] + n_3\bar{x}_{st}(v_{st}/v_2)\left[(v_1 - n_2v_2)/v_1\right]$
	$\qquad\qquad\qquad\qquad + n_2n_3\bar{x}_{st}(v_{st}/v_1)\,(\text{FPC-2})$
2	$\bar{x}_{st}(v_{st}/v_3)\left[(v_2 - n_3v_3)/v_2\right] + n_3\bar{x}_{st}(v_{st}/v_2)\left[(v_1 - v_2)/v_1\right]$
3	$\bar{x}_{st}(v_{st}/v_3)\left[(v_2 - v_3)/v_2\right]$

These equations can be expanded or contracted to include any number of stages. For less than three levels, the equations are eliminated sequentially from the top; for more than three levels, additional equations are added, each one containing within it the variance from the lower levels (with appropriate FPC adjustments) plus an additional term for the variability at the new level.

For any specific subsampling regime, the formulae may often be simplified. The finite population corrections can be ignored when the subsamples represent

---

1. In an earlier paper on this subject (Venrick, 1972a) this distinction was not recognized and FPC-2 was used throughout.

less than one-tenth of the larger sample. Furthermore, when the data are analysed in terms of the smallest subsample volume (the volume counted), and when the highest level is between field samples so that an FPC is not applicable, the FPCs at lower levels cancel against the variance adjustment factors and the expected mean square at the highest level remains equal to the mean. However, the proportion of this variance contributed by each level depends upon the subsampling design.

Example

Consider the case in which there are four replicate 1,000 ml samples collected from the field. From each is preserved a single 500 ml sub-sample from which 100 ml is settled and counted. Data are analysed as number of cells per 100 ml and the populations are assumed random at each level. The expected mean square between field samples is given by:

$$MS_1 \approx \bar{x}_3 \left[(500 - 100)/500\right] + \bar{x}_3 (100/500) \left[(1,000 - 500)/1,000\right] +$$

$$+ \bar{x}_3 (100/1,000)$$

$$\approx 0{\cdot}8\,\bar{x}_3 + 0{\cdot}1\,\bar{x}_3 + 0{\cdot}1\,\bar{x}_3$$

$$\approx \bar{x}_3$$

The final expected mean square is still equal to the mean, but only 10 per cent of this is contributed by the variance of the population in the field. On the other hand, if the cells are enumerated in 100 ml aliquots but the data standardized to cells per litre before calculation of the variance components, the expected MS becomes $100\,\bar{x}_3 = 10\,\bar{x}_1$. Further aspects of this have been considered (Venrick, 1972*a*).

COUNTING BY FIELDS

It frequently happens that the final settled aliquot contains too many cells of one or more species to be completely enumerated. It is then common practice to count only a fraction of the aliquot and estimate the total aliquot count from this. This is the final stage in the subsampling design and the statistical principles are the same, the only difference being that the different species within a sample may be counted in different sized fractions, although all counts are standardized to a single volume (often the volume of the aliquot). Any error introduced by counting only a fraction of the aliquot will be included in the error between replicate aliquots, increasing the variance to be expected on the basis of random distribution within and between aliquots.

82

As a consequence of this additional handling error, when different fractions of the same sample are enumerated for different species, the species estimated from the smaller fraction will be more variable, relative to its mean, even though the species are each distributed according to a Poisson.

If a settled aliquot (in a settling chamber or a Sedgwick-Rafter counting cell) is divided into N contiguous, non-overlapping fields of which n are chosen at random to be enumerated and pooled into a single estimate of the abundance in the aliquot, then the expected mean square between aliquots is increased by the factor $n(N^2/n^2) = N^2/n$ over the mean square expected when the identity of the fields is preserved and the data expressed as number of cells per field. The factor (N^2/n^2) is the correction of volume counted to aliquot volume, while the factor n is the correction for pooling fields within aliquots. If the fields within aliquots are designated as subsampling level 4, and the aliquots as level 3, then,

$$MS_3 \approx (N^2/n_4) \left[\sigma_4^2 \, (\text{FPC-1}) + n_4 \sigma_3^2 \, (\text{FPC-2}) \right]$$

Example

The following numbers (actually drawn from a random numbers table) can be used to represent counts within five replicate fields in each of 4 chambers (out of a total of 50 fields/chamber):

Chamber no.	1	2	3	4	
Field no. 1	0	9	1	8	$n = 5, N = 50$
2	9	0	0	4	
3	7	3	1	8	
4	7	5	7	6	
5	0	2	5	9	
Σ	23	19	14	35	

Estimated chamber total 230 190 140 350

The analysis of variance on the counts/field gives:

Source	Sum of squares (SS)	Degrees of freedom (df)	Mean square (MS)	Estimates
Between chambers	48·15	3	16·05	$\sigma_4^2 + 5\sigma_3^2$
Within chambers (between fields)	172·80	16	10·8	σ_4^4

83

On the other hand, if the data are first pooled into an estimate of total abundance in the chamber, the precision of replicate chambers as measured by the variance between 230, 190, 140 and 350 = 8,025, and

$$16 \cdot 05 \left(\frac{50^2}{5} \right) \equiv 8,025.$$

As seen by the formula for the decomposition of variance, the precision between replicate aliquots estimated from partial counts includes the variance between fields, even though these have been pooled. Thus, if one wishes to express the mean square between replicate aliquots estimated from partial counts when there is a random distribution of cells between and within aliquots, the appropriate formula, including finite population corrections, would be:

$$\mathrm{MS}_3 \simeq \frac{N^2}{n_4} \left[\bar{x}_4 \left(\frac{N - n_4}{N} \right) + \frac{\bar{x}_3}{N^2} \mathrm{FPC\text{-}2} \right]$$

$$\simeq \frac{N}{n_4} \left[\bar{x}_3 \left(\frac{N - n_4}{N} \right) + \frac{n_4}{N} \bar{x}_3 (\mathrm{FPC\text{-}2}) \right]$$

$$\simeq \left(\frac{N - n_4}{n_4} \right) \bar{x}_3 + \bar{x}_3 \, (\mathrm{FPC\text{-}2}).$$

If the finite population correction between aliquots can be ignored, this reduces further to:

$$\mathrm{MS}_3 \approx (N/n_4) \, \bar{x}_3$$

which may be compared to $\mathrm{MS}_3 \approx \bar{x}_3$ expected when the entire aliquots are enumerated.

ALLOCATION OF EFFORT AND COST

A given number of replicate data points may have arisen from any of several possible subsampling designs. One might, for instance, take a single sample in the field and replicate it n times, or one might take n replicate field samples and make one determination on each. The more levels of subsampling, the more potential designs. The ultimate precision of the data is a function of the sub-sampling strategy, and it is advantageous to select a design which will maximize this.

Considerations of cost, in terms of time or money, are also important. Optimal allocation of resources generally refers to designs which maximize the precision of the estimate and minimize costs. Formulae are available for two- and three- and higher-level subsampling designs (Marcuse, 1949; Brooks, 1955; Cochran, 1963; Sokal and Rohlf, 1969). These are based on the determination of a cost function in which total cost is equal to the cost of a sample at any level times the number of samples at that level summed over all levels. Solution of the formulae depend upon knowledge of these costs as well as knowledge of the variance introduced at each level. Optimal allocation formulae for two-stage sampling problems have been applied to phytoplankton by Uehlinger (1964), McAlice (1971) and Woelkerling et al. (1976). A slightly different approach was used by Gillbricht (1962).

For a three-level sampling design, the number of subsamples at levels 1, 2 and 3 may be calculated from the relative variances and costs per sample:

$$n_3 = \sqrt{\frac{\sigma_3^2 \, c_2}{\sigma_2^2 \, c_3}}$$

$$n_2 = \sqrt{\frac{\sigma_2^2 \, c_1}{\sigma_1^2 \, c_2}}$$

These are independent of either total cost or total variance and may be presented in graphical form (Fig. 16). The number of primary samples to be collected depends upon whether one has a maximum total cost, or a maximum allowable variance. In the first case:

$$n_1 = \frac{\sigma_1}{\sum\limits_{i=1}^{3} \sqrt{\sigma_i^2 \, c_i} \, \sqrt{c_1}} \cdot C.$$

In the second:

$$n_1 = \frac{\sigma_1}{VA} \frac{\sum\limits_{i=1}^{3} (\sigma_i^2 \sqrt{c_i})}{c_1}$$

where C = total cost
$$= n_1 c_1 + n_1 n_2 c_2 + n_1 n_2 n_3 c_3$$

and VA = maximum allowable variance
$$= \frac{MS_1}{n_1 n_2 n_3}.$$

The variances appropriate for these formulae are the corrected variances (corrected for standardization of volumes and for FPCs), which are best calculated from the mean squares given by a standard analysis of variance. (The various corrections are inherent in the calculation of the mean squares and can be happily ignored at this point.)

The above formulae show that once the optimal n_2 and n_3 have been achieved, the most economical means of increasing precision is to increase effort at the primary sampling level.

In phytoplankton work the greatest variance is either s_3^2, at the counting level, or s_1^2, the field variance. Although the latter is generally much greater than

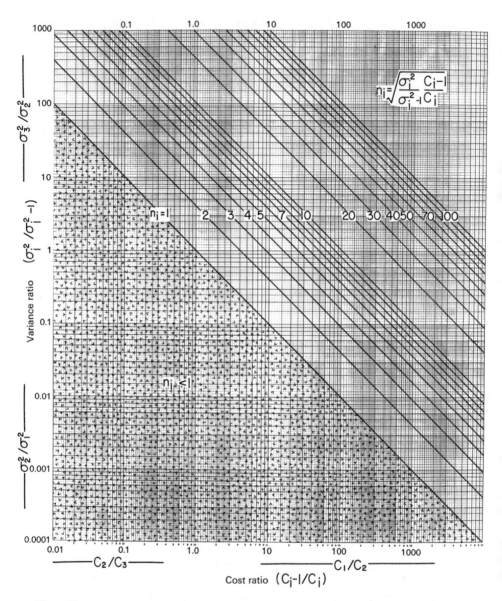

Figure 16
Values of optimal sample number (n_i) as a function of the cost ratio c_{i-1}/c_i and the variance ratio $\sigma_i^2/\sigma_{i-1}^2$ in a nested sampling programme (modified from Uehlinger, 1964). Realistic ranges for cost and variance ratios for samples/subsamples (c_1/c_2 and σ_2^2/σ_1^2) and for subsamples/aliquots (c_2/c_3 and σ_3^2/σ_2^2) are indicated.

random expectation (Cassie, 1968), its magnitude will be reduced according to the relative volumes of the field samples and the aliquot counted. Likewise, the

86

major cost of phytoplankton sampling is either in the collection of the primary field samples (c_1) or the settling and counting of the final volume (c_3). The secondary level of subsampling generally contributes little in the way of either variance or cost. Most reasonable ratios of cost and variance lead to values of $n_3 = 1$ and $1 \leqslant n_2 \leqslant 3$ (Fig. 16); counting more than one aliquot per subsample may not increase precision sufficiently to warrant the extra cost. Thus, as a very general rule of thumb (in the absence of specific information of cost and variance), to obtain a precise estimate of abundance in the field, it is probably most efficient to take as many replicate field samples as time and money allow, and to take one or two subsamples for each, counting only one aliquot per subsample.

There are, however, a few specialized situations in which minimizing the error mean square may not be the optimum strategy. In studies of spatial distributions in the field, for instance, dilution of the field variance through successive subsampling may be undesirable. For maximum sensitivity the initial field sample should be no larger than that ultimately enumerated (Venrick, 1972a).

The dearth of specific recommendations in this discussion is unavoidable since the desirability of any specific subsampling strategy depends upon the goals of the programme, the facilities available and the amount of time and money to be spent. The intent of this section is to explain the statistical implications of subsampling so that the researcher may better evaluate his own programme. It is hoped that the following points are clear:
1. The effect of multiple-stage subsampling is to add an error component at each stage and to dilute the relative contribution of the initial field variance.
2. It is not necessarily true that a nested series of subsamples drawn from a Poisson distribution will have an expected MS equal to the mean.
3. Two or more series of samples cannot be directly compared for heterogeneity unless they are based on the same subsampling design. Otherwise the individual variance components must be removed from the observed mean square adjusted to a common volume before comparison.
4. When complete data are reported, they should be accompanied by the subsampling design, including numbers and volumes of each subsampling level, to facilitate comparison with data from different designs.

5.2   Settling

5.2.1   The inverted-microscope method

Grethe Rytter Hasle

The inverted-microscope method, or Utermöhl method, was introduced in the early 1930s. Another, less correct, name is the sedimentation method. Utermöhl (1931, p. 594) characterized the inverted-microscope method as a combination of the Kolkwitz (1907) chamber method and the sedimentation method of Volk (1906), or in his own words: 'Die hier mitgeteilte Untersuchungsweise stellt eine Erweiterung und zum Teil Abänderung des Kolkwitzschen Kammerverfahrens dar, zugleich aber auch eine Vereinfachung des Volkschen Sedimentierverfahrens.' Volk's sedimentation method involves stepwise concentration to successively smaller volumes of water by transfer from larger to smaller containers. (Somewhat modified, the method is now used with success to obtain non-quantitative material for morphologic, taxonomic and distributional studies of coccolithophorids, small dinoflagellates and diatoms, and other plankton organisms too small to be collected adequately by nets; semi-quantitative results may also be expected: see Chapter 5.2.2.) The simplification of Volk's sedimentation method made by Utermöhl is the combined use of the same chamber for sedimentation and counting without any transfer and consequent loss of material.

The Kolkwitz chamber was designed for a standard compound microscope. Because of its height an objective with a long working distance and correspondingly low power must be used. By using an 'inverted' microscope and chambers with a glass bottom of coverslip thickness, Utermöhl made it possible to use objectives with shorter working distance and higher power.

COUNTING CHAMBERS

Utermöhl (1931) designed an assortment of counting (= sedimentation) chambers or cylinders in order to provide chambers for specific sample characteristics. A sample with a dense phytoplankton population or a lot of detritus is most easily examined in a chamber in which the bottom (floor) area is large relative to the volume. A sample with a sparse population and less detritus is best examined in a chamber in which the bottom area is small relative to the volume, reducing the time wasted scanning the empty bottom. The chambers constructed by Utermöhl (1931) were too tall to be used with a condenser;

however, he later (1958) described a chamber with a detachable bottom plate which is now manufactured by optical firms. The principle is to remove the upper part of the chamber (the sedimentation cylinder) after sedimentation, leaving organisms in the bottom part, which has a height less than the working

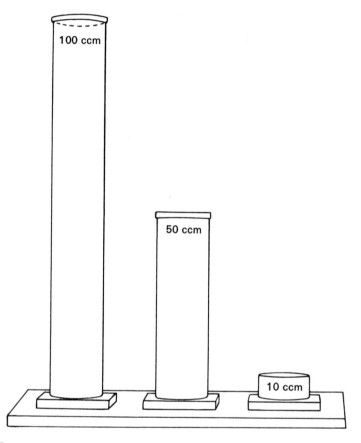

Figure 17
Top (sedimentation) cylinders of combined plate chambers (after Wild Ml, Catalogue 140d XII.69).

distance of the condenser. The model now in common use is a combined plate chamber consisting of a top cylinder (sedimentation cylinder) of 10, 50 or 100 ml capacity (Fig. 17) (5 and 25 ml may also be available) and a bottom-plate chamber (Fig. 18). The parts of the plate chamber are a rectangular perspex plate (Fig. 18a), a ring (18b) and a circular bottom (base) plate of coverslip thickness. The plate fits into the mechanical stage of the inverted microscope of the type for which it is constructed. It has a small opening close to one end,

D*

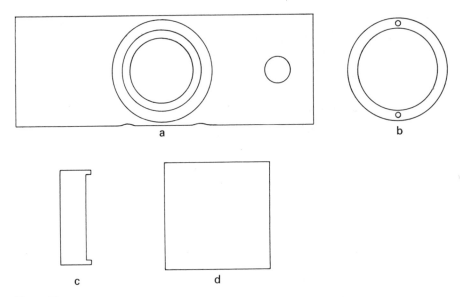

Figure 18
Bottom part of combined plate chamber: (a) perspex plate with larger opening for top cylinder
and smaller drainage hole; (b) ring for support of the bottom (base) plate; (c) key to fasten ring
to underside of perspex plate; (d) top plate of sedimentation chamber, also to be used to remove
top cylinder after sedimentation.

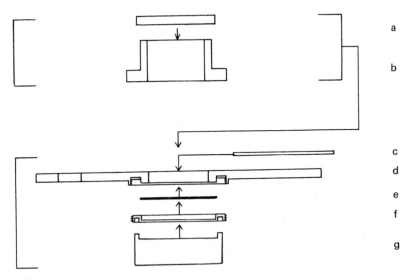

Figure 19
Vertical cross-section of combined plate chamber (partly after Wild Ml, Catalogue 140d XII.69):
(a) top plate of sedimentation cylinder; (b) sedimentation cylinder; (c) top plate of bottom plate
chamber; (d) perspex plate; (e) bottom (base) plate; (f) ring; (g) key.

and a circular central opening of 26 mm diameter, which is slightly larger than the diameter of the sedimentation cylinders. The ring can be fastened to the underside of the rectangular plate by a key (Fig. 18c). The bottom plate is placed between the ring and the large opening of the perspex plate and thus forms the floor of the plate chamber (Fig. 19). The combined chamber is ready for use when the cylinder of the desired capacity is placed on top of the plate chamber. When the well-shaken preserved water sample has been poured into the combined chamber to overflow, a top plate is placed in position to eliminate dust and evaporation. Care should be taken to remove all water outside the chamber to keep, in particular, the thin glass bottom clean. After sedimentation the top cylinder is slowly pushed away from the plate chamber by using the square top plate of the plate chamber (Figs. 18–19). Pushing stops when the cylinder reaches the small opening near the end of the perspex plate of the plate chamber. As soon as the circular top plate of the sedimentation cylinder is removed, water is drained out of the cylinder through the small hole below.

In contrast to the assortment of counting chambers he originally recommended, Utermöhl (1958) ultimately found standardization of the bottom area of the new chambers a convenience. He recommended examination of only a portion of the bottom area when high magnification is needed and when the population is dense enough so that such a procedure will yield reliable results. This implies that the sediments generally are randomly distributed on the chamber floor. When dealing with marine phytoplankton, non-random (aggregated) distributions of the settled organisms are often encountered, particularly when chain-forming species are present. Instead of examining a small part of the bottom of the commercial chambers, a one-piece chamber shallow enough to permit the use of a condenser, and with a small volume and a bottom area of reasonable size, can be constructed (see below).

Utermöhl (1958) warned against the use of tall cylinders because of attachment of organisms to the chamber wall. Paasche (1960) made comparisons which indicated that this might be the case with chain-forming species of the setae-bearing genus *Chaetoceros*. Counts of other phytoplankton forms may be unaffected by this source of error (Margalef, 1969c). Personal investigations of the spindle-shaped marine dinoflagellate *Ceratium fusus* and the lens-shaped marine dinoflagellate *Prorocentrum micans* showed no significant difference in cell densities estimated from a 2-ml (15-mm high) chamber and a 10-ml (40-mm high) chamber. Another objection to tall chambers was given by Nauwerck (1963, p. 16) who observed that convection currents could not be avoided in chambers higher than five times the chamber diameter. The problem of convection currents in these chambers could apparently not be overcome by a longer settling time since it was emphasized that a considerable amount of plankton did not settle at all. The height of the 100 ml commercial chamber greatly exceeds this relative size and should therefore be used cautiously, since sedimentation of phytoplankton will be hindered by convection currents.

SEDIMENTATION TIME

Variations in preservation method and species composition probably account for the diversity of sedimentation time recommended in the literature. Utermöhl (1931) assumed that all organisms would have settled by the day after preparation of the sample, while in 1958 he wrote more explicitly that a settling time of at least 24 hours was needed. Lund *et al.* (1958) recommended 18 hours for 100-ml chambers, 3 hours for 10-ml and 1 hour for 1-ml, while Willén (1976) used about 8 hours for 10-ml and 48 hours for 50- and 100-ml chambers. In an investigation by Nauwerck (1963) a sedimentation time of 4 hours per centimetre chamber height resulted in a complete sedimentation of even the smallest organisms 'von μ-Grösse'. These apply to freshwater phytoplankton preserved by an iodine solution. Steemann Nielsen (1933) recommended 24 hours as sedimentation time for marine phytoplankton preserved with formaldehyde, and Margalef (1969c) as a general formula suggested the sedimentation time in hours to be at least three times the height of the sedimentation chamber in centimetres.

Since iodine preservation apparently precludes a proper investigation of coccolithophorids (Chapter 4), neutralized formaldehyde is often used. Formaldehyde lacks the advantage of iodine, which increases the weight of the cells, thereby decreasing the settling time.

A test performed on formaldehyde-preserved marine phytoplankton stored for 12 to 13 years and counted in 2-ml (about 15-mm high) chambers showed that in particular cases a period of more than 24 hours was necessary to ensure sedimentation of the algae (Hasle, 1969, p. 19). The error due to incomplete sedimentation never exceeded a factor of two, and it varied from one species to the other. In addition to colony formation, shape, size and silicification appeared to be decisive factors. No similar test was performed on other sizes of chambers but, based on experience with the 2-ml chambers, a settling time of at least 40 hours was recommended for formaldehyde-preserved marine phytoplankton independent of chamber size.

SPECIAL DEVICES

Home-made bipartite chambers have been made to be used with a standard compound microscope as well as with an inverted microscope (e.g. Lund, 1951; Dawson, 1960), the principle being the same as for the chambers discussed above. A similar two-piece chamber with a sliding-shutter assembly manufactured from nylon and cemented to the bottom of the sedimentation tube was constructed by Tungate (1967) to be used under a standard microscope. The tube is slightly larger at the bottom than at the top and this has been shown to reduce the number of diatoms sticking to the chamber walls.

The Throndsen (1970a) chamber is one of the small chambers which can be used instead of examining a small portion of the bottom of commercial chambers. It holds 2 ml, has a diameter of 16 mm and a height of 10 mm (Fig. 20). The main part of the chamber is a perspex cylinder. The model shown on the figures has two different basal diameters to fit two different plankton microscopes. The bottom of the chamber is a circular coverslip with a thickness of

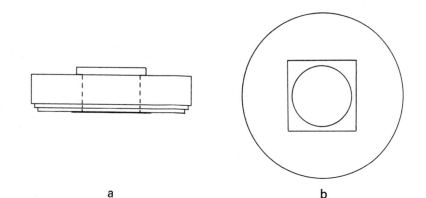

a   b

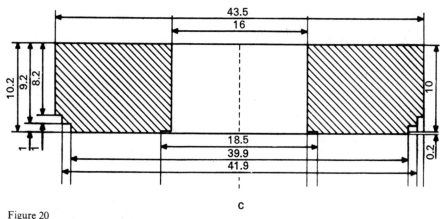

c

Figure 20
Throndsen's sedimentation/counting chamber: (a) side view; (b) top view; (c) vertical cross-section (redrawn from Throndsen, 1970a). Height and diameters are in mm.

about 0·2 mm or slightly less if an immersion objective will be used. To fit the bottom glass a 0·2 mm deep groove is cut along the margin of the inner cylinder. The bottom may be conveniently fixed to the cylinder by nail polish, which will last for a fairly long time and which has the advantage that it is easily removed when the coverslip breaks. More permanent glue may also be used.

Utermöhl (1958) recommended the use of a special filling chamber which had a sieve bottom plate, in order to produce a homogeneous distribution of the sediments on the chamber floor. Nauwerck (1963) found that this procedure made little difference and a homogeneous distribution of the plankton could never be guaranteed.

Tangen (1976) constructed a sedimentation table (Fig. 21) which provides a firm base for chambers during filling, sedimentation and drainage of the supernatant water. A circular spirit level is fixed in the table plane to set it

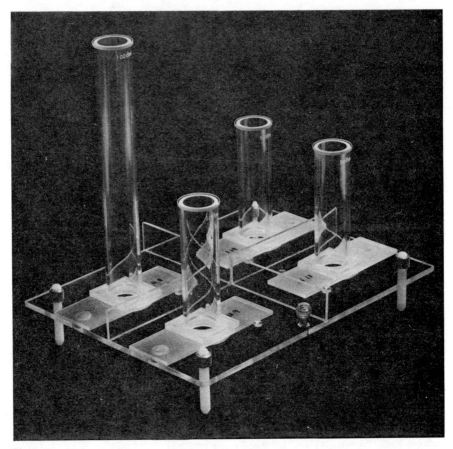

Figure 21
Sedimentation table with combined plate chamber set for sedimentation (after Tangen, 1976).

horizontally, which assures a more homogeneous distribution of the sediment on the chamber floor. Moreover, experience has shown that leakage rarely occurs when the plate chamber is exactly horizontal (Tangen, 1976). The table has openings into which the chambers fit and smaller openings corresponding to the smaller drain holes of the plate chamber. When the sedimentation cylinder is removed, the water is easily collected in a petri dish placed under the drainage hole.

94

SPECIAL PRECAUTIONS

The first step in order to obtain a representative subsample drawn from the storage bottle is a thorough, but gentle, shaking of the sample before it is poured into the sedimentation chamber. Shaking 100 to 200 times by hand is considered necessary to loosen organisms attached to the bottom of the storage bottle. Utermöhl (1931) emphasized that the shaking should be done in such a manner that circulating water movements are avoided. (Other planktologists prefer to subsample by a large-bore pipette which is slowly emptied in the chamber while progressively raised from chamber bottom to opening, the lower end of the pipette remaining barely immersed.)

Care should be taken that the sample is brought to room temperature, before filling the chamber, to prevent formation of air bubbles on the chamber walls. It is also important that the temperature be kept fairly constant during the sedimentation period to prevent convection currents. Sun-ray exposure must particularly be avoided.

Evaporation will often occur, causing formation of air bubbles under the top plate of the counting chambers after removal of the sedimentation cylinder. Air bubbles hinder a clear optical image and should be avoided. This may be done by covering the sample with the lid of a petri dish lined with wet filter paper. If air bubbles arise under the top plate of a 2-ml chamber during the settling period, the plate is gently pushed aside and a drop of distilled water is added.

The chambers should be cleaned immediately after use. It is particularly important to clean the glass bottom properly. This is easily done when dealing with commercial chambers where the glass bottom can be unscrewed from the metal ring, although the glass breaks easily and must be replaced frequently. Chambers with more permanently attached bottoms can be cleaned by a wooden stick or pincers coated with cotton to avoid scratching the glass.

Grease, elastic bands and clamps have been used to keep the top cylinder and the plate chamber tightly together to ensure that no leakage will take place when the sample is left for sedimentation. One should bear in mind that grease may easily damage the sample.

SIZE OF SUBSAMPLES

Choice of adequate volume depends on the material to be examined and the goal of the investigation.

When the phytoplankton concentration can be characterized as dense or medium-sized (e.g. in coastal waters, fjords, bays), sedimentation of two subsamples, one smaller and one larger, is usually an advantage. The smaller may be 2 ml and the larger 10 or 50 ml. Although the phytoplankton population may be sparse in inshore waters at some seasons, concentration of larger subsamples is not recommended because of the heavy load of detritus usually present in such waters. In studies of the seasonal distribution of cell concentration and species composition it may be an advantage to use the same volume throughout the year. This is particularly applicable if emphasis is laid on number of species, e.g. by calculation of species diversity, since the number of species often varies with the size of subsample (Lohmann, 1920; Hasle, 1959 and Chapter **8.3** herein). If, however, the goal of the investigation is an estimation of total phytoplankton volume, only a few of the most abundant species are of interest (Willén, 1976). These species are usually found in such numbers that concentration of a small, e.g. 2 ml, subsample will suffice.

When dealing with sparse phytoplankton populations (e.g. from greater depths or oceanic, barren waters) subsamples of 50 or 100 ml or even more may be needed to answer the problems being investigated. Since 100-ml chambers are not to be recommended (Nauwerck, 1963), 50-ml chambers should be used. This may be done by settling the organisms in successive 50-ml cylinders into the same plate chamber. The organisms present in a multiple of 50 ml will in the end be settled on the floor of the plate chamber. If there are small species present in such small cell numbers that the whole of the large bottom has to be examined by a high magnification to get a reliable cell count, it may be desirable to pre-concentrate the material before it is introduced into a 2-ml chamber for final sedimentation. The pre-concentration may be done by centrifugation (Chapter **5.3**) or in a 50-ml chamber by passive settling (Hasle, 1959). In the latter case the loss by pipetting off the supernatant water and transferring the sediment from one chamber to the other was estimated at less than 10 per cent (7 per cent as the mean of eight observations). Care must be taken to rinse the walls of the larger chamber with the supernatant water and also to use this water to fill up the chambers if needed. This is done to include within the concentrate organisms which are possibly attached to the walls, as well as to prevent a possible dissolution of coccoliths.

5.2.2 Settling without the inverted microscope

I. N. Sukhanova

The settling method, first proposed by Volk, has been given wide use in oceanological studies in the U.S.S.R. (Morozova-Vodianitskaya, 1954; Usachev, 1961; Kusjmina, 1962; Semina, 1962, 1974; Vinogradova, 1962; Sukhanova, 1973, 1976, etc.). Water from the sampler is poured into a 1-litre vessel, which has a height of about 20 cm. The sample is immediately fixed with 40 per cent formalin in the proportion 1:100 and covered with a lid. The sample is allowed to settle for a week and then slowly decanted (by drops) through a glass tube—a siphon with an elongated end bent upwards 2 or 3 cm. A rubber tube of appropriate diameter is fitted on to the external end of the siphon. The siphon is placed into the vessel so that its distance from the bottom of the vessel is 1 to 1·5 cm and the distance of the elongated end from the bottom is about 3 to 4 cm. The water is siphoned off until the water surface is just a few millimetres above the siphon end. At that moment, the siphon is carefully removed, and the rest of the sample, some 150 to 200 ml, is poured into a glass of corresponding volume. The original vessel is rinsed twice with filtered seawater. Samples in this form are easy to transport. For subsequent processing, a sample is settled again and decanted down to a volume of 50 to 100 ml.

A phytoplankton-rich sample needs no further preliminary processing. A less rich sample is centrifuged using 20–25 ml conical tubes in a cup-type centrifuge at 3,000 rev/min (centrifugation methods are fully described in Chapter **5.3**).

The primary advantage of this method is that it requires simple glassware only and no special equipment at the initial stage of field sampling. The shortcomings of the method are: (a) the sample has to be fixed, which leads to a partial or complete decay or deformation of some phytoplankters (Hasle, 1969), although good preservation of diatoms and most dinoflagellates is ensured; (b) samples should be of a standard volume; (c) algae having a specific weight of less than unity are not concentrated; (d) cells which adhere to the walls of settling or centrifuging containers get lost, as well as those which roll under the sediment when the centrifuge stops, and this probability increases with the number of concentration stages.

5.3   Centrifugation

Jahn Throndsen

Centrifuges may be used for concentrating living or preserved material in the laboratory. Provided that the boat or ship is stable enough to avoid difficulties from the gyro effect of the rotating head, the centrifuge is a suitable aid for shipboard concentration work as well.

Centrifuges used in phytoplankton surveys are of two main types: the cup-type centrifuge with the samples enclosed in special tubes (Fig. 22A-B), and the continuous centrifuge in which the fluid containing the algae is passed through the centrifugation chamber (Fig. 22C-D). Another type, hitherto not used in plankton research, has a continuous flow through radially arranged chambers (Beckman Elutriator rotor) and may separate cells of different sedimentation rate.

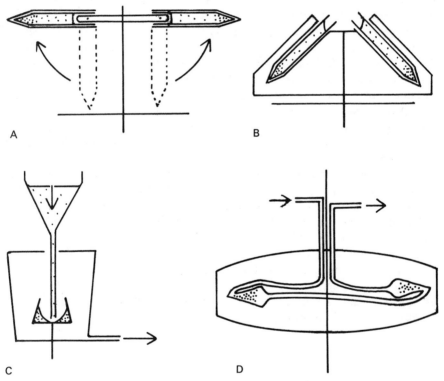

Figure 22
(A) Cup-type centrifuge, swing-out type. (B) Cup-type centrifuge, fixed-angle type. (C) Continuous centrifuge, simple-rotor type. (D) Continuous centrifuge, counter-flow rotor type.

CUP-TYPE CENTRIFUGES

The tubes containing the material are usually made of glass or plastic. The shape of the centrifuge tube is important for the efficiency and the convenience of the method, and a narrow lower part is usually to be preferred. During centrifugation the tubes may be allowed to swing out to adjust themselves in the plane of the centrifugal force (Fig. 22A) or they may remain in a fixed oblique position (Fig. 22B).

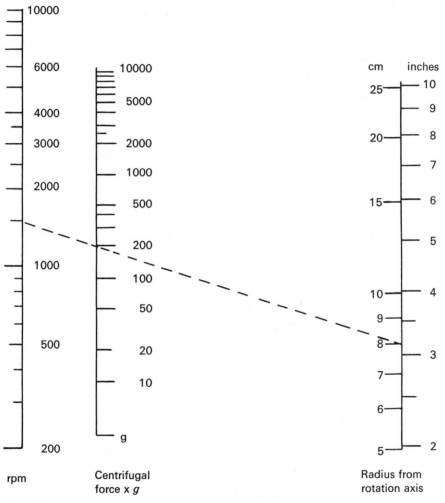

Figure 23
Nomogram to show relation between speed of the centrifuge in rev/min (revolutions per minute), relative centrifugal force and distance of the sample from the axis of rotation; a straight line between rev/min value and the value for the radius of rotation will cut the relative centrifugal force rule at the corresponding value.

For practical estimation of the centrifugal force acting on the sample, the relation between rotation speed (rev/min) and radius of the spin is shown in Figure 23.

The advantage of the swing-out type is that the centrifugal force will be directed along the length axis of the tube, the organisms settling in the basal end of the tube. Its disadvantage is the formation of currents in the fluid when the centrifuge is about to stop and the tube moves from a horizontal to a vertical position (Fig. 24A-B). These currents may stir up the sediment and hence resuspend the material. In order to avoid this, Steemann Nielsen (1933) sealed the centrifuge tubes and fixed them in a horizontal position, but he failed to achieve any significant improvement of the method.

The fixed-position rotor head may, on the one hand, cause settling of cells on the side of the centrifuge tube but, on the other hand, the problems with currents being created during the end of the run are minimized.

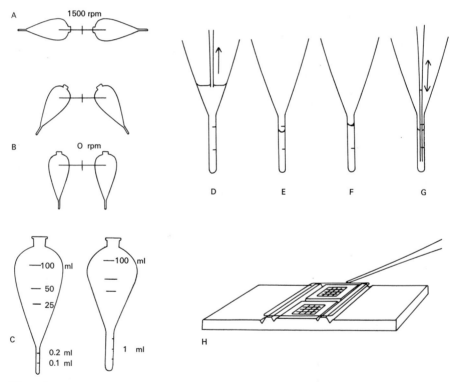

Figure 24
Cup-type centrifuge method: (A) position of the centrifuge tubes at working speed; (B) the last part of the slowing-down phase should be as short as possible to reduce the resuspension of precipitated material; (C) two versions of pear-shaped centrifuge tubes; (D) siphoning off the excess water; (E) waiting for the water of the centrifuge-tube walls to drain off; (F) adjusting the volume exactly to the 0·1 ml mark; (G) resuspending the material settled, by use of an empty pipette; (H) filling the counting chambers.

The following method using a swing-out centrifuge is applicable for ship-board counting of phytoplankton samples in oceanic waters, but it may be used in other areas or in the laboratory as well, provided that the amount of alien particles is low.

Equipment

A centrifuge taking 100-ml pear-shaped centrifuge tubes of Kimax (No. 45220, Kimble Products), or ASTM D96 (American Standard Test Methods) type, counting slide, microscope, simple siphoning or suction device for taking off excess fluid after centrifugation, Pasteur-type pipettes for handling the concentrated sample, 2 per cent aqueous solution of OsO_4.

Procedure

1. The 100-ml pear-shaped centrifuge tubes are filled with the water samples (Fig. 24c). Tubes may have a relatively wide basal part of 1 to 1·5 ml (ASTM D96) or a narrow one of 0·1 to 0·2 ml (Kimax No. 45220). These large-volume tubes should be adjusted to the same weight and arranged in pairs as usual.
2. Three to four drops of 2 per cent aqueous osmic acid (OsO_4) solution are added and the centrifuge is set to spin at 1,200 to 1,500 rev/min (200 × g) for 20 to 30 minutes.
3. When centrifugation is finished, water is siphoned off to slightly below the 0·1- or 0·2-ml mark (Kimax) or the 1-ml mark (ASTM). Let the water drain off the walls and adjust the fluid level exactly to the mark with supernatant water. Then, with a dry pipette, resuspend the sediment thoroughly (Fig. 24D-G).
4. A ruled counting chamber of 0·1- to 0·2-mm depth, e.g. an Improved Neubauer or a Fuchs–Rosenthal haemacytometer cell (see Chapter **7.2.2** and Figs. 32–34 for details) is filled with the concentrated plankton suspension (Fig. 24H). Let the material settle for 1 to 2 minutes and then count the cells at about × 200 magnification under a standard compound microscope with phase-contrast illumination.

When a concentration of 100 ml to 0·1 ml has been achieved through steps 1 to 3 above, counting of the whole ruled area of the Fuchs–Rosenthal cell will correspond to 3·2 ml of the original sample (0·9 ml for Improved Neubauer). With the use of a larger chamber such as the Palmer-Maloney chamber (see Chapter **7.2.2**, Fig. 32), the total volume of which is 100 μl, the volume will correspond to the whole concentrated sample, i.e. 100 ml.

Reliability

The reliability of the method depends on the efficiency of the settling of the cells, accuracy of the centrifuge-tube ruling, precision by which the final volume is achieved, homogeneity of the resuspended sample, precision in filling the counting chamber, the counting (and identification) of the cells, and the statistical variation between the subsamples due to their relatively small size.

Experience on the reliability of the centrifugation method is varied. Lohmann (1908) found that the number of organisms lost by his version of the centrifuge method was negligible ('meistenteils ohne Bedeutung'), whereas Steemann Nielsen (1933) found a relatively poor (11 to 64 per cent) recovery when he compared Lohmann's centrifuge method with Utermöhl counts (on diatoms and coccolithophorids). Ballantine (1953) made comparisons with direct counts, membrane-filter-concentrated material and serial dilution cultures, and she concluded that, of the methods tested, 'centrifugation of a living sample, followed by counts of the numbers of organisms in the concentrate, appears to be the most satisfactory method from all points of view'. The centrifuge method described above (i.e. with osmic-acid fixation) was used during the SCOR Working Group 15 Cruise in May 1970, and the counts could be compared with counts on formaldehyde-preserved samples by the inverted-microscope method. The three examples in Table 5 show that, except for the coccolithophorids, cell-number estimates from the centrifuge samples were generally the highest. Small diatoms (A–C) may have dissolved in the weakly alkaline water of the sedimented sample.

TABLE 5. Cell-concentration estimates from three oceanic stations in the Caribbean Sea (stations 05 and 17) and the eastern Pacific Ocean (station 09), sampled during SCOR Working Group 15 Cruise in May 1970; concentrated by cup-type centrifugation (fixed with osmic acid) and by sedimentation (preserved with neutralized formaldehyde solution, inverted-microscope method)

Station	Centrifugation	Sedimentation
A. STATION 17 (16°52' N, 76°04' W, 100 m)		
Dinoflagellates	4,800	1,020
Coccolithophorids	12,600	8,200
Diatoms	10,800	180
Monads, flagellates		
<2·5 μm	30,000	Not counted
2·5 to 5 μm	7,200	1,440
B. STATION 05 (9°49' N, 79°41' W, 10 m)		
Emiliania huxleyi[1]	15,000	30,040
Gephyrocapsa oceanica	3,600	2,100
Gymnodinium simplex	900	300
Katodinium rotundatum	300	140
Chaetoceros diversus	900	340
Pennate diatoms	5,400	4,660
Centric diatoms	3,300	2,860
C. STATION 09 (3°19' S, 85°04' W, 0 m)		
Dinoflagellates	13,200	1,900
Coccolithophorids	3,600	10,280
Centric diatoms	11,000	420
Pennate diatoms	28,800	1,360

1. Emiliania huxleyi = Gephyrocapsa h. = Coccolithus h. = Pontosphaera h.

102

The advantage of the cup-centrifuge method is that it can be applied to studies on either living, narcotized or fixed samples as the procedure takes only 20 to 30 minutes (enumeration not included). It may be used at sea as well as in the laboratory and a standard phase-contrast microscope only is needed for the enumeration work (in addition to suitable counting chambers).

CONTINUOUS CENTRIFUGES

Continuous centrifuges may be of two types, either with a simple rotor head in which the material is collected on the chamber wall (Fig. 22c), or with a counter-current flow rotor in which the cells are concentrated in suspension (Fig. 22D). The Kimball and Wood (1964) continuous centrifuge belongs to the first type, as does a similar instrument, Kahlsico, commercially available from Kahl Scientific Instrum. Corp.; these are probably the only continuous centrifuges designed especially for phytoplankton studies (counter-flow centrifuges are still awaiting application to plankton research). Kimball and Wood (1964) assumed their centrifuge to be useful in qualitative phytoplankton work, and it has been used in combination with cup-type centrifugation (e.g. Wood, 1968). Quantitatively it was found to retain cells better than GF/C (Whatman-glass filter pads) as far as flagellates (*Dunaliella* and *Gymnodinium*) were concerned; however, poorer results were obtained for diatoms such as *Phaeodactylum* (Kimball and Wood, 1964). G. Vargo (unpubl.) again tested the Kimball and Wood centrifuge for a variety of cultured species and natural populations; retention, as measured by chlorophyll concentrations in the effluent, was found to range from 18 to 98 per cent according to delivery rate and species composition.

Material concentrated by the continuous centrifuge is thoroughly re-suspended in a known amount of seawater (of the same salinity and temperature as the sample, filtered if necessary). A subsample of the suspension is then counted in a chamber, e.g. of Sedgwick-Rafter type (cf. Chapter 7.2.2). If the sample is to be kept for some time, addition of a fixing and preserving agent may prove necessary.

The advantage of the continuous centrifuge is, most of all, that a large sample volume can be used. The efficiency in retaining the phytoplankton cells, however, depends on the flow rate through the centrifuge, and the flow rate to give sufficiently good retention has to be determined for each centrifuge and type of material. Kimball and Wood (1964) found 1 litre/2 minutes suitable and 80 to 90 per cent of natural phytoplankton was then retained by their centrifuge. Littleford *et al.* (1940) used a flow rate of 1 litre/12 minutes in a Foerst centrifuge, and found only a 5 per cent increase in the number of organisms retained when the sample was recirculated. Comparison with a cup-type centrifuge showed that the latter gave about 30 per cent lower numbers of organisms (Littleford *et al.*, 1940).

5.4   Filtration

5.4.1   Reverse filtration

Anne N. Dodson
and William H. Thomas

When phytoplankters are required in a live and undamaged condition, they may be concentrated gently by removal of water that flows upward through a filter. The reverse concentrator consists essentially of: (a) a vessel for containing the sample; and (b) an insert of smaller circumference having some type of filter on the lower end and open at the upper end. The water seeps upward through the filter and is drawn off, leaving the organisms more concentrated in the sample vessel. This method was developed to reduce the damage to cells which results when a sample is poured down through a filter. Gravimetric filtration entraps cells to a much greater extent, and some cells become distorted by the downward pull and squeeze through even though the filter is rated at a pore size smaller than the cells. With reverse filtration, the upward force of the water through the filter is so small that cells smaller than the pore size are not pushed through.

DEVICE AND PROCEDURE

The following description is taken with slight modifications from Dodson and Thomas (1964). Figure 25 is a diagram of a basic concentrator. The sample vessel is a plastic or glass 2·5-litre beaker, 15·5 cm in diameter, 17 cm deep, and the insert is a clear plastic tube, 12·5 cm in diameter, 15 cm deep. Large-pore (60 μm) nylon plankton netting is pressed into the plastic after it has been temporarily softened with ethylene dichloride. Excess netting is trimmed away after rehardening. This netting serves to support 15-cm diameter filter paper which is stretched over the tube and held in place with a wide rubber band. The paper is pressed to conform to the tube with many small creases, and the edge is held so that the wide rubber band extends beyond the filter on to the tube, sealing the edge so that water and organisms cannot enter the filter tube through the creases.

The filter tube is placed, filter side down, into the sample container. Filtrate seeps up through the paper and is removed by siphoning, using small-bore flexible plastic tubing clamped to the side of the filter tube; 1·5 to 2·0 mm tubing carries off filtrate at approximately the same rate that it seeps through and is self-priming through capillary action in case the siphon should break during the process. The weight of the filter tube is adequate to exert downward pressure on the sample. The type of filter paper used determines the extent to which phytoplankton become embedded, the firm grades having fewer trapped cells than soft, porous paper. The exposed area of the filter also affects the

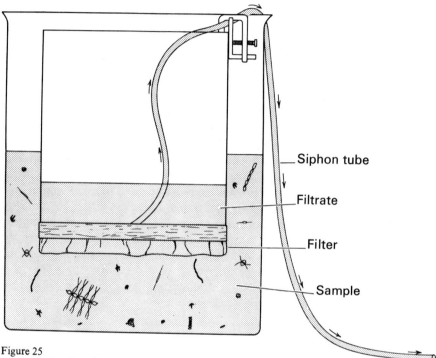

Figure 25
Schematic drawing of a reverse-filtration device.

flow rate. With this size concentrator and No. 42 Whatman paper, 2 litres of sample can be reduced to 20 ml in approximately one hour, although the rate would vary somewhat depending on the richness and species variation in the sample.

MODIFICATIONS

Refinements on the basic concentrator have proved useful for specific purposes. Pomeroy and Johannes (1968) cemented a known pore-size membrane (0·8 µm) directly to the bottom of a 15-cm diameter × 15-cm-high filter tube, maintained a constant head in the sample vessel, and used an aspirator to draw off the filtrate. Using 8 concentrators of this size, 200 litres were reduced to 10 ml in 2 to 3 hours. Griffiths and Fee (unpubl.) used a variety of mesh sizes of Nitex (Kressilk Products Inc.) cemented to filter tubes to separate size fractions of

105

preserved zooplankton, and reported a minimum of entrapment of delicate spiny Crustacea. Holm-Hansen *et al.* (1970), in their study of the limitations of reverse-flow filter technique, fabricated a refined version using plexiglas (Rohm & Haas) which incorporated a constant head in the sample, aspirated filtrate, and also a teflon pressure ring to hold a stainless-steel support screen and filter against the bottom of the filter tube. With this apparatus and using 142-mm-diameter membrane filter (pore size not given), the concentration of 20 litres to a volume of 300 ml in about 3 hours was reported.

A continuous reverse-flow concentrator developed by Hinga and Sieburth (in prep.) has been used gently to concentrate organisms greater than 3 µm in size (Sieburth *et al.*, 1976). It uses a Nuclepore (Nuclepore Corp.) membrane held sandwich-like between thin upper and lower chambers. The water in a reservoir passes through the lower chamber where the concentrating particles are kept from clogging the membrane by a gently rotating mixer or magnetic stirring bar, and the effluent with the plankton of less than 3 µm passes into the upper chamber and out of an effluent tube the height of which, along with that of the reservoir, dictates processing speed. Two-litre chambers with a 68 × 68 cm supported membrane can process at rates up to 1 litre per minute and process 200 litres in less than two hours in a double-decked processor. Volumes from the large concentrator can be reduced in a 70-ml, 10 × 10 cm chamber.

Recently, another reverse-filtration technique has been described by I. N. Sukhanova (in Sorokin *et al.*, 1975). It implies a several-step procedure using a 3-cock filtering device; the filter is a nylon sieve of 15 to 20-µm pore size. It can be used on board ship and allows two categories of cells (those carried away in the filtrate as well as those retained on the filter) to be subsequently studied or enumerated.

EVALUATION

This method of concentrating the phytoplankton in a seawater sample has intrinsic limitations. The pore size of the filter paper or membrane defines approximately the lower size limits of the organisms retained, and the filters may entrap varying numbers and kinds of organisms. Holm-Hansen *et al.* (1970) suggest an easy method using chlorophyll *a* analysis for arriving at a correction factor. Replicate samples preserved and settled for microscopic examination could also be used as a check on species percentages if a more quantitative differentiation is desired.

The original purpose for which the concentrator was designed was to reduce the amount of water in a sample to be examined for phytoplankton in the size range above 5 µm. With the water removed, the phytoplankton, originally in a large volume, could be examined in a few pipetted aliquots; for example, 1,000 ml reduced to 10 ml could be examined in two 5-ml aliquots. Very few phytoplankters are damaged by this method, and those required for isolation are collected in a small amount of water, saving long hours spent looking at many aliquots in order to find enough cells for isolation. Loosely entrapped organisms can be washed into the concentrate with a small amount of filtrate sprayed on the bottom of the filter from a fine-tipped wash-bottle. For collecting undamaged phytoplankton with a minimum of sample handling in a short period of time, concentration using the reverse-filtration method is most satisfactory, and with attention to the inherent limitations, the principle may be adapted satisfactorily for other uses, such as the size fractionation of zooplankters previously mentioned.

5.4.2 Membrane filtering

Robert O. Fournier

Since 1952, when Goldberg, Baker and Fox introduced membrane filters into marine phytoplankton research, numerous attempts have been made to broaden their usefulness. Today membrane filters are used for concentration and direct microscopic observation of phytoplankton (Holmes, 1962), for concentration and resuspension of living cells (Clark and Sigler, 1963), for gravimetric or elemental analysis (Banse et al., 1963; Menzel and Dunstan, 1973) and as a preparative step in scanning and transmission electron microscopy (Fournier, 1970; Bisalputra et al., 1973; Paerl, 1973). The basic assumptions and methodology employed in all of the above are similar.

The advantages of using membrane filters over other forms of phytoplankton concentration include: high filtration rates coupled with positive retention of all particles above a stated pore size; a wide range of pore sizes suitable for various filtration or screening procedures; organisms concentrated on to filters may be examined directly with high-magnification light microscopy, scanning electron microscopy, or embedded for transmission electron microscopy. Finally, the apparatus involved is light, compact and easy to use in the field even in the absence of electrical power (a bicycle or automobile tyre pump with the valve reversed makes a suitable vacuum pump; De Noyelles (1968) used a syringe with a weight attached). The disadvantages, however, include: difficulties in taxonomic characterization of organisms viewed on filters since they cannot be physically manipulated; the question whether or not cells settle on to filters in a random distribution has never been adequately resolved. Finally, and most important, many cells are distorted and unrecognizable or totally destroyed by the rigours of filtration.

Attempts have been made to develop techniques using membrane filters that will retain all organisms, even the most fragile, in an identifiable state (Goldberg et al., 1952; Holmes, 1962; De Noyelles, 1968; Bisalputra, 1973). Each author has claimed that his respective technique will work on fragile forms although only Holmes specifically excluded the so-called µ-flagellates. The fact remains that no one has yet undertaken a survey of the various planktonic genera. Diatoms, dinoflagellates, coccolithophorids, coccoid blue-greens and some cryptomonads have been shown to withstand filtration under the proper circumstances. However, most marine Chrysophytes and Haptophytes possess little more than a periplast (or a few non-structural scales) to contain the organelles and retain the taxonomically critical form. Many of these organisms are so delicate that the application of a coverslip or a vital stain to a drop of suspension on a microscope slide is sufficient to promote loss of flagella and/or haptonema plus irreversible loss of shape. Even the best fixatives, such as Lugol,

108

can render many of these unrecognizable (Bernhard *et al.*, 1967), and when followed by filtration the latter are destroyed (Holmes and Anderson, 1963).

Therefore, until it can be conclusively demonstrated that the small flagellate component of the plankton can be filtered and remain intact and recognizable, this technique has a clear bias towards the more robust members of the plankton and it should not be used as a general method of assessing phytoplankton abundance. Rather it should be used when a specific component of the plankton is being monitored or if the losses of unknown quantities of small delicate forms do not have serious consequences.

FILTERS

The term membrane filter was used by Goldberg *et al.* (1952) to describe membranes composed of incompletely cross-linked molecules of cellulose acetate and cellulose nitrate. Today the term has a slightly more generic connotation which sometimes includes silver filters, glass-fibre filters and nuclepore-like filters. Silver and glass-fibre filters will not be mentioned further since they do not possess all of the advantages ascribed earlier to membrane filters. Nuclepore filters, unlike the cellulose-acetate variety, have a smooth flat surface containing uniform cylindrical pores formed by chemical etching following neutron impingement. The nuclepore offers the easiest filter from which concentrated cells may be removed for resuspension. However, this relatively minor advantage is offset by the fact that they possess much less filtering capacity, less resolution when viewed at high magnifications, and personal experience suggests that they occasionally allow small, easily distorted cells to squeeze through the pores.

Cellulose-acetate filters have a microscopically coarse surface with many minute peaks and depressions which undoubtedly increase the difficulty of removing adhering cells. Resuspension of filtered cells is probably not quantitative with either filter except for robust forms such as those studied by Clark and Sigler (1963). Holm-Hansen *et al.* (1970) found that even with the very gentle reverse-filtration technique small cells adhered to the filter surface and were not easily dislodged.

For the remainder of this section only cellulose-acetate membrane filters will be considered since they are clearly the most versatile filters at present available, lending themselves to filtration, separation, direct microscopic examination and various cytological procedures. Although several companies sell them (e.g. Millipore, Gelman and Sartorius) they appear to differ only in terms of price, availability and personal preference rather than performance. A pore size of 0·45 to 0·8 μm is suggested since the smallest known phytoplankter, *Micromonas pusilla*, has 0·8 μm as its smallest dimension (Manton and Parke, 1960).

109

STANDARD PROCEDURE

Keeping in mind the limitations expressed above, the recommended procedure for concentrating phytoplankton is as follows:

1. Collect a water sample of a few ml to 500 ml, with 25 ml recommended for estuaries, 50 ml for inshore and 150 ml for offshore. The most appropriate volume for a particular location should be empirically determined, keeping in mind that too many organisms will be difficult to examine since they will rest one upon the other.
2. The sample should be placed in a container on which the desired volume has been marked and which contains 3 to 6 per cent neutralized glutaraldehyde or neutralized formalin freshly made up from paraformaldehyde.
3. Agitate the sample gently but firmly by inverting the collecting container at least 10 times.
4. Pour entire contents into collecting funnel above a 25 mm-wide membrane filter; the tubes available from the filter manufacturers usually have a 15 to 25 ml capacity so the addition of any funnel extension ensures that the sample will remain mixed while filtering (Moore, 1963).
5. Apply gentle vacuum at no more than 1/6 atmosphere (125mm Hg). Dozier and Richerson (1975) recommend a vacuum of no more than 1/15 atmosphere (50mm Hg) as a means of preventing destruction and minimizing distortion of fragile forms. Clearly, the gentler the filtration the better the results but this can be accommodated empirically to individual needs. Filter entire contents through filter until dry, then add 5 ml of distilled water and filter once again to remove salt.
6. Remove filter and place in a desiccator until ready for mounting on slide.
7. Obtain microscope slide and inscribe relevant identifying information at one end with diamond or hard metal scribe, then clean slide thoroughly.
8. Place dried filter on to a drop of immersion oil previously spotted in centre of slide, then add 2 to 3 drops on top side of filter. Filter should become transparent.
9. Place 25 mm round or square coverslip on top of filter and press downward. Usually the oil and coverslip keep the filter flat but occasional, persistent bowing upward can be reduced by the application of a small weight overnight. This procedure is harsh but quite acceptable for the most robust forms such as diatoms, coccoid blue-greens, some coccolithophorids and greens. It would also be satisfactory as a preparative step for scanning electron microscopy (Paerl, 1973).

GENTLE FILTRATION

If a mixture of cells, both fragile and robust, is assumed, then considerably more caution is called for. This technique differs principally from the previous one in that the cells are not allowed to dry out on the filter prior to mounting

110

on a slide. When 1 to 2 ml of sample remains to be filtered, then the technique of Holmes (1962) is employed:

1. Wash the filter with successively more dilute seawater (75, 50, 25 and 10 per cent) and finally with distilled water made basic (pH *c.* 7·1 to 7.5) by the addition of ammonium hydroxide.
2. The material on the filter is next dehydrated by washing successively with 10, 30, 50, 75 and 95 per cent aqueous ethanol. The amount of reagent used in each of the washings depends upon the filter diameter—generally 10 to 15 ml is adequate.
3. The filter disc is then covered with alcoholic Fast Green (0·1 per cent in 95 per cent ethanol, or alternatively Fast Green may be dissolved in water and added prior to alcohol washes) and allowed to stand for about 20 minutes.
4. The Fast Green solution is then passed through the filter disc (with suction), and the disc is rinsed with 10 to 15 ml of absolute ethanol. Staining with Fast Green, although an optional step, makes it easier to locate organisms on the filter disc and to distinguish small detrital particles from small flagellates.
5. The filter disc is carefully removed from the filter assembly, and is placed, filtering surface up, on a few drops of a clearing agent, e.g. Canada Balsam, on a glass microscope slide.
6. After the filter disc is cleared (usually less than 10 minutes), the excess clearing agent is removed from the bottom of the filter disc by dragging the filter across a clean microscope slide with forceps.
7. The filter is transferred with forceps on to a few drops of toluene or xylene on another clean microscope slide.
8. A few additional drops of xylene or toluene are added to the upper surface of the disc, and a clean coverglass is carefully lowered on the preparation.
9. The slide is dried at room temperature on a warming table or in an oven at about 40° C. Once again a small weight on the cover glass will help flatten out any wrinkles in the filter. After drying, the preparation is ready for enumeration.

SPECIAL PROCEDURE FOR CYTOLOGICAL STUDIES

An alternative approach is that described by Bisalputra *et al.* (1973):

1. The phytoplankton is filtered prior to fixation.
2. When several ml remain in the filter funnel, the material is fixed by adding 5 to 10 ml of a solution containing 2·5 per cent glutaraldehyde and 0·5 per cent paraformaldehyde in a buffer solution composed of 0·15 M sodium cacodylate and seawater in the ratio 2:3 (v/v), and enriched with 6 per cent sucrose.
3. The filter funnel is then covered with a watch glass and allowed to stand for 1 to 2 h.
4. At the end of the fixation period the fixative that has not drained through should be removed by gentle suction.

5. Fixation is followed by three rinses of approximately 5 to 6 ml each of the above buffer, the first containing 6 per cent sucrose and the remaining two without. The buffer should be allowed to drain by gravity or gentle suction such that each rinse lasts for 5 to 10 min.

6. Post-fixation is carried out by adding 3 to 5 ml of 1 per cent osmium tetroxide (OsO_4), made up in the above buffer minus the sucrose, for 1 to $1\frac{1}{2}$ h.

7. Following post-fixation the fixative is gently drawn through and rinsed as before except that this sucrose-free buffer should be made up in distilled water, *not* seawater.

8. The filter and adhering material may now be placed in a vial and dehydrated through a graded series of aqueous methanol solutions (e.g. 50, 70, 90, 95 and 100 per cent).

9. Finally, after dehydration is complete and *all* water has been removed from the filter, it can be embedded either directly in Spurr's (1969) resin or the filter can first be dissolved away in a solution of equal parts alcohol and propylene oxide. If the filter is left intact it will provide a useful guide to trimming and orientation of specimens, while if it is removed it will yield thin sheets of cellular material that can be easily manipulated.

10. In either case polymerization of the resin should take place in a vacuum oven at 70°C for 12 to 18 h. Beyond this point standard electron microscopical procedures for trimming, sectioning, staining and examination can be employed.

5.5 Permanent records

5.5.1 Preliminary treatment of the sample

F. M. H. Reid[1]

The general principles of fixing and staining samples are given in Chapter 4. The different techniques for preparing permanent records may involve more specific procedures which will be indicated below in the relevant section. Fixatives are usually added prior to concentration but some of the methods to be described involve post-concentration fixation (Bisalputra *et al.*, 1973; Dozier and Richerson, 1975).

Removal of the dissolved salts of seawater is usually necessary before making permanent mounts. This is achieved by several washes in water or in a graduated series of water/seawater solutions. The pH must be carefully controlled if coccolithophorids or other calcareous organisms are present, and in some areas it is better to use soft tapwater rather than distilled water. Washing with a solution of ammonium formate isotonic with the seawater medium is recommended in some instances (Conover, 1966). This helps to reduce osmotic effects in delicate organisms; however, it has an undesirable reaction with some fixatives when heated, e.g. glutaraldehyde.

The separation of cells from accompanying material can be achieved by suspension in solutions of varying specific gravity but quantitative results are not possible. Bromoform or a saturated solution of zinc chloride have been used. Carefully controlled centrifugation or the use of filters of selected pore size are also suggested (see Chapters **5.3** and **6.2.1** respectively, for more detail).

Removal of organic matter from the cells is desirable where the morphology of the external structural elements is important (see Chapter **6.3** for specific methods for diatoms and dinoflagellates). Organic contents may also be removed by subjecting the sample to ultraviolet (UV) irradiation (Holmes, 1967; Swift, 1967). This method is useful for diatoms (either lightly or heavily silicified), coccolithophorids and thecate dinoflagellates though, sometimes, the integrity of the cell is destroyed. A suspension of living or fixed cells is placed in a quartz tube to which several drops of 30 per cent H_2O_2 are added just prior to irradiation. After 1 to 2 hours the irradiated material is allowed to settle, centrifuged gently, rinsed in distilled water, centrifuged, dried and mounted. For resistant cells it may be necessary to add more H_2O_2 and to repeat the process. Hasle (1972) modified the method for *Nitzschia* species by first adding a small amount

1. This contribution, together with Chapters **5.5.2** and **5.5.4** by the same author, was supported by U.S. Energy Research and Development Administration Contract No. AT (11-1) GEN 10, P.A. 20 to the Food Chain Research Group, Institute of Marine Resources, University of California, San Diego.

113

E

of 2·5 per cent HCl to samples to remove calcium salts. Distilled water is used for washing by centrifuging and decanting and for subsequent resuspension of the cells. H_2O_2 is added and exposure to UV irradiation for 6 hours follows. A final wash with distilled water precedes mounting.

Reimann *et al.* (1965) describe a method for enzymatic removal of organic cell contents. The cells are treated with acetone to remove lipids and pigments. After washing with diluted acetone and distilled water, a 5 per cent solution of pancreatin in a phosphate buffer (to pH 7·5) is added. The suspension is incubated for four days at 40° C, with daily washing and addition of fresh enzyme solution. According to B. E. Volcani (pers. comm.), lightly silicified diatoms are more likely to remain intact with this method, but success depends on the quality of the pancreatin enzyme solution.

5.5.2 Permanent slides

F. M. H. Reid

The production of permanent phytoplankton slides involves several steps depending on the method of concentration, the organisms of interest and the reagents used. For qualitative work strewn mounts can be simply made by pipetting a suspension of cells on to a coverslip, drying and mounting. Concentration of phytoplankton by sedimentation or filtration techniques makes it possible to retain a quantitative sample, but methods for producing permanent slides from these techniques are as yet experimental and there is disagreement as to whether there is random distribution of cells on such preparations (McNabb, 1960; Holmes, 1962; Moore, 1963; Sanford *et al.*, 1969). These are discussed further in the next two sections.

Steps involved in preparing permanent mounts include dehydration, clearing, staining, mounting and sealing (ringing) but they may not all be necessary for any given method.

DEHYDRATION

Removal of free water from the cells is of great importance unless a water-soluble mountant is used (see Table 6). However, use of the latter means that a strong seal is not formed, air pockets may develop from evaporation, and sealing is therefore necessary.

The dehydration procedure may be very simple, as in the case of diatoms. A No. $1\frac{1}{2}$ coverslip ($0 \cdot 17 \pm 0 \cdot 02$ mm) with strewn cells is dried over gentle heat or overnight and then inverted on a slide with a mounting medium of suitable refractive index, and gently warmed (see Chapter **6.3.1**). A more elaborate treatment is required for flagellates and cells whose internal structure is of interest. In this case dehydration should be a gradual process to prevent distortion. Removal of free water by substitution with another liquid such as alcohol is accomplished by using a series of gradually increasing concentrations of solutions leading to absolute alcohol. Ethyl, isopropyl or butyl alcohol may be used depending on availability and cost; however, the first is more likely to shrink and harden cells. Dioxan is a dehydrating agent which produces minimal shrinkage. The alcohol-vapour chamber technique (Bridges, 1937; Bradley, 1948) provides a method for gentle dehydration without the use of washes. A rack placed in a vessel containing about 150 ml of absolute ethyl alcohol can be used for this. Eddy and Banse (1972) suggest exposure to silica gel to evaporate the liquid down to a damp film prior to using the alcohol chamber.

CLEARING

This step is not often required for whole mounts of phytoplankton since the cells are usually already translucent. It is necessary for embedded materials. If concentration was by membrane filtration, the filter pad must be cleared before mounting, usually after thorough dehydration. If delicate cells are involved, gradual clearing may be accomplished by using mixtures of alcohol and the clearing agent. The most widely used clearing agents are xylene, benzene, toluene, cedar oil, trichloroethylene and clove oil. The first three must be used with care because of undesirable toxic or inflammatory properties. They are faster-acting than most others but xylene, especially, tends to distort cells and fade stains. Cedar oil is expensive, but needs less careful dehydration and can be used as a mountant.

Glutaraldehyde (50 per cent), which is primarily used as a fixative for flagellates, is used as a clearing agent in a method described by Dozier and Richerson (1975). Being water-soluble, the glutaraldehyde is applied to a wet filter on a slide which is then placed in a 60° to 70° C oven for about 20 minutes until clearing occurs. A coverslip is placed on the filter with a few drops of any mounting medium with refractive index (n) of about 1·5. This is a rapid process and cell walls of dinoflagellates are well defined.

STAINING

Staining techniques for distinguishing special features are frequently incorporated into the procedures. For the staining of permanent slides there are many possibilities which are described in detail in texts on microtechnique (Johansen, 1940; Clark, 1973). One of the most commonly used phytoplankton stains is Fast Green FCF (Conn, 1961) dissolved in alcohol, water or glacial acetic acid depending on the other reagents used. This stains cellulose walls and cytoplasm and is very resistant to fading. The use of acetocarmine is advocated for nuclear staining (Weiler and Chisholm, 1976) as it is a simple method requiring only a rinse with distilled water after staining. Fleming (1954) used haematoxylin and eosin or methylene blue and eosin for differential staining. Von Stosch (1974) details a method for emphasizing the thecal sutures of dinoflagellates through the simultaneous use of Pleurax (Hanna, 1949) and trypan blue (see Chapter **6.3.2**). However, Mohr (in Clark, 1973) expresses the opinion that 'exploitation of ambient fluids of different refractive index or of special optical systems . . . or combinations of these, reveal some structures more satisfactorily than stains' in the case of protozoans, and this no doubt applies to many phytoplankters.

The problem of fading of stained mounts can be mitigated by keeping the preparations out of the light and by the use of benzene instead of xylene when the latter is called for. Some artificial mounting resins have less tendency to fade preparations than does Canada Balsam. Mounting directly in Euparal after

TABLE 6. Some mounting media in general use

Media	Refractive index (n)	Solvent	Other information	Manufacturer
Natural media				
Canada Balsam	1·53	Xylene, benzene, toluene, trichloro-ethylene, dioxan	Yellows with age; bleaches stains; dries slowly but can be combined with other resins to remedy this.	
Dammar Balsam	1·53	Xylene, benzene	Superior to Canada Balsam; faster drying if dissolved in benzene.	
Karo (mixture of dextrose, dextrin, maltose)	?	Water, alcohol	Hardens so that no sealing is necessary except in moist climates.	
Styrax	1·62	Xylene, benzene	Very expensive and no longer easily obtained.	
Synthetic media				
Aroclor	1·63	Xylene	Good for diatoms (N. I. Hendey, pers. comm.).	Monsanto
Clearax	1·67	Xylene, acetone	Good for diatoms.	G. T. Gurr
Clearmount	1·51	Xylene, benzene, toluene, alcohol, dioxan, etc.	Conserves stains.	G. T. Gurr
CMCP-10	?	Water, alcohol	Quick drying but stains fade.	Turtox-Cambosco
Euparal	1·48	Xylene, alcohol	Mixture of natural and synthetic resins; can use directly after 95% alcohol; intensifies haematoxylin stains.	Flatters & Garnett
Glycerine jelly	1·47	Water	Semi-permanent; store slides flat; sealing mandatory. Add 1 g phenol crystals to 100 g jelly for stock solution.	
Hyrax	1·63	Xylene, benzene, toluene	Very expensive; good for diatoms. (Hanna, 1930)	Custom Research and Development
Naphrax	1·72	Xylene, toluene, acetone	Good for diatoms. (Fleming, 1943, 1954)	Northern Biological Supplies
Permount	—[1]	Toluene	Conserves stains; does not yellow.	Fisher Scientific
Pleurax	1·75	Alcohol	Good for delicate diatoms (von Stosch, 1974). Use recipe in Hanna (1949).	

1. 'Slightly higher than Canada Balsam.'

alcohol dehydration helps to preserve stains, but this mountant has a low refractive index.

The choice of a mountant depends on previous treatment of the cells and on a suitable refractive index. Before mounting, the preparation must have been subjected to a reagent miscible with the mountant. The mounting medium must have a refractive index as different as possible from that of the cells to be viewed. Diatom cells have a refractive index of about 1·15 and dinoflagellate cells of 1·21. Very gentle heating on a hotplate set at about 50° to 70° C is often necessary to remove bubbles and to produce a hard mount.

There is a large choice of mounting media available from many suppliers. Table 6 lists only those which are readily obtainable and more generally used in botanical work.

Glutaraldehyde is used as a mountant in the settling chamber technique described by Coulon and Alexander (1972), being added to the water sample before settling.

An inexpensive medium described by Czarnecki and Williams (1972) utilizes a polystyrene source such as styrofoam cups. Solution in toluene is followed by mixture with methylene iodide. The refractive index of the resulting medium is 1·75 and the slight stain produced by the iodide improves the contrast. Air drying overnight is mandatory.

Less permanent preparations can be simply made using glycerine jelly ($n = 1·5$).

Reimann and Lewin (1964) describe a technique for permanent preparations of *Cylindrotheca* species in which cleaned cells are air dried on a coverslip and heated on an aluminium plate to 350° C for 10 min. The coverslip is then sealed to a glass slide by ringing its edges with enamel paint. The mounting medium is simply air, and the refractive index in this case is lower than that of the cells.

Ringing is always necessary when aqueous mountants are used, but is less necessary for the other types. Commercial ringing media are available and waterproof glue, shellac, asphaltum and nail polish are also used. In any case a circular coverslip and a special turntable will simplify the process and the seal should extend 1·5 mm over the coverslip.

5.5.3 Membrane filters as permanent records

Robert O. Fournier

Once a phytoplankton sample has been collected, filtered and processed in one of the three ways described in Chapter **5.4.2**, then this filter can easily be retained as a permanent record for future examination or exchanged with other scientists. Filters placed immediately on a microscope slide and then covered with Canada Balsam, following dehydration and staining, are quite permanent. Immersion oil may be substituted for Canada Balsam in the final step without any loss of permanence. The immersion oil permits removal of the coverslip at any future time for easy subsampling of the filter, if the need arises. A diamond-tipped scribe is useful for recording useful information directly on the slide.

Filters which have been rinsed with distilled water and were then stored in desiccators (Chapter **5.4.2**) are just as permanent a record as those mounted on slides. Preferably, each filter should be stored in a small container such as a cardboard pill box or plastic petri dish. Whatever container is used, it should have one or two holes in the bottom to permit easy egress of moisture. Every container should then be inscribed with the requisite identifying information. The advantage of this approach is that the filters may be examined at any time in the future in any number of ways other than by just light microscopy. For example, a small portion of a filter can be mounted directly on to a scanning electron microscope stub, followed by shadowing and examination (Bisalputra *et al.*, 1973; Paerl, 1973).

One final way in which a permanent record can be achieved using membrane filters is to embed them in plastic for future transmission electron microscopy. Bisalputra *et al.* (1973) recommend a procedure (Chapter **5.4.2**) whereby the filter is dissolved away just prior to embedding. Thus concentrated phytoplankton can be mixed with an embedding medium and placed on a microscope slide for light microscopy, and embedded directly prior to thin sectioning for transmission electron microscopy. Lastly it should be pointed out that these embeddings can at any time be *thick* sectioned using an ultramicrotome and then placed on a slide for light microscopy.

5.5.4 Settling techniques for permanent records

F. M. H. Reid

The Utermöhl technique for concentrating plankton does not lead to a permanent record unless the settled materials are photographed. Three methods which produce slides for use with a conventional microscope are available; however, they were devised for freshwater phytoplankton, and the problem of removing dissolved salts arises when marine samples are involved.

In the settle-freeze method (Sanford *et al.*, 1969) organisms are settled on to a microscope slide, the bottom few millilitres of liquid are frozen with dry ice, and the supernatant poured off. Dehydration and mounting follow. The superior preservation of delicate organisms and of flagellae seen in freshwater samples is partially negated in marine samples as rinsing is necessary to remove dissolved salts. It is also unlikely that a random distribution is maintained.

Dickman (1968) describes a method in which the sample water is settled into a well at the bottom of which is a coverslip spread with a water-soluble mounting medium such as CMCP-10. The settling tube and supernatant can be slid off and the well contents allowed to evaporate. The coverslip is freed and inverted on to a slide with a drop of mounting medium. For marine material it is necessary to add the preliminary steps of sedimentation, removal of supernatant and a distilled-water wash before proceeding as above, thus introducing the potential for cell loss and disturbance.

Coulon and Alexander (1972) described a technique which probably causes less cell damage and retains a more natural distribution with marine samples. They use a three-part settling chamber constructed from three glass slides and a length of glass tubing. Holes equal to the inner diameter of the tubing are drilled in two of the slides and the tubing is cemented to the top slide. The top slide can be moved back and forth on the others, and after settling a sample in the tube, the supernatant can be drained off, leaving the settled material on the other slides. This is then evaporated to dryness or the alcohol-vapour technique (see Chapter **5.5.2**) can be used. The mounting medium (glutaraldehyde is recommended) is added before draining to prevent disturbance of the settled material. With this method it is possible to preserve pigments so that fluorescent microscopy can be carried out. It is also valuable in the field of autoradiography, since after incubation, the organisms can be fixed, settled and finally washed using the sliding chamber.

Eddy and Banse (1972) tested the three methods and concluded that each gives problems in marine work. They have devised a combination of the methods which is fairly successful especially if the silica gel/alcohol chamber dehydra-

120

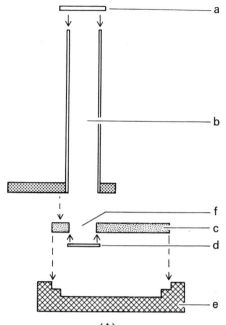

(A)

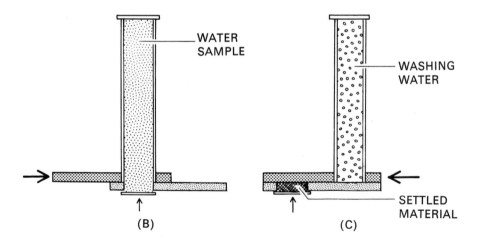

WATER
SAMPLE

WASHING
WATER

SETTLED
MATERIAL

(B)   (C)

Figure 26
Diagram of apparatus needed to make permanent mount from settled material (redrawn from Eddy and Banse, 1972): (a) top coverslip: (b) cylindrical settling chamber; (c) baseplate; (d) bottom coverslip; (e) baseplate rack; (f) baseplate well.

tion system is used (Fig. 26). The organisms are settled on to a bottom coverslip (d) attached to a baseplate (c) by stopcock grease (or by an O-ring and bolts). The supernatant can be removed by sliding the chamber (b), as in Figure 26B–C, and, if necessary, can be replaced with water prior to repositioning the cylinder for desalting. The mounting medium can be added to the original sample or can be smeared on the coverslip (d). The remaining water in the baseplate well (f) is evaporated to a damp film with silica gel and final dehydration is carried out in an alcohol chamber. The coverslip is detached and inverted on to a slide with more mounting medium to produce a permanent mount.

5.5.5   Perspectives on holography

John R. Beers

Although of limited usefulness at present, continuing developments in the field of holography, i.e. three-dimensional image recording, and their application to plankton study may eventually provide a valuable means of making permanent records of phytoplankton. Holograms are recordings on film or photographic plates of interference patterns produced by diffracted and un-diffracted laser-generated coherent light waves. When making a hologram, all seston particles, including phytoplankton, in the path between the laser and the recording medium will variously diffract the light waves in the beam, whereas those light waves that travel from the source to the film without encountering any particles will remain undiffracted. When a hologram is reconstructed, images of the particles are produced. Being truly three-dimensional, any and all depths within the scene recorded on the hologram can be brought into focus on a microscopic scale in the reconstruction, and the images can be viewed using a magnifying lens system comparable to a conventional microscope.

Holographic techniques have the theoretical potential of providing records of the total size spectrum of phytoplankters and can be made without concentrating or even fixing the samples. However, they are currently restricted by the limited resolution generally attained. This results from a combination of diverse factors such as the quality of the laser illumination, the photographic emulsions available on which to record holograms, and the large amount of diffracting materials (from colloidal size upwards) other than phytoplankton in most seawater samples.

Beers *et al.* (1970) proposed that holograms of the contents of settling chambers used for inverted microscope study could provide a permanent record of the sample. The simple procedure described is limited in practice to use with the larger phytoplankters ($> 30 \, \mu m$ in at least one dimension) in samples settled on nearly a single plane. An on-axis mode of holography was employed; any use of holograms for recording the materials in non-settled samples would undoubtedly be best accomplished by the more demanding, in terms of laser coherence, off-axis holography.

6

Identification problems

6.1   General recommendations

Grethe Rytter Hasle

Plankton investigations involving cell counts usually involve identification of organisms as well. The goal of the investigation determines the appropriate taxonomic level for such identification.

Cell numbers of the main groups—blue-green algae, diatoms, coccolitho-phorids and 'other flagellates'—offer some ecological information since to a certain degree spatial and seasonal distribution patterns do exist. For example, predominance of blue-green algae is found in some brackish and in some sub-tropical or tropical waters only. Diatoms appear abundantly and as the pre-dominant group in arctic and antarctic waters, in temperate waters during the cold season and generally in inshore waters of all climatic zones. Coccolitho-phorids (particularly in number of species) may outnumber all other well-investigated groups in oceanic waters. The greatest number of dinoflagellate species is found in warm waters. 'Other flagellates' may be found everywhere, but are more or less abundant.

Regardless of these gross distributional features, abundant dinoflagellates may appear in the autumn plankton at high latitudes in the northern hemisphere, and mass occurrence of either dinoflagellates (several taxa) or the coccolitho-phorid *Emiliana huxleyi* (= *Coccolithus* or *Gephyrocapsa huxleyi*), which can cause discoloration of the sea, has been observed frequently in inshore waters of the northern temperate zone.

It is generally assumed that centric diatoms are better represented in marine plankton than are pennate diatoms. But, in fact, the pennate genus *Thalassiothrix*, the section *Pseudonitzschia* and the bicapitate species of the pennate genus *Nitzschia* are very well represented in warm, oceanic waters.

Identification at the genus level will in some cases offer rather precise information, as many genera, particularly of dinoflagellates and coccolitho-phorids, are found to be restricted to, or have their main distribution in, parti-cular geographic and climatic zones. It should be kept in mind however that, for example, some *Ceratium* species are cosmopolitan, others are restricted to sub-tropical and tropical waters and *C. arcticum* inhabits arctic waters. The diatom

125

genera *Nitzschia* and *Thalassiosira* are extremely heterogeneous in spatial and seasonal distribution. Records of *Thalassiosira* without reference to specific epithet are not very useful since the genus is represented in all types of waters; however, individual species have definite distribution patterns and thus definite ecologies (e.g. the arctic species *T. hyalina*, the bipolar species *T. antarctica*, and the probably cosmopolitan species *T. gravida*).

Although everyone should be encouraged to identify organisms to the lowest taxonomic level which time and capability permit, it has to be realized that limitations are set by methods used for preservation, concentration and microscopical examination.

As mentioned elsewhere (Chapter 4) flagellates without a rigid cell covering or an internal skeleton will usually be unrecognizable after preservatives in common use have been added to the water sample. Concentration by filtration may also damage such organisms. Specimens not completely distintegrated should be included in cell counts, however, under the term 'monad'. This term stands for a small cell (about 10 μm or less) with a more or less spherical outline. If flagella are attached to the cell, it may be classified as 'flagellate' although such records seem to be no more significant than just 'monads and flagellates'. In spite of such an approximate 'identification', experience has shown that certain trends are thus discernible in the distribution of these groups.

Coccolithophorids are characterized by their calcified external plates, the coccoliths, which are extremely sensitive to any level of acidity. Moreover, morphological details of the coccolith structure which are essential for determination are often discernible only under the best optical conditions. An immersion objective (oil or water) and a total magnification of about 800 to 1,000 times will be needed to distinguish between *Emiliana huxleyi* and *Gephyrocapsa oceanica*, approximately 5 to 10 μm in cell diameter. However, characteristic and larger coccolithophorid species such as *Syracosphaera pulchra, Anthosphaera robusta* and *Coccolithus pelagicus* can be identified using a dry objective and a total magnification of 400 to 500 times. In many cases, however, electron microscopy will be required (see Chapter **6.3.3**).

Although the classification of dinoflagellates with a rigid cell covering (the 'armoured' or thecate dinoflagellates) is principally based on tabulation of the theca, planktologists usually have to use the gross morphology for identification. This includes presence and absence of apical horns, wings and other extensions, cell shape and displacement of the girdle. Rough characters, such as the generally more spiny and coarser theca of *Gonyaulax* and its left-handed displacement of the girdle as compared with *Protoperidinium*, are routinely used to distinguish these two genera. An outline of the various plates is readily seen in some cases; but often more sophisticated treatment is necessary to reveal this feature (see Chapter **6.3.2**).

Identification of planktonic diatoms at generic level usually offers fewer problems since so many centric genera have a characteristic gross morphology. Identification at the species level, for instance, of *Chaetoceros* and most *Rhizo-*

solenia species can readily be made during routine analysis of water samples. Difficulties may be encountered, however, if *R. fragilissima, R. delicatula, Leptocylindrus danicus* and narrow specimens of *Cerataulina pelagica* are present in the same sample, which sometimes happens. Use of 400 to 500 times magnification and phase contrast should allow clear distinction between the species. *Lauderia annulata* (incl. *L. borealis*), *Detonula pumila* (= *Schroederella delicatula*) and *Bacterosira fragilis* may be confused with each other as well as with *Thalassiosira* spp. Only a few of the 70 to 80 *Thalassiosira* species described can be identified when observed in girdle view and in colonies. Valve structure is diagnostic for these genera and species as for all diatoms, but this can only be seen in water mounts under favourable conditions. Effort should therefore be made to observe them in valve view. This is also the case with species of the group *Pseudonitzschia*; identification of these species based on their girdle view only is not reliable, except perhaps for *Nitzschia delicatissima* and *N. turgidula*. Pennate diatoms in ribbon-shaped colonies also must be examined in valve view to permit identification of genus as well as species.

The use of more sophisticated methods than those employed in routine counting is mostly necessary for species identification of small flagellates (see Chapter **6.3.4**). This is also often the case for identification at higher taxonomic levels when dealing with this group.

Most of the phytoplankton literature in use for identification purposes deals with the flora of a particular area. Since the most comprehensive publications available concentrate on North Atlantic waters, a newcomer to the field using them uncritically may easily be misled. The many records of *Nitzschia seriata* from areas where more critical examinations have shown that it does not belong exemplify such a situation. *Thalassiosira subtilis* is another planktonic diatom which apparently has been confused with other species of the genus (the common character is appearance in large mucilage colonies, and without examining the structure of these colonies and the structure of the single frustule, three or four species may be misidentified as *T. subtilis*).

Identification based on illustrations alone should be avoided. A self-trained planktologist will often pay attention to details of the illustrations which actually are not significant. If the accompanying text is properly written, it will include more information on variation and salient morphological features of a given species. Moreover, most authors willl discuss similar species and clarify how they are distinguished.

Species contributing to the greatest cell numbers or biomass are usually regarded as the important ones in a quantitative phytoplankton investigation. If the important species cannot be correctly identified in the preparation used for cell enumeration, and if isolation of single cells for further treatment is impracticable, a concentrate of any remaining water sample may be used for examination of the quantitatively important species. It should be realized, however, that loss of material can scarcely be prevented, and that a species occurring in a concentration of thousands or millions of cells per litre may not be found again in the concentrate. Net hauls collected at the same stations as the water samples are very useful in such cases as additional material for species identification. Examination of net hauls prior to analysis of water samples is also recommended in order to be acquainted with some of the species present.

Species which are unimportant quantitatively may yield interesting ecological and hydrographical information, and some effort should therefore be made to identify them if this is within the scope of the investigation. If net hauls corresponding to the water samples are available, the chances of having rare species examined are greatly increased. This is particularly important if electron microscopy is necessary for identification (e.g. minute diatoms).

Preparation of marine phytoplankton organisms for electron microscopy does not differ much from methods used for other biological specimens except that seawater and preservatives must be removed to prevent formation of crystals. A gentle way to do this is to use a dialysis tube immersed in distilled water or buffered distilled water (see Chapter **6.3.3**). Scanning electron microscopy is now much used for examination of whole cells of many of the groups present in marine phytoplankton. General techniques are given in textbooks on scanning electron microscopy, and more specific techniques used for groups in question are dealt with below. The 'critical point drying method', as detailed by Cohen (1974), may be used with success for all groups of plankton algae. It prevents the collapse and distortion which often follows simple drying of specimens for electron microscopy.

Electron microscopy has revolutionized the concepts of morphology and taxonomy of most of the phytoplankton groups. Distinction between taxa of all levels has been elucidated, and new ideas about classification systems have arisen. Its use for identification in routine phytoplankton counting is more questionable, however, except in particular cases. The present cost of instruments and the time-consuming techniques make electron microscopy of today a luxury that most institutions cannot afford for routine work. Moreover, it often introduces new problems in the identification, since only a minute— though increasing—number of all species described from phytoplankton have been studied by electron microscopy, and comparisons between what is discernible in the electron and the light microscopes cannot always be made readily. This is most obviously the case for organisms whose finer morphological structures are not resolved under the light microscope. When working with organisms slightly coarser in structure, details first discovered in the electron microscope are often seen later in the light microscope also, the electron microscopy thus being of indirect importance for identification.

6.2 Isolating the cells

6.2.1 Separating phytoplankton components

Robert R. L. Guillard

The utility of separating phytoplankton components varies from absolute necessity to mere convenience. Certain nanoplankters can hardly be assigned even to an algal class without observations of pigments, storage products, or ultrastructural features; these observations are best made from cultures. Methods for obtaining cultures by pipette isolation of single cells or by dilution-culture techniques, in each case with or without previous selective enrichment, are given elsewhere (Chapters **6.2.2** and **7.6**; see also the relevant chapters in Stein, 1973). At the other extreme of necessity, it may simply be convenient to eliminate the mass of organisms that are not of interest in some particular circumstance. The treatments used to clean the silica parts of diatoms and certain flagellates destroy other organisms and some detritus, but there is no comparable treatment for other systematic groups.

In some samples there may be an advantage in separating components, alive or preserved, according to size or density; electric charge may also offer possibilities. No system consistently separates components quantitatively according to any scheme of interest, taxonomic or otherwise, nor is any, save size fractionation, ready for routine application. They are nevertheless worth discussion for potential application. Possibilities of sorting cells electronically are included in Chapter **7.5.1**.

SEPARATION ACCORDING TO SIZE

Stainless-steel screens are available with mesh sizes down to a few μm and nylon netting is made with mesh as small as 0·5 μm (see Chapter **3.3** for manufacturers). Metal-membrane filters (e.g. Selas Flotronics) and perforated polycarbonate filters (e.g. Nuclepore Corp.) have pores spanning the size range 8 to 0·1 μm. Sheldon (1972) has shown that 'Flotronics' and 'Nuclepore' membranes, though called 'filters', act as 'screens', in that they separate two size fractions more or less quantitatively, rather than simply removing as many particles as possible regardless of size. Both types have particle-size median retentions close to the manufacturers' nominal pore sizes. The membranes can be overloaded easily and there-

129

fore serve for quantitative separation only from small samples with moderate turbidity. They select for shape of particle also (Sheldon, 1972). One can separate small drum-shaped diatoms (e.g. *Thalassiosira pseudonana*) from similarly sized subspherical green algae (e.g. *Nannochloris atomus*) by means of 'Nuclepore' filters with 5 μm pores, about the same size as the algae. Diatoms are retained by the filter, while most of the green algae pass through it (pers. observation).

Metal and nylon screens retain significant proportions of particles larger than about half their nominal pore size, and they retain about half the particles having the nominal mesh dimension (Sheldon and Sutcliffe, 1969). Colony formation and cell protuberances, as well as cell size and shape, also influence retention, as is well demonstrated in the study by Durbin *et al.* (1975) of Narragansett Bay phytoplankton separated into four size classes (< 20, 20 to 60, 60 to 100, > 100 μm) by nylon netting.

SEPARATION ACCORDING TO DENSITY

Preserved zooplankton components can be separated usefully by centrifugation of samples in density gradients made using a colloidal silica preparation ('Ludox AM', DuPont Instruments; Bowen *et al.*, 1972). The selectivity of separation can be improved for specific plankton components—notably fish eggs and larvae—by adding suitable polyanions or polyvinyl alcohol (St. Onge and Price, 1975). The technique has not been applied systematically to preserved phytoplankton components.

Living phytoplankters from cultures have been concentrated and to some extent separated by continuous centrifugation into 'Ludox AM' gradients in a zonal rotor (Price *et al.*, 1974). The same method, when used to collect algae from 200 l of seawater (in 2 h) was strongly biased against dinoflagellates in the population. An improved gradient medium (containing sorbitol) has permitted the collection of viable cells of three delicate dinoflagellate species from cultures (Price and Guillard, unpubl.). None of these methods has been used at sea or tested routinely ashore.

SEPARATION BY ELECTROPHORESIS

A commercial continuous-flow particle electrophoresis system ('CPE', Beckman Instruments Inc.) has been used with some success to separate components of natural samples of freshwater phytoplankton after preliminary concentration by centrifugation and resuspension in buffer solutions of *c.* 1 mM (Bayne and Lawrence, 1972). Heat-killed flagellates migrated much as did living ones, suggesting that preserved samples might also be separable, but no data are available.

If the same electric fields (10 to 90 V/cm) are necessary for migration of marine phytoplankters as for freshwater ones, the power requirements and heat generation of the system would seem to be excessive. This should be investigated. Alternatively, preserved samples could be concentrated and re-suspended in a medium of suitable composition and low ionic strength to permit use of the system described. This possibility has not been tested.

6.2.2   Isolation of single cells

Jahn Throndsen

Identification of many species met with during phytoplankton enumeration work with the inverted microscope is often impeded by the optical conditions in the counting chamber. In such cases, the cells in question may be picked up by means of a simple micropipette and transferred to an ordinary slide for closer examination by a compound microscope (see Throndsen, 1969a and Fig. 27).

A more complicated procedure is needed when single cells are to be prepared for study in the electron microscope. The following technique may be employed when the very same specimen is to be examined in the light microscope (LM) and the electron microscope (EM) as well. This technique conforms partly to that described by Halldal et al. (1954).

The material to be studied should be fixed with an appropriate fixative, e.g. formaldehyde-acetic acid for diatoms, neutralized formaldehyde for coccolithophorids, osmic acid for naked flagellates.

EQUIPMENT

Compound microscope for identification and microphotography; inverted microscope with micropipette device for isolation of cells; grids for electron microscopy; 'hole slide' with Formvar film (Merck); object slides and cover glasses; preparation needle; marking ink; brass cylinder (support for EM grids); forceps; soft brush; container for EM grids.

PROCEDURE

1. Place the material on a microscope slide under a coverslip and select a typical specimen by use of a compound microscope. Make light micrographs and take essential measurements.
2. Indicate the position of the cell on the edges of the slide, transfer the slide to an inverted microscope fitted with a micropipette, and localize the cell in question by use of the position marks (Fig. 28A). Uncover the specimen by pushing the coverslip aside with a preparation needle while watching the cell in the microscope (B). When the cell is exposed, lower the pipette until it encircles it and remove the specimen by gentle suction (C). Raise the pipette. (If critical light microscopy is not required, step 1 may be omitted and the cell picked up directly from a slide on the inverted microscope.)

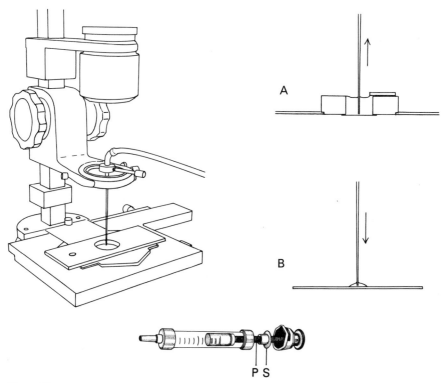

Figure 27
Left: Arrangement of a simple micropipette in the condenser holder on an inverted microscope. Bottom: modified syringe (suction device); (P) threaded piston rod; (s) stopper with threads for the piston rod. Right: (A) removing a cell from the sedimentation chamber; (B) ejecting the cell on to a microscopic slide for closer examination by a compound microscope. (Partly redrawn from Throndsen, 1969a).

3. Replace the object slide with a Formvar film-covered 'hole slide', lower the pipette until it just touches the film (on one of the holes) and eject the specimen (Fig. 28D). Let the cell settle and withdraw excess fluid into the pipette.
4. Raise the pipette, empty it on to a piece of filter paper and refill it from a drop of clean water (pH 7 for coccolithophorids) on an object slide. Lower the pipette near the specimen and flush it gently with water and then withdraw the water into the pipette (Fig. 28E). Be careful not to wash away the cell. Repeat the rinsing procedure until no salt crystals are formed in the residual water evaporating on the film.
5. Note the position of the dry cell, transfer the slide to a simple standard microscope and relocate the specimen. Lower the condenser and place a brass cylinder, with an EM-grid on it, in the centre of the top lens. Raise the condenser till the grid can be recognized under the specimen. Use the mechanical stage to place the cell just over one of the central holes in the

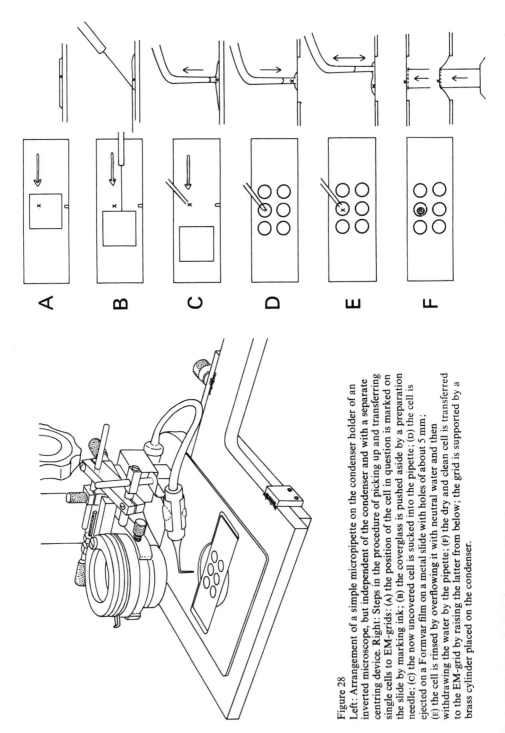

Figure 28
Left: Arrangement of a simple micropipette on the condenser holder of an inverted microscope, but independent of the condenser and with a separate centring device. Right: Steps in the procedure of picking up and transferring single cells to EM-grids: (A) the position of the cell in question is marked on the slide by marking ink; (B) the coverglass is pushed aside by a preparation needle; (C) the now uncovered cell is sucked into the pipette; (D) the cell is ejected on a Formvar film on a metal slide with holes of about 5 mm; (E) the cell is rinsed by overflowing it with neutral water and then withdrawing the water by the pipette; (F) the dry and clean cell is transferred to the EM-grid by raising the latter from below; the grid is supported by a brass cylinder placed on the condenser.

EM-grid, and raise the condenser until grid and Formvar film are in contact (while watching the process in the microscope). The cylinder is then raised further, until the film breaks (Fig. 28F), and it is removed by means of a pair of forceps. The grid is cut free by using a preparation needle, and then transferred to a specimen grid box by means of a soft brush or a fine pair of forceps.

The EM preparation can either be studied directly with the transmission electron microscope (TEM) or be coated for scanning electron microscopy (SEM), or shadowed with gold-palladium to reveal surface details in TEM. A carbon replica may be produced if the specimen is too dense to reveal details with direct TEM.

6.3. Some specific preparations

6.3.1 Diatoms

Grethe Rytter Hasle

The reader of the older literature may easily get the impression that diatoms are the predominant algae in marine plankton in general, independent of locality and time of year. Although this is certainly not true, the planktologist engaged in routine phytoplankton counting will seldom encounter a sample which does not contain at least one diatom cell. Fortunately, some of the most common and largest diatom genera in marine plankton are identifiable even to species level without too many problems during routine counting. But other diatoms also important in marine plankton (e.g. those in ribbon-shaped colonies and the smallest and most weakly silicified ones) are usually not identifiable to species, and often not to genus during routine counting. As discussed in the general recommendations (Chapter 6.1), net hauls corresponding to the water samples can provide material for examination of the diatoms in question by more sophisticated methods. This is possible because, as pointed out elsewhere (Chapter 3.3), not only the larger and colony-forming diatoms, but also the smallest single-celled ones (e.g *Minidiscus trioculatus*, diameter 2 to 5 μm), can under certain circumstances be caught by nets of mesh-size 30 to 50 μm.

Usually the planktologist wants to include all algal groups in the water-sample counts. The water sample must therefore not be exposed to a too low pH (see Chapters 4 and 6.3.3). If, however, the goal is merely an examination of diatoms, either in the water samples or in the associated net hauls, a pH lower than 7 is preferable because it hinders dissolution of the siliceous diatom structures. A mixture of acetic acid and formalin (equal volumes of 40 per cent formaldehyde and 100 per cent acetic acid; 20 ml to 70 ml net sample) has been used as a fixing and preservative agent with good results. Lugol solution (including acetic acid) may serve the same purpose. The latter preservative has its general limitations (Chapter 4), but for diatoms it has been found to preserve colonies better than formaldehyde does (Derek Harbour, pers. comm.).

No conclusive experiments on preservation and storage of diatoms are known. Experience has shown that use of formaldehyde and acetic acid, or buffered or unbuffered formaldehyde, may result in diatoms with well-preserved colonies, cell contents and siliceous structures even after twenty or fifty years of storage. In other cases diatoms treated more or less in the same manner have been damaged in less than one year, or even sooner.

Although methods in current use for estimation of cell numbers will not

always suffice to ascertain reliable identification, it is believed that light microscopy is the method to be used for counting diatoms (Chapter **6.1**). Whether
this can best be done in water mounts (Chapters **7.2.2, 7.3, 7.4**), in permanent
mounts (Chapter **5.5.2**) or on membrane filters (Chapter **7.2.3**) may be ques-
tioned, however. Embedding diatoms in a medium of higher refractive index
than that of silica and water, partly in order to increase the contrast, has been
done ever since serious diatom studies were first undertaken. Although it is the
author's opinion that this method is still preferable for identification of diatoms
in general, electron microscopy is occasionally needed to distinguish between
morphologically closely related species, e.g. *Skeletonema subsalsum* and *S.
costatum* in brackish waters, and *Nitzschia delicatissima* and *N. pseudodelica-
tissima*. Distinction of this type is important in routine work since the species
have different ecology and distribution. The main value of electron microscopy
in diatom studies seems to be on the basic level, that is to study the morphology
and fine structure of the diatom frustule and to use this information to determine relationships and distinctions between taxa.

LIGHT MICROSCOPY (LM)

Raw material in water mounts

This method of examination usually gives sufficient information to identify the
following genera and most of their species: *Bacteriastrum, Cerataulina, Chaetoceros, Corethron, Ditylum, Guinardia, Leptocylindrus, Lithodesmium, Odontella,
Planktoniella, Rhizosolenia, Skeletonema, Stephanopyxis, Thalassionema*, and
Thalassiothrix. These diatoms are readily recognized by the shape of colonies,
gross morphology of the whole cell, chloroplasts and specific structures such
as *Bacteriastrum* and *Chaetoceros* setae, *Cerataulina* elevations, and *Corethron*
and *Skeletonema* processes. Other planktonic, marine diatoms can only be
identified properly by examination of valve structure. The valve structure of a
series of species is revealed in water mounts, e.g. *Actinoptychus* spp., *Asterom-
phalus* spp., some *Coscinodiscus* spp., *Nitzschia kerguelensis*, *N. pungens*,
Roperia tesselata, but not so for the more weakly silicified and finely structured
Coscinodiscus, Navicula, Nitzschia and *Thalassiosira* species.

Raw material in permanent slides

The resin Pleurax (refractive index about 1·7) is particularly well suited for
mounting raw diatom material. This technique is described in detail by von
Stosch (1974). The whole, intact diatom frustule and colonies can be studied,
and also the nucleus can be stained as an indication that the cell was live when
fixed. The method has thus all the advantages of water-mount examination,
and in addition offers the possibility of studying the finer details of the diatom
siliceous wall and the composition of the girdle. However, examination of

137

areola array of, for instance, *Coscinodiscus* and *Thalassiosira* species will be hampered by examination of whole cells, as the pattern of the two valves will interfere and thus confuse the picture.

Cleaned material in permanent slides

Methods to free the siliceous diatom cell of organic material require use of UV radiation, enzymes, combustion or chemicals. UV and enzyme techniques are described elsewhere (Chapter **5.5.1**), and a technique involving chemicals is found at the end of this section. As noted elsewhere (Chapter **5.5.2**), many resins of refractive index 1·6 to 1·7, suitable for mounting diatoms, are available; a method for preparing permanent slides using Hyrax as the resin is also described below.

By varying the amount of hydrogen peroxide added and the time of exposure to UV radiation, the degree of removal of organic substance can be regulated. A short time of exposure to UV (less than 6 hours) will usually not remove the organic substance which is apparently active in keeping the various elements of the siliceous frustule together. Cleaning by enzyme treatment may also result in whole frustules free of cell content. These techniques are thus comparable to the use of Pleurax for embedding raw material.

The purpose of cleaning diatoms with chemicals is to remove the cell contents (which confuse the image of valve structure), and to separate the single components of the siliceous frustule. Then, special structures such as valve processes and septa of bands as well as areolation of valves and bands will be more easily available for examination.

The most delicate siliceous structures may be damaged by chemical cleaning if done too energetically. It should be kept in mind, however, that when dealing with a preserved diatom sample stored for some time, a distinction can hardly be made between the damage taking place before cleaning and that caused by the cleaning. The method described at the end of this section must be regarded as fairly gentle, since the finest structures such as the areola velum are seldom destroyed. It should be noted that HCl rather than H_2SO_4 is used, thus avoiding any precipitation of calcium salts.

Whenever possible within the limitation of LM, cell dimensions, number of marginal processes, areolae and striae should be measured by LM, either on water or permanent mounts or on photographs made from these mounts.

SCANNING ELECTRON MICROSCOPY (SEM)

Raw material

When working with thick-walled diatoms the only preparation required is the removal of seawater and preservatives by rinsing (see below under 'Note: Cleaning Diatoms'), and to let the material dry either on the stub itself or on a small piece of a coverslip. The time for drying can be reduced by transferring the material to alcohol. The coverslip is then cemented to the stub by colloidal silver. The raw sample may be used as a strewn mount, or selected specimens may be used if the diatom under study is too rare in the raw sample to be found easily in the strewn mount.

Lightly silicified specimens will often collapse with this treatment and should be prepared by the critical-point drying method (e.g. Cohen, 1974). In this technique the material is concentrated on a Nuclepore filter or on a grid coated with a supporting film (see under 'Transmission Electron Microscopy (TEM)' below), which after drying is mounted to the stub. Coating with gold-palladium (<500 Å by evaporation or 'sputtering') is used to avoid charging (see also Chapter **6.3.3**).

Since SEM gives a three-dimensional image, valuable information is obtained on cell morphology, such as curvature of valve face, structure of processes, possible extrusions through processes and composition of the girdle. It should be kept in mind, however, that SEM shows only the structure of that surface of the diatom wall which faces the beam. In some cases this is an advantage, in others a disadvantage (see below).

Cleaned material

Mounting and coating are done as for uncleaned thick-walled whole cells. The advantages of this method as compared to the examination of cleaned material in LM lies in the higher resolution power of SEM and the possibility of tilting and rotating the specimen and thus obtaining an image of its morphology in three dimensions. Some of the diatom processes can be distinguished only by their internal parts, and the siliceous wall of valves, and often also of the bands, is structured differently on the two surfaces. To understand the complete structure of a given diatom it is therefore important to examine some valves— and bands—from the inside and others from the outside. This can be done when dealing with separated valves and bands, but not with whole cells. It should be noted that occasionally valve and bands do not completely separate during the cleaning. The valve then often falls down into its girdle, which may lead to misinterpretation of the pictures. This relates to the connection between valve mantle and valvocopula (the band next to the valve) as well as to which side of the valve, internal or external surface, is shown on the picture.

TRANSMISSION ELECTRON MICROSCOPY (TEM)

Raw material

Material free of preservative and seawater is mounted on a grid coated with a supporting Formvar film slightly thicker than commonly used for mounting cleaned diatom material (0·4 per cent Formvar solution in ethylene dichloride instead of 0·3 per cent). The grid is placed in a carrier constructed for the critical-point drying of grids (e.g. E. Syvertsen, in prep.) and then processed as is usual for this technique. Since this method leaves the cell intact and fully expanded, it offers the possibility of examining the same details as by examining raw material in LM or SEM. The advantage of TEM is its much higher resolution power than for LM, and also for SEM when dealing with older instruments. Moreover, in contrast to SEM, the two surfaces of the diatom wall are seen at the same time, which means that, for instance, the internal part of processes is discernible through the valve wall. The inner surface of the bands in their correct position is also revealed. Some experience is needed however to figure out the composition of the girdle.

Cleaned material

The material is transferred from distilled water to the grid covered by a Formvar film (in critical cases the Formvar may be coated with carbon (approx. 200 Å) to produce a stronger supporting film), and allowed to dry. The specimens are then ready for TEM, or they may be coated with carbon before the examination starts, and also shadowed with carbon/platinum. Compared to most of the SEM currently in use, TEM gives a better resolution and its use is therefore necessary to give sufficient information on the smallest species and the most delicate diatom structures such as the finest perforations of plates. Since some models have devices for tilting the specimen (a goniometer stage), the possibility of viewing a diatom valve from different angles is present in TEM as well as in SEM. This is particularly important in order to study the valve mantle, which often includes essential diagnostic structures such as strutted processes, costae and a specific areola array.

Thick-walled, heavily silicified diatoms are seldom studied with success in TEM since the electron beam is absorbed by the dense silica. Replicas can then be used, prepared in the same way as for coccolithophorids (Chapter **6.3.3**) except for the use of HF to dissolve the silica.

SCANNING ELECTRON MICROSCOPY (SEM) VERSUS
TRANSMISSION ELECTRON MICROSCOPY (TEM)

Examples showing how these two methods can be used for different purposes in the study of diatom morphology and taxonomy were recently given by Hasle (1977). These examples are already partly outdated relative to what is presented in this chapter. Whole diatom cells may be examined in TEM under angles commonly used in SEM, namely 40° to 45°, if the instrument is equipped with a goniometer stage. Observations of gross morphology of the frustule and of internal and external process structures, otherwise obtained by SEM, can thus be achieved by TEM. With the high magnification and resolution (typically in the order of 100 Å) of modern SEM, fairly small diatoms and delicately structured morphological details of the diatom frustule can be studied with this instrument. It should be noted that the resolution announced for a particular instrument is generally based on material giving a higher resolution than obtainable on biological specimens. Moreover, maximum magnification announced is usually of little practical interest since it is obtained only under optimal conditions. Well-focused diatom pictures at a magnification of about 100,000 can be produced nowadays in SEM, although most SEM instruments still in use are not capable of such results. Most diatom SEM examinations are thus performed at maximum magnifications of 20–30,000. It should be emphasized, however, that well-focused pictures can hardly be made even at these magnifications when using the smaller SEM models most useful for certain biological and geological studies.

NOTE: CLEANING DIATOMS

One of the many methods in use for oxidation of the organic material of diatom cells, used with success for planktonic marine diatoms, is as follows (after Simonsen, 1974):
1. Rinse the sample with distilled water by centrifugation, or passive settling, or in a dialysis tube (for the latter, see Chapter **6.3.3**).
2. Add an equal amount of saturated $KMnO_4$. Agitate. Leave for 24 hours.
3. Add an equal amount of concentrated HCl (be careful, because of the danger) to the sample and $KMnO_4$ mixture. The solution turns dark brown. Heat gently over an alcohol lamp until it becomes transparent and colourless or light yellow-green.
4. Rinse with distilled water until sample is acid-free.

Cleaned diatom material should be stored in distilled water to which are added a few drops of the mixture of formaldehyde and acetic acid to hinder growth of bacteria and fungi and the dissolution of silica. Glycerin may be added to prevent desiccation. Storage in alcohol or in a mixture of distilled water and alcohol is preferable if the material is to be used for scanning electron microscopy later on (the stored material can then be used without rinsing).

NOTE: PREPARING DIATOM SLIDES

One of the many methods used to prepare permanent diatom slides is (Hasle and Fryxell, 1970):
1. Clean coverslips (0·17 ± 0·02 mm) with alcohol to remove any oil.
2. Lay cleaned coverslips on a labelled tray.
3. With a new, disposable pipette, place one to four drops of cleaned sample on each coverslip, depending upon the density of the sample. It is desirable

to have specimens distributed evenly so that one can be viewed or photo-graphed alone. With a dense sample, use one small drop with three drops of distilled water to spread it. Use the pipette to spread the water evenly on the coverslip, and use that pipette for only one sample to avoid mixing.

4. Dry over gentle heat or leave overnight. Protect from dust.
5. Add two to four drops (depending on viscosity) of resin with a high refrac-tive index, e.g. Hyrax (see Chapter **5.2.2**), to give contrast to the silica of the frustules.
6. Dry over gentle heat or leave overnight, until the resin becomes firm. Protect from dust.
7. Adjust a hotplate to moderate heat (calculated at about 200° C) and clean and mark microscope slides. Heat slides, leaving one end off the plate for easy handling.
8. Place one slide face down on a prepared coverslip. (Some mark under the slide glass helps to centre the cover slip.) Turn it over quickly when the resin melts enough to stick, and replace the slide on the hotplate.
9. Heat until resin has spread under the entire coverslip. Do not boil. Gently tap the coverslip with a wood stick to remove bubbles.
10. Cool the slide. Trim excess resin with a knife or use a solvent. Seal with nail polish. Affix permanent label.

6.3.2 Dinoflagellates

F. J. R. Taylor

The choice of a generalized procedure for the handling of this group is complicated by the physiological and structural diversity of its members, a problem
which applies to flagellates generally (Taylor, 1976*b*).

Routine fixation and preservation of those species with cell walls consisting of heavily developed cellulose plates is readily accomplished with weak
(1 to 2 per cent) formaldehyde added to samples from a 20 per cent (or weaker)
stock solution (Chapter 4). However, forms with very thin plates, or no wall at
all, are only poorly fixed by this method, many disintegrating rapidly or
deforming unrecognizably on contact with the fixative. Lugol's iodine solution
(Chapter 4) fixes a greater proportion of a natural sample, but the naked forms
are often distorted. Most typically they 'round-up', showing less accentuated
girdle and sulcal features and may be difficult to identify even to genus or
group. Nevertheless, Lugol's iodine seems to be the best choice for quantitative
estimates.

The flagella are shed very readily in this group, being thrown off in the
presence of most fixatives and also in living cells under stress. However, when
appreciable numbers are present some may be found which still possess one or
both flagella. Uranyl acetate, which can be used to stop flagellar motion in most
flagellate groups, causes flagellar loss in those dinoflagellates with which we
have experimented.

Fixation with osmium tetroxide has been used on naked dinoflagellates
(since Pouchet, 1883), either in the form of exposure to the vapour of a 2 per
cent solution or by the addition of drops to the sample. However, the danger
of its use outside a fume-hood, its expense, and the black staining of the oils
found so plentifully in many dinoflagellates, limit its value. This is perhaps the
most widely effective fixative, but there are some dinoflagellates which burst in
its presence. Glutaraldehyde (2 to 5 per cent solution) is also fairly effective for
some naked species, especially when fixation is carried out in a refrigerator.
Although not as dangerous as osmium tetroxide, glutaraldehyde is unpleasant
to smell and should not be inhaled for the long periods usually involved in
light-microscopic examination.

In ecological studies, especially those relating to primary productivity, data
referring to photosynthetic members should be distinguished from those
referring to non-photosynthetic species. Unfortunately, formalin eventually
bleaches the pigments and it may be difficult to determine if cells subjected to
lengthy preservation were originally pigmented or not. The starch reserve of
dinoflagellates stains bluish with iodine, but this is usually masked by the dark
reddish-brown general staining of the photosynthetic forms which produce it,

143

and it is therefore not very useful as an indicator of group affiliation under routine conditions.

Athecate ('naked') forms are identified principally by the shape of their cells, and the detailed observation of the girdle and sulcal furrows is often difficult. Various types of shallow-focus light microscopy, such as Nomarski Interference Microscopy, are useful for such purposes, but regular bright-field systems can be manipulated unconventionally to good effect, either by opening the iris diaphragm much more than usual, or by offsetting the condenser laterally by a small amount.

Thecate (cellulose-plated) cells require the determination of the complete 'plate pattern' for critical taxonomy, although in routine practice this is often not possible and identifications are based on size, shape and the sutural patterns of some areas. Optical interference from the cell contents often obscures the thecal sutures and, therefore, empty thecae provide the greatest ease of analysis (with due caution to avoid image reversal when focusing through the theca). When the contents are present the sutures may be observed by shallow-focus techniques or staining (or scanning electron microscopy: see below). The cell contents may also be plasmolysed by the addition of sucrose or salt to see thecal details more clearly.

Cells can be suitably positioned in water mounts by light coverslip pressure (a procedure calling for much patience) or by placing isolated cells in warmed glycerine jelly, the aspect being fixed once the jelly hardens (Graham, 1942).

Staining

Staining the cellulose plates can be accomplished by several techniques. Combinations of iodine with oxidizing agents will stain cellulose bluish or red-brown. Von Stosch (1969) has recommended the following procedure (preferably for application with non-preserved material).

Prepare two small test tubes, one containing a few ml of hydroiodic acid, the other with 3·5 g chloral hydrate dissolved in one ml of hydroiodic acid with a few crystals of iodine added to it. The most effective combination of these two varies with the material at hand, and therefore experiments can be carried out by starting with a weak dilution of the chloral hydrate/iodine with the HI run under the coverslip, increasing the strength of the former as necessary. As the plates tend to be thinner at their edges, this results in the sutures appearing as lightly stained zones between reddish-brown plates.

Iodine staining of the plates may not be very effective when applied to formalin-preserved material. If so, Trypan blue (also known as Diamine blue 3B) in a 0·2 to 0·3 per cent solution can be tried. This dye is also effective on live material and in a lower concentration can act as a vital stain having a low

144

toxicity for the cells. Apparently its effectiveness can vary according to the supplier (von Stosch, 1974). Trypan blue stains the theca a pale blue, the sutures holding the stain more strongly. It was used by Lebour (1925), and has recently been combined with the mountant Pleurax (see Chapter **5.5.2**) to provide stained, permanently mounted material (von Stosch, 1974). It is important that permanent mounts should permit the subsequent movement of the cells to observe full thecal detail.

If there is access to a vacuum evaporator of the type used for 'shadowing' in electron microscopy, coating of washed thecae can be carried out with examination by light microscopy. R. E. Norris (pers. comm.) has coated thecate dinoflagellates directly on glass slides, using carbon/platinum or chromium as a coating material, finding that thecal surface features were readily observed with a light microscope when mounted in immersion oil.

The amphiesmal vesicles (the peripheral vesicles in which plates may or may not be present) can be prepared for light microscope examination by 'silver line' staining by methods identical to those used to observe the pellicular alveoli of ciliates. Biecheler (1934, 1952) has described patterns exhibited by this technique in both thecate and athecate forms. By this method she demonstrated small vesicles in the vicinity of the flagellar pores of *Prorocentrum*, this indirect demonstration of cryptic periflagellar plates being neglected by most subsequent authors. (Frankel and Heckmann, 1968, have described a simplified version of this technique.)

The internal features of dinoflagellates can be examined by conventional cytological techniques (e.g. Jensen, 1962). It may be noted that the prominent nuclei, with chromosomes usually condensed and visible in interphase, particularly with acetocarmine or Feulgen-staining, can be a useful diagnostic feature when cells of unusual or deformed shape, or suspected dinoflagellate cysts, are encountered.

Thecal dissociation

Another approach to the examination of the plates, and particularly the small platelets in the ventral area, is to dissociate the theca in a controlled manner. Sodium hypochlorite (commercial 'bleach') is effective for this purpose, although the preparation may look messy as a result of the partial oxidation of the cell contents. A drop of a concentrated solution is run under the coverslip and allowed to act for a few minutes, at which time the plates will separate with light pressure on the coverslip. This should be done very gradually and tentatively, with sketches or photographs made at various stages of dissociation before plates become lost or disoriented. The papers of Balech (e.g. 1959a) show the fruits of this technique.

Holmes (1967) has used a combination of strong ultraviolet radiation and hydrogen peroxide to clean dinoflagellates, the technique requiring a very powerful UV lamp (1500 W or more) and an efficient cooling system.

145

F

ELECTRON MICROSCOPY

Although external morphology is the principal basis for the identification of most members of the group, the number of species which have been examined by transmission electron microscopy is rapidly growing and some require this type of observation for critical taxonomic purposes (e.g. zooxanthellae and many parasitic forms) and it is essential for the recognition of very weakly developed thecal plates. A simple, effective method of preparation is described by Bisalputra *et al.* (1973).

The scanning electron microscope (SEM) has proved to be a considerable aid in the examination of thecal surface features, particularly for delicate hyaline extensions of the surface ('lists'). These can achieve considerable complexity and the SEM permits an unequivocal interpretation (Taylor, 1971, 1973). Many thecate forms can be washed with distilled water and then air-dried, their preparation being extremely simple (cf. Taylor, 1976*c* or any recent manual of SEM methodology). However, thin-walled, naked or cyst forms will collapse on air-drying and require freeze- or critical-point drying (details of the latter are given by Cohen, 1974; see also Chapter **6.3.1** herein). The SEM does not solve all observational problems in dinoflagellates, however. Often sulcal plates are hidden by the sulcal lists, or the sutures may be obscured by other surface ornamentation. Formalin-preserved cells frequently lack the outer amphiesmal membranes and provide a clearer view of the thecal plate surfaces when viewed with the SEM.

The 10 to 30 nm coating of gold/palladium commonly used for SEM is light-transparent and, provided the cells are first dried on small coverslip surfaces and then temporarily attached with metallic paint to the metal 'stubs', can be gently removed and mounted in Hyrax or Pleurax on slides (MacAdam, 1971). The principal difficulties experienced with this method of retaining permanent slides of SEM material are coverslip breakage and the occasional presence of air bubbles within the cells.

NOTE ON DINOFLAGELLATE CYSTS[1]

Dinoflagellate cysts vary from ephemeral, mucoid-coated cells, to those with strongly acid-resistant walls commonly present in marine sediments and studied by palynologists. The latter types of cysts are readily concentrated from sediments by acid treatment, their walls containing a substance similar to sporo-pollenin. Records of cyst occurrence in plankton samples, with accompanying environmental data, should be made whenever they are observed. Because cysts may look very different from the cells which produce them it is valuable to describe the associations when they are observed and cysts with viable contents can serve as hardy 'seeds' from which to initiate cultures.

1. The author would like to acknowledge useful discussions with Barrie Dale and Barbara Whitney when preparing this subsection.

146

Wall and Dale (1968) and Dale (1976) have profitably developed the application of palynological techniques to recent plankton, with results of interest to palynologists and planktologists alike. B. Dale (pers. comm.) recommends the following mild acetolysis treatment for net samples of plankton.

Dilute the liquid of the sample with approximately the same volume of distilled water. Spin down up to 1 ml of concentrated plankton in a 15 ml centrifuge tube, suck off the supernatant and soak in 5 ml glacial acetic acid for at least 12 h. Centrifuge and pour off the acid. Add 5 ml acetic anhydride, then add 10 to 15 drops of concentrated H_2SO_4 (drop by drop *with caution* since great heat may be generated!). Place the tube in a cold-water bath, and heat to boiling point. Allow to stand hot for 5 to 10 minutes, then centrifuge, and pour off liquid (again, *with great caution*, since spitting occurs on contact with water!). Add a few drops of alcohol, and top up with distilled water; centrifuge, pour off the liquid and add fresh distilled water repeatedly until neutralized. Gently cleaned diatom frustules are present together with the dinoflagellate cysts. The former can be removed by pre-treatment with hydrofluoric acid, as in the standard palynological treatment described by Barss and Williams (1973) or Sarjeant (1974).

147

6.3.3   Coccolithophorids

Berit Riddervold Heimdal

The classification of coccolithophorids is based almost entirely upon the structure and arrangement of the coccoliths, with only minor emphasis on other features such as the cell shape. Identification to genus and species level is hampered in most cases by the minuteness of the cells and their individual coccoliths; electron microscopy is thus required to bring out essential details. As a matter of fact, coccolithophorids can only be satisfactorily described through the combined use of light microscopy (LM), scanning electron microscopy (SEM) and transmission electron microscopy (TEM). In order to obtain the most complete information for identification and to avoid taxonomic uncertainties, it is recommended that the same specimen be studied under both the light and the electron microscopes whenever possible.

LIGHT MICROSCOPY

LM observations of water mounts give information on cell shape and on possible spines. Larger and characteristic species (e.g. *Coccolithus pelagicus, Cyclococcolithus leptoporus, Rhabdosphaera claviger, Scyphosphaera apsteinii* and *Syracosphaera pulchra*) can be identified using a dry objective and a total magnification of 400 to 500 times, while an immersion objective and a total magnification of about 800 to 1000 times is necessary to distinguish between *Emiliania huxleyi* and *Gephyrocapsa oceanica*, 5 to 10 µm in diameter. Also, small cells of *Umbilicosphaera hulburtiana* (diameter 8·5 to 24 µm) may easily be confused with *E. huxleyi* on water mounts at lower magnifications.

The slides are generally examined with bright-field illumination, the resolution being better in oblique light. Species which are only weakly calcified can often be studied better in phase contrast/dark field (using a Heine phase-contrast condenser). Photomicrographs taken with such illumination should preferably be accompanied by bright-field photomicrographs for better comparison of specimens and more accurate measurements. The specimen being studied closely should be photographed with different levels of focus.

ELECTRON MICROSCOPY

Before examination in the electron microscope, salts and preservatives must be removed from the sample to prevent formation of crystals. A gentle way to do this is to use bags made of dialysis tubing (glycerol impregnated regenerated

148

cellulose, e.g. Arthur H. Thomas & Co.) immersed in membrane-filtered tap water, at nearly neutral pH, or buffered distilled water. (Pure distilled water would rapidly dissolve the coccoliths.)

If salts are not removed before the transfer of identified single cells from the LM to grids for electron microscopy, the cells must be rinsed by over-flowing them with neutral water (as described in Chapter **6.2.2**) before examination with the TEM. For examination under the SEM, the cleaned cocco-lithophorid sample is mounted as for examination with the TEM or on a small piece of coverslip. Then the grid or coverslip is cemented to a metal stub as described below.

Simple strewn mounts may help if loss of any small, fragile or sparse cells is expected. An untreated suspension of the sample is sprayed directly on a Formvar (Merck) film mounted on a metal slide in the same way as described for single isolated cells (see Chapter **6.2.2**), washed and allowed to dry in air. It is advisable to stir the drop occasionally to prevent floating material from becoming concentrated at the end of drying. Then the film should preferably be mounted on a '200-mesh' electron microscope grid (i.e. 200 holes per linear inch), the centre of which is marked to permit exact location of the individual specimens.

When studying the same specimen under both the light and the electron microscopes, a sketch of the area surrounding each specimen photographed under the LM is recommended to ensure relocation. The coccolithophorids usually disintegrate as they dry on the grid; some of the coccoliths then become detached and it is only these that can be examined more closely.

During the last decade the study of coccolithophorids has been carried out mainly with the SEM; the resolution of the instruments most currently available does not yet approach that of the TEM. In most cases, however, the coccoliths are too opaque to be penetrated by the rather weak electron beam of the latter instrument, so carbon replicas must be prepared. For replication the grid should be placed with the mounted material in a vacuum evaporator. The coccoliths are lightly shadowed by evaporating platinum–palladium under vacuum at an angle of 30° or 45° to the horizontal, depending on the coccolith morphology, and coated with carbon at an angle normal to the surface. The Formvar film is removed by soaking in dioxane for 6 to 8 h, and then the thin replica film remaining on the grid is placed in a bath of dilute HCl to dissolve $CaCO_3$, washed in distilled water with added alcohol and dried in air. Great care must be taken during the different steps to avoid making mirror-images of the specimen studied, since classification is often based on imbrication and suture direction of the elements.

For SEM the grids or the coverslips are mounted on a specimen stub with colloidal silver and coated with a thin film of carbon and metal (e.g. gold–palladium). Thin coatings are desirable so that fine surface details will not be obscured and so that the specimen can be observed again under the TEM if required. However, good electrical contact between specimens and stub is

149

necessary. Coccolithophorids are mostly spherical or oval, so they often rest on one or more small protuberances, where coating is difficult to apply. Specimens can thus become charged when exposed to the electron beams of the SEM unless rather thicker metallic coatings are applied. The application of carbon somewhat reduces this problem.

An accelerating beam voltage of 30 kV is generally used at high magnification (× 20,000 to 50,000) because significantly better resolution is thus achieved than at lower voltages. At lower magnifications accelerating voltages of either 20 kV or 10 kV are used. The lower voltages produce electron micrographs of a more even density by eliminating the bright areas caused by the penetrating effect of the electron beam at high voltages, but at the same time small holes may be obscured or poorly defined.

The depth of focus is much greater in the SEM than in the LM. The three-dimensional effect of the image produced by this instrument can be further enhanced by stereoscopy, i.e. by making two exposures of the specimen at different angles of tilt. As the change of angle tends to alter the working distance, and hence the magnification, it is necessary to move the specimen vertically back to its original position.

Some valuable recommendations for the study of calcareous nannoplankton are provided by Farinacci (1971), and practical information about electron microscopy may be found in Meek (1976).

6.3.4   Other flagellates

B. S. C. Leadbeater

Excluding the dinoflagellates and coccolithophorids, the most commonly occurring marine flagellates are those forming part of the nannoplankton (Lohmann, 1911; Leadbeater, 1972, 1974). In this category, pigmented flagellates belonging to the algal classes Haptophyceae, Chrysophyceae (including Silicoflagellata) and Prasinophyceae (*sensu* Christensen, 1962) are probably the most common, although a limited number of flagellates attributable to the Cryptophyceae, Euglenophyceae and Chlorophyceae are usually present. Nonpigmented flagellates include choanoflagellates and bodonids. All groups are apparently ubiquitous in distribution.

For collection of flagellates, 2-litre water samples obtained by any of the standard methods are required. After collection, the samples should be kept as close to the temperature recorded at collection as possible. On return to the laboratory the samples should be filtered through plankton netting with mesh size *c*. 25 μm to remove the larger components of the plankton, and allowed to stand in suitable containers in a cool place to permit settling of suspended particles and detritus.

Concentration of flagellates can be achieved by gentle centrifugation or membrane filtration followed by centrifugation. To obtain sufficient cells it is usually necessary to process between 1 and 2 litres of seawater. For immediate centrifugation of water samples a centrifuge with swing-out head is essential. However, this method of concentration is tedious if large volumes of liquid have to be processed and the centrifuge tubes are small. Concentration by membrane filtration is quicker and for this a membrane filter-funnel which holds *c*. 100 ml of liquid is required. Membranes with a standard pore size between 0·5 and 3·0 μm should be used, the smaller pore-size membranes being used for seawater samples containing few cells and suspended particles. A piece of tubing approximately 1 m in length is attached to the lower end of the filter funnel. During filtration the tubing is allowed to hang vertically below the funnel and the descent of the column of liquid through the tube is sufficient to draw further liquid through the filter. During filtration, the level of the liquid in the funnel should always be at least 1 cm above the filter and at all times the upper surface of the membrane must be covered with seawater. When sufficient water has been filtered, about 20 ml of liquid should be left in the funnel and this is gently agitated to dislodge cells from the surface of the membrane. The cell suspension is then concentrated into a pellet by centrifugation at low speed using a centrifuge with swing-out head and conical centrifuge tubes.

The satisfactory description of nannoplankton flagellates requires correlated light and electron microscopical observations, although for some species

151

with one or more easily identifiable morphological features (e.g. the scaly cell wall of haptophycean cells) electron microscopy of shadow-cast whole mounts may be all that is necessary for positive identification.

LIGHT MICROSCOPY

A good-quality microscope with an oil immersion objective of 100 magnitude is essential for the observation of small flagellates. Bright-field microscopy is necessary for observation of pigmentation. For minute flagellates, where there may be difficulty in observing small quantities of pigment, chlorophyll content can be ascertained using fluorescence microscopy. For general cell morphology, phase- and anoptral-contrast microscopy have been used extensively. Cellular appendages (e.g. flagella, haptonemata and cell-wall projections) can then be clearly resolved, although halos of light obscure details of protoplast contents. Interference-contrast microscopy overcomes this problem and both cell appendages and organelles can be clearly resolved.

For observation of cells, a small drop of the sample, just enough to reach the edges of the coverslip when mounted, should be placed on a slide. If the cells are living, the investigator will have to find quickly a suitable cell, using a low-power objective, and then change rapidly to the oil-immersion objective. Flagellates do not survive for long in the confined conditions under a coverslip and swimming movements may be affected. However, under these conditions cell organelles can usually be seen clearly. If it is desirable to fix cells, a small drop of 5 per cent glutaraldehyde or 1 per cent osmium tetroxide solution may be used (see below).

ELECTRON MICROSCOPY

Three aspects of electron microscopy are important for the identification of flagellates: (a) shadow-cast whole mounts observed in transmission electron microscopy (TEM), (b) sectioned material observed in TEM and (c) scanning electron microscopy (SEM).

Shadow-cast whole mounts permit satisfactory identification of species with distinctive extracellular structures such as flagellar appendages and cell walls composed of detachable units (e.g. cell-wall scales of the Haptophyceae, Prasinophyceae and Chrysophyceae). For preparation of whole mounts, samples must be concentrated by gentle centrifugation usually in 10-ml tubes. Once a visible pellet of cells has been produced, samples are fixed for 5 minutes with 1 per cent osmium tetroxide in 0·1 M cacodylate buffer at pH 7.0. Subsequently, cells require a thorough washing in distilled water. To ensure clean preparations, the cells should be concentrated and resuspended in at least three separate washes of distilled water. Drops of the final concentrate may then be placed on Formvar (Merck)-coated or carbon-coated copper grids and allowed to dry. Drops should also be placed on freshly cleaned slides or coverslips and these can subsequently be used for light microscopy or SEM. Shadow-casting, usually with chromium or an alloy of gold and palladium, is carried out in a vacuum coating unit.

Some characters of importance in the identification of flagellates (e.g. the number of microtubules in the haptonema or, in some species, pyrenoid form) can only be determined from material embedded and sectioned for electron microscopy. This involves fixation of cells in 5 per cent glutaraldehyde followed by a wash in buffer solution and post-osmication in 2 per cent osmium tetroxide solution. After dehydration in an ethanol series, cells are usually embedded in an epoxy resin and sections cut with a glass or diamond knife. Full details of fixation and embedding procedures may be found in Leadbeater (1972) or Manton and Leadbeater (1974).

Whole mounts on slides, coverslips or aluminium stubs may be uniformly coated with gold/palladium and used for SEM. If cells are dried on to slides, there is the advantage that a cell can be observed with the light microscope prior to SEM. If specimens for SEM can be dried in a critical-point drying apparatus (see e.g. Cohen, 1974), then distortion caused by collapse of the cell during drying can be minimized. Although the resolving power of the current SEM is not as good as that of the transmission electron microscope, valuable information on flagellar insertion, arrangement of cell-wall scales, costae, etc., can nevertheless be obtained.

F*

6.4　A guide to the literature for identification[1]

There is no 'plankton flora' of the world ocean. Although some group of experienced taxonomists may well undertake such a conspectus and could succeed in producing it, this would represent the upper limit of scientific devotion and human courage as well.

Neither does there exist a world review of any taxonomic group such as diatoms or dinoflagellates. In this connection, the more comprehensive treatments which are available encompass a more or less restricted geographical area; many of them are already somewhat old and may mislead the beginner with obsolete names and acceptations.

Even monographs at the generic level are desperately needed. Recent efforts to synthesize and update the knowledge at this level may be exemplified by the works of Hasle (1964, 1965) on the diatom genus *Nitzschia*, Dodge (1975) on the dinoflagellate genus *Prorocentrum*, and Gaarder and Heimdal (1977) on the coccolithophorid genus *Syracosphaera*.

Recommending some references for the identification of marine[2] phytoplankton is thus a difficult choice. The following may be much too selective on the one hand by the omission of older works, even if of major significance in the field, and on the other hand by deletion of shorter contributions, even if recent and excellent. The reason for this choice is that accepting broader limits would have soon increased the list tremendously.

The eight subdivisions are self-explanatory. It should be pointed out, however, that one or another 'check-list' is particularly helpful in providing an extensive bibliography for a given area.

Once a reference is known, obtaining the book is often another problem. Inquiries to the author or his or her institution are the first logical possibility and should not be neglected. Some major works which have been out of stock for years have been reprinted and may then be bought at any bookseller or directly from the publisher. Such publications will be indicated below by an asterisk (∗); for availability and cost, the reader is referred to the catalogues of such publishers as: Antiquariaat Junk, Asher & Co., Cramer, Johnson Reprint, or Otto Koeltz (see page 329 for full addresses). The investigator may also ask the librarian of his or her institution to obtain photostat or microfilm reproduction from national documentation centres.

---

1. When planning the authorship of the manual, it had not been made clear by the editor who should assume responsibility for the present chapter. Thus B. S. C. Leadbeater and B. R. Heimdal kindly sent lists for 'Other flagellates' and 'Coccolithophorids' respectively, and the extensive list used for a phytoplankton course for experienced participants was provided by the Institute for Marine Biology at Oslo. Eventually the editor decided to select references from this material, add others and divide them all into categories, under his own responsibility.
2. Some useful indications of relevant limnological literature may be found in Chapter 9.3.

INTRODUCTORY BOOKS TO
MARINE PHYTOPLANKTON

BRUNEL, J. 1962. *Le phytoplancton de la baie des Chaleurs.* 2ème éd., 1970. Montreal, Presses Univ. Montreal. 365 p., incl., 66 pl. (Contrib. Minist. Chasse Pêche 91.)

DREBES, G. 1974. *Marines Phytoplankton. Eine Auswahl der Helgoländer Planktonalgen (Diatomeen, Peridineen).* Stuttgart, Georg Thieme. 186 p.

MARUMO, R.; KAWARADA, Y.; TAKANO, H. 1966. *Marine plankton of Japan.* I: *Cyanophyceae: Xanthophyceae; Chrysophyceae; Bacillariophyceae.* Tokyo, Soyosha. 69 p.

MASSUTI, M.; MARGALEF, R. 1950. *Introducción al estudio del plancton marino.* Barcelona, Patronato Juan de la Cierva de Investigación Técnica. 182 p.

PANKOW, H. 1976. *Algenflora der Ostsee.* II. *Plankton (einschl. benthischer Kieselalgen).* Stuttgart, Gustav Fischer. 493 p., incl. 26 pl.

PROSHKINA-LAVRENKO, A. I.; MAKAROVA, I. V. 1968. (*Plankton algae of the Caspian sea.*) Leningrad, Izdat. Nauka. 291 p., 9 pl. (In Russian.)

TRÉGOUBOFF, G.; ROSE, M. 1957. *Manuel de planctonologie méditerranéenne.* I, II. Paris, Centre National de la Recherche Scientifique. 587 p., 207 pl. (*).

WIMPENNY, R. S. 1966. *The plankton of the sea.* London, Faber & Faber. 426 p.

DIATOMS

ANON. 1975. Proposals for a standardization of diatom terminology and diagnoses. (In: R. Simonsen (ed.), *Third Symp. Recent and Fossil Mar. Diatoms.* Kiel, 1974.) *Nova Hedwigia (Beih.),* vol. 53, p. 323–54.

ALLEN, W. E.; CUPP, E. E. 1935. Plankton diatoms of the Java sea. *Annls Jard. bot. Buitenzorg,* vol. 44, no. 2, p. 101–74 + figs. 1–127.

CLEVE-EULER, A. 1951–55. Die Diatomeen von Schweden und Finnland, I–V. *K. Svenska Vetenskakad. Handl.,* ser. 4, vol. 2, no. 1, p. 1–163 + 294 figs., pl. 1–6 (1951); vol. 4, no. 1, p. 1–158 + figs. 292–483 (1953); vol. 4, no. 5, p. 1–255 + figs. 484–970 (1953); vol. 4, no. 4, p. 1–232 + figs. 971–1306 (1955); vol. 3, no. 3, p. 1–153 + figs. 1318–1583, pl. 7 (1952). (*)

CUPP, E. E. 1943. Marine plankton diatoms of the west coast of north America. *Bull. Scripps Instn Oceanogr.,* vol. 5, no. 1, p. 1–237 incl. pl. 1–5. (*)

GAARDER, K. R. 1951. Bacillariophyceae from the 'Michael Sars' North Atlantic deep-sea expedition 1910. *Rep. scient. Results Michael Sars N. Atlant. Deep-sea Exped.,* vol. 2, no. 2, p. 1–36, tabl. 1–3.

GRAN, H. H.; ANGST, E. C. 1931. Plankton diatoms of the Puget Sound. *Trans. Puget Sound mar. Stn Univ. Washington,* vol. 7, p. 417–516.

HASLE, G. R. 1964. *Nitzschia* and *Fragilariopsis* species studied in the light and electron microscopes, I. Some marine species of the groups *Nitzschiella* and *Lanceolatae, Skr. norske Vidensk.-Akad., Mat. nat. Kl.,* n. ser., no. 16, p. 1–48, pl. 1–16.

——. 1965. *Nitzschia* and *Fragilariopsis* . . . , II. The group *Pseudonitzschia. Skr. norske Vidensk.-Akad., Mat. nat. Kl.,* n. ser., no. 18, p. 1–45, pl. 1–17.

HENDEY, N. I. 1937. The plankton diatoms of the southern seas. *Discovery Rep.,* vol. 16, p. 151–364, pl. 6–13.

——. 1964. An introductory account of the smaller algae of British coastal waters, V. Bacillariophyceae (Diatoms). *Fish. Invest. Lond.,* ser. 4, p. 1–317, pl. 1–45. (*)

HUSTEDT, F. 1927–66. Die Kieselalgen Deutschlands, Österreichs und der Schweiz mit Berücksichtigung der übrigen Länder Europas sowie der angrenzenden Meeresgebiete. In: L. Rabenhorst, *Kryptogamen-Flora* (Leipzig, Akad. Verlag), vol. 7: Teil 1, p. 1–920 (1927–30); Teil 2, p. 1–845 (1931–59); Teil 3, p. 1–816 (1961–66). (*)

LEBOUR, M. V. 1930. *The planktonic diatoms of northern seas.* London, Ray Soc. 244 p., 4 pl.

PROSHKINA-LAVRENKO, A. I. 1955. (*Plankton diatoms of the Black sea.*) Moscow, Leningrad, Izdat. Nauka. 222 p., 8 pl. (In Russian.)

——. 1963. (*Plankton diatoms of the Azov sea.*) Moscow, Leningrad, Izdat. Nauka. 190 p., 9 pl. (In Russian.)

RAMPI, L.; BERNHARD, M. 1978. *Key for the determination of Mediterranean diatoms.* Roma, Comit. Naz. Energia Nucleare. 71 p. (RT/B10 (78)1.)

SAUNDERS, R. P.; GLENN, D. A. 1969. Diatoms. *Mem. Hourglass cruises,* vol. 1, no. 3, p. 1–119, incl. pl. 1–16.

SIMONSEN, R. 1974. The diatom plankton of the Indian Ocean Expedition of R.V. 'Meteor' 1964–65. *Meteor Forschungsergeb. (D. Biol.),* vol. 19, p. 1–66, pl. 1–41.

SOURNIA, A. 1968. Diatomées planctoniques du Canal de Mozambique et de l'île Maurice. *Mém. ORSTOM,* vol. 31, p. 1–120, pl. 1–13. (Also from: *Colld repr.,* Int. Ind. Ocean Exped., vol. 7.)

SUBRAHMANYAN, R. 1946. A systematic account of the marine plankton diatoms of the Madras coast. *Proc. Ind. Acad. Sci.,* ser. B, vol. 24, no. 4, p. 85–197, pl. 2.

DINOFLAGELLATES

BALECH, E. 1971. Microplancton de la campaña oceanográfica Productividad III. *Rev. Mus. Argent. Cienc. Nat. Bernardino Rivadavia, Hidrobiol.,* vol. 3, no. 1, p. 1–202, pl. 1–39.

——. 1971. Microplancton del Atlantico ecuatorial oeste (Equalant I). *Publ. Serv. hidrogr. naval B. Aires,* vol. 654, p. 1–103, pl. 1–12.

——. 1976. Clave ilustrada de dinoflagelados antárticos. *Publ. Inst. Antárt. Argent. B. Aires,* vol. 11, p. 1–99.

CHATTON, E. 1952. Classe des Dinoflagellés ou Péridiniens. In: P.-P. Grassé (éd.), *Traité de zoologie* (Paris, Masson), vol. I, no. 1, p. 309–406, pl. 1.

DODGE, J. D. 1975. The Prorocentrales (Dinophyceae), II. Revision of the taxonomy within the genus *Prorocentrum. Bot. Linn. Soc.,* vol. 71, no. 2, p. 103–25, pl. 1–4.

GAARDER, K. R. 1954. Dinoflagellatae from the 'Michael Sars' North Atlantic deep-sea expedition 1910. *Rep. scient. Results Michael Sars N. Atlant. Deep-sea Exped.,* vol. 2, no. 3, p. 1–62, tabl. 1–5.

GRAHAM, H. W. 1942. Studies in the morphology, taxonomy and ecology of the Peridiniales. *Scient. Results Cruise VII Carnegie 1928–29,* Biol., vol. 3, p. 1–209.

GRAHAM, H. W.; BRONIKOVSKY, N. 1944. The genus *Ceratium* in the Pacific and North Atlantic oceans. *Scient. Results Cruise VII Carnegie 1928–29,* Biol., vol. 5, p. 1–209.

KISSELEV, I. A. 1950. (*Dinoflagellates of marine and fresh waters of USSR.*) Moscow, Leningrad, Izdat. Nauka. 280 p. (In Russian.)

KOFOID, C. A.; SKOGSBERG, T. 1928. The Dinoflagellata: the Dinophysoidae (Reports on the scientific results of the expedition to the eastern tropical Pacific, in charge of Alexander Agassiz . . ., 35). *Mem. Mus. comp. Zool. Harvard Coll.,* vol. 51, p. 1–766, pl. 1–31.

KOFOID, C. A.; SWEZY, O. 1921. The free-living unarmored Dinoflagellata. *Mem. Univ. Calif.,* vol. 5, p. 1–562, pl. 1–12. (∗)

LEBOUR, M. V. 1925. *The dinoflagellates of northern seas.* Plymouth, Marine Biological Association. 250 p. incl. 35 pl.

SCHILLER, J. 1931–37. Dinoflagellatae (Peridineae) in monographischer Behandlung. In: L. Rabenhorst, *Kryptogamen-Flora* (Leipzig, Akad. Verlag), vol. 10 (3): Teil 1, p. 1–617 (1931–33); Teil 2, p. 1–590 (1935–37). (∗)

SILVA, E. S. 1957. Dinoflagelados do plâncton marinho de Angola. *An. Jta Invest. Ultramar,* vol. 10 [1955], no. 2, p. 107–91, pl. 1–11, tabl. (Also in: *Trabalhos Missão Biol. marit.,* 15.)

SOURNIA, A. 1968. Le genre *Ceratium* (Péridinien planctonique) dans le Canal de Mozambique. Contribution à une révision mondiale. *Vie milieu,* sér. A, vol. 18 [1967], no. 2–3, p. 375–499, incl. pl. 1–3. (Also from: *Colld repr.,* Int. Ind. Ocean Exped., vol. 6.)

STEIDINGER, K. A.; WILLIAMS, J. 1970. Dinoflagellates. *Mem. Hourglass cruises,* vol. 2, p. 1–251, incl. pl. 1–45.

SUBRAHMANYAN, R. 1968. *The Dinophyceae of the Indian seas,* I. *Genus* Ceratium *Schrank.* Mandapam Camp, Marine Biological Association of India. 129 p., 9 pl. (Mem. II.)

——. 1971. *The Dinophyceae of the Indian seas,* II. *Family Peridiniaceae Schütt emend. Lindemann.* Cochin, Marine Biological Association of India. 334 p. incl. 79 pl. (Mem. II.)

TAYLOR, F. J. R. 1976. Dinoflagellates from the International Indian Ocean Expedition. A report on material collected by the R.V. 'Anton Bruun' 1963–64. *Bibthca bot.,* vol. 132, p. 1–234, pl. 1–46.

WOOD, E. J. F. 1954. Dinoflagellates in the Australian region. *Austr. J. mar. freshwat. Res.*, vol. 5, no. 2, p. 171–351.

COCCOLITHOPHORIDS

DEFLANDRE, G. 1952. Classe des Coccolithophoridés. In: P.-P. Grassé, (éd.), *Traité de zoologie* (Paris, Masson.), vol. I, no. 1, p. 439–70.

DEFLANDRE, G.; FERT, C. 1954. Observations sur les Coccolithophoridés actuels et fossiles en microscopie ordinaire et électronique. *Ann. Paléontol.,* vol. 40, p. 115–76, pl. 1–15.

FARINACCI, A. 1971. Round table on calcareous nannoplankton, Roma, September 23–28, 1970. In: A. Farinacci (ed.), *Proc. II Planktonic Conf., Roma 1970*, vol. II, p. 1343–60. Roma, Tecnoscienza.

GAARDER, K. R.; HASLE, G. R. 1971. Coccolithophorids of the gulf of Mexico. *Bull. mar. Sci.*, vol. 21, no. 2, p. 519–44.

GAARDER, K. R.; HEIMDAL, B. R. 1977. A revision of the genus *Syracosphaera* Lohmann (Coccolithineae). *Meteor. Forschungsergeb. (D. Biol.),* vol. 24, p. 54–71, incl. pl. 1–8.

KAMPTNER, E. 1941. Die Coccolithineen der Südwestküste von Istrien. *Ann. Naturhist. Mus. Wien*, vol. 51, p. 54–149, pl. 1–15.

OKADA, H.; MCINTYRE, A. 1977. Modern coccolithophores of the Pacific and North Atlantic oceans. *Micropaleontology*, vol. 23, no. 1, p. 1–55, incl. pl. 1–13.

REINHARDT, P. 1972. *Coccolithen. Kalkiges Plankton seit Jahrmillionen.* Wittenberg Lutherstadt, A. Ziemsen Verlag. 99 p.

SCHILLER, J. 1930. Coccolithineae. In: L. Rabenhorst, *Kryptogamen-Flora* (Leipzig, Akad. Verlag), vol. 10 (2), p. 89–273. (∗)

OTHER FLAGELLATES AND NANNOPLANKTERS

BUTCHER, R. W. 1959. An introductory account of the smaller algae of British coastal waters, I. Introduction and Chlorophyceae. *Fish. Invest. Lond.,* ser. 4, p. 1–74, pl. 1–14.

——. 1961. An introductory account . . . , VIII. Euglenophyceae (Eugleninae). *Fish. Invest. Lond.,* ser. 4, p. 1–17, pl. 1–3.

——. 1967. An introductory account . . . , IV. Cryptophyceae. *Fish. Invest. Lond.,* ser. 4, p. 1–54, pl. 1–20.

CARTER, N. 1937. New or interesting algae from brackish water. *Arch. Protistenkd.*, vol. 90, no. 1, p. 1–68, pl. 1–8.

GEMEINHARDT, K. 1930. Silicoflagellatae. In: L. Rabenhorst, *Kryptogamen-Flora* (Leipzig, Akad. Verlag), vol. 10 (2), p. 1–87. (∗)

LEADBEATER, B. S. C. 1972. Identification, by means of electron microscopy, of flagellate nanoplankton from the coast of Norway. *Sarsia*, vol. 49, p. 107–24, pl. 1–4.

LEEDALE, G. F. 1967. *Euglenoid flagellates.* Englewood Cliffs, Prentice-Hall. 242 p.

MANTON, I.; LEADBEATER, B. S. C. 1974. Fine-structural observations on six species of *Chrysochromulina* from wild Danish marine nanoplankton, including a description of *C. campanulifera* sp. nov. and a preliminary summary of the nanoplankton as a whole. *Biol. Skr.*, vol. 20, no. 5, p. 1–26, pl. 1–12.

MARSHALL, S. M. 1934. The Silicoflagellata and Tintinnoinea. *Scient. Rep. Gt. Barrier Reef Exped.*, vol. 4, no. 15, p. 623–64.

SCHILLER, J. 1925. Die planktonischen Vegetationen des adriatischen Meeres, A. Die Coccolithophoridien-Vegetation den Jahren 1911–14. (Nach den Ergebnissen der österreichischen Adriaforschung in den Jahren 1911–14.) *Arch. Protistenkd.*, vol. 51, no. 1, p. 1–130, pl. 1–9.

——. 1925. Die planktonischen Vegetationen . . . , B. Chrysomonadina, Heterokontae, Cryptomonadina, Eugleninae, Volvocales. 1. Systematischer Teil. *Arch. Protistenkd.*, vol. 53, no. 1, p. 59–123, pl. 3–6.

THRONDSEN, J. 1969. Flagellates of Norwegian coastal waters. *Nytt Mag. Bot.*, vol. 16, no. 3–4, p. 161–216.

TSUMURA, K. 1963. A systematic study of Silicoflagellatae. *J. Yokohama munic. Univ.*, vol. 45, no. 146, p. 3–85, pl. 1–28.

BLUE-GREEN ALGAE

GEITLER, L. 1932. Cyanophyceae von Europa unter Berücksichtigung der anderen Kontinente. In: L. Rabenhorst, *Kryptogamen-Flora* (Leipzig, Akad. Verlag), vol. 14, p. 1–1196. (*)
SOURNIA, A. 1968. La Cyanophycée *Oscillatoria* (= *Trichodesmium*) dans le plancton marin: taxinomie, et observations dans le Canal de Mozambique. *Nova Hedwigia*, vol. 15, no. 1, p. 1–12, pl. 1–2. (Also from: *Colld repr.*, Int. Ind. Ocean Exped., vol. 6.)
——. 1970. Les Cyanophycées dans le plancton marin. *Année biol.*, vol. 9, no. 1–2, p. 63–76.

REGIONAL CHECKLISTS

CONGER, P. S.; FRYXELL, G. A.; EL-SAYED, S. Z. 1972. Diatom species reported from the gulf of Mexico: In: S. Z. El-Sayed *et al.* (ed.), *Chemistry, primary productivity, and benthic algae of the gulf of Mexico*, p. 18–23. New York, N.Y., American Geographical Society. (Serial atlas of the marine environment, folio 22.)
DOWIDAR, N. M. 1974. The phytoplankton of the Mediterranean waters of Egypt, I. A check list of the species recorded. *Bull. Inst. Oceanogr. Fish.*, Cairo, vol. '1974', no. 4, p. 319–44.
DREBES, G.; ELBRÄCHTER, M. 1976. A checklist of planktonic diatoms and dinoflagellates from Helgoland and List (Sylt), German Bight. *Bot. Mar.*, vol. 19, no. 2, p. 75–83.
HALIM, Y. 1969. Plankton of the Red sea. *Oceanogr. mar. Biol. ann. Rev.*, vol. 7, p. 231–75 + table.
HEIMDAL, B. R.; HASLE, G. R.; THRONDSEN, J. 1973. An annotated checklist of plankton algae from the Oslofjord, Norway (1951–72). *Norw. J. Bot.*, vol. 20, no. 1, p. 13–19.
HENDEY, N. I. 1974. A revised check-list of British marine diatoms. *J. Mar. Biol. Assoc. U.K.*, vol. 54, no. 2, p. 277–300.
KELL, V.; KÜHNER, E. 1975. Untersuchungen am Phytoplankton der nördlichen Zentralatlantiks, IV. Zusammenfassende Artenliste. *Wiss. Z. Univ. Rostock*, vol. 24, no. 6, p. 775–80.
KERŽAN, I. 1976. *Katalog pelagicnih alg Jadrana*. 34 p. (Mimeo.)
MARSHALL, H. G. 1976. Phytoplankton distribution along the eastern coast of the USA, I. Phytoplankton composition. *Mar. Biol.*, vol. 38, no. 1, p. 81–9.
PARKE, M.; DIXON, P. S. 1976. Check-list of British marine algae—third revision. *J. Mar. Biol. Assoc. U.K.*, vol. 56, no. 3, p. 527–94.
PAVILLARD, J. 1937. Les Péridiniens et Diatomées pélagiques de la mer de Monaco de 1907 à 1914. Observations générales et conclusions. *Bull. Inst. océanogr.*, Monaco, no. 738, p. 1–56.
SEMINA, H. J. 1974. (*Pacific phytoplankton.*) Moscow. Izdat. Nauka. 239 p. (In Russian.)
SKOLKA, V. H. 1960. Espèces phytoplanctoniques des eaux roumaines de la mer noire. *Rapp. P.-v. Réun. Commn int. Explor. scient. Mer Méditerr.*, vol. 15, no. 2, p. 249–68.
SOURNIA, A. 1970. A checklist of planktonic diatoms and dinoflagellates from the Mozambique Channel. *Bull. mar. Sci.*, vol. 20, no. 3, p. 678–96.
STEIDINGER, K. A. 1972. Dinoflagellate species reported from the gulf of Mexico and adjacent coastal areas (compiled 1971). In: S. Z. El-Sayed *et al.* (ed.), *Chemistry, primary productivity, and benthic algae of the gulf of Mexico*, p. 15, 23–5. New York, N.Y., American Geographical Society. (Serial atlas of the marine environment, folio 22.)
SUBRAHMANYAN, R. 1958. Phytoplankton organisms of the Arabian sea off the west coast of India. *J. Ind. bot. Soc.*, vol. 37, no. 4, p. 435–41.
TAASEN, J. P.; SAUGESTAD, T. 1974. A list of plankton algae, collected in net hauls from Raunefjorden, western Norway with some remarks on the seasonal variation of the dominant species. *Sarsia*, vol. 55, p. 121–8.
TAYLOR, F. J. 1970. A preliminary annotated check list of diatoms from New Zealand coastal waters. *Trans. r. Soc. N.Z.*, Biol. Sci., vol. 12, no. 14, p. 153–74.
——. 1974. A preliminary annotated check list of dinoflagellates from New Zealand coastal waters. *J. r. Soc. N.Z.*, vol. 4, no. 2, p. 193–201.
——. 1974. A preliminary annotated check list of micro-algae other than diatoms and dinoflagellates from New Zealand coastal waters. *J. r. Soc. N.Z.*, vol. 4, no. 4, p. 395–400.

TRAVERS, A.; TRAVERS, M. 1975. Catalogue du microplancton du golfe de Marseille. *Int. Rev. gesamt. Hydrobiol.*, vol. 60, no. 2, p. 251–76.

WOOD, E. J. F. 1963. *Check-list of dinoflagellates recorded from the Indian ocean*, p. 1–58. Div. Fish. Oceanogr., CSIRO. (Rep. 28.)

——. 1963. Checklist of Diatoms recorded from the Indian ocean, p. 1–311. Div. Fish. Oceanogr., CSIRO. (Rep. 36.)

TAXONOMIC CATALOGUES

LOEBLICH, A. R. JR; LOEBLICH, A. R. III. 1966. Index to the genera, subgenera, and sections of the Pyrrhophyta. *Stud. trop. Oceanogr.*, vol. 3, p. 1–94, pl. 1.

——. 1968. Index to the genera . . . , II. *J. Paleont.*, vol. 42, no. 1, p. 210–13.

——. 1969. Index to the genera . . . , III. *J. Paleont.*, vol. 43, no. 1, p. 193–8.

——. 1970. Index to the genera . . . , IV. *J. Paleont.*, vol. 44, no. 3, p. 536–43.

——. 1970. Index to the genera . . . , V. *Phycologia*, vol. 9, no. 3–4, p. 199–203.

——. 1972. Index to the genera . . . , VI. *Phycologia*, vol. 10 [1971], no. 4, p. 309–14.

——. 1974. Index to the genera . . . , VII. *Phycologia*, vol. 13, no. 1, p. 57–61.

LOEBLICH, A. R. JR; TAPPAN, H. 1966. Annotated index and bibliography of the calcareous nanoplankton. *Phycologia*, vol. 5, no. 2–3, p. 81–216.

——. 1968. Annotated index . . . , II. *J. Paleont.*, vol. 42, no. 2, p. 584–98.

——. 1969. Annotated index . . . , III. *J. Paleont.*, vol. 43, no. 2, p. 568–88.

——. 1970. Annotated index . . . , IV. *J. Paleont.*, vol. 44, no. 3, p. 558–74.

——. 1970. Annotated index . . . , V. *Phycologia*, vol. 9, no. 2, p. 157–74.

——. 1972. Annotated index . . . , VI. *Phycologia*, vol. 10 [1971], no. 4, p. 315–39.

——. 1973. Annotated index . . . , VII. *J. Paleont.*, vol. 47, no. 4, p. 715–59.

SOURNIA, A. 1973. Catalogue des espèces et taxons infraspécifiques de Dinoflagellés marins actuels publiés depuis la révision de J. Schiller, I. Dinoflagellés libres. *Nova Hedwigia (Beih.)*, vol. 48, p. 1–92.

——. 1978. Catalogue des espèces . . . , III. (Complément.) *Rev. algol.*, n. sér., vol. 13, no. 1, p. 3–40.

SOURNIA, A.; CACHON, J.; CACHON, M. 1975. Catalogue des espèces . . . , II. Dinoflagellés parasites ou symbiotiques. *Arch. Protistenkd.*, vol. 117, no. 1–2, p. 1–19.

——. *Catalogue of the fossil and recent genera . . .*, Part IV. Neidium *through* Pyxilla. (In press.)

VAN LANDINGHAM, S. L. 1967. *Catalogue of the fossil and recent genera and species of Diatoms and their synonyms. A revision of F. W. Mills' 'An index to the genera and species of the Diatomaceae and their synonyms'*, Part I. Acanthoceras *through* Bacillaria, p. 1–493. Lehre, J. Cramer.

——. 1968. *Catalogue of the fossil and recent genera . . .*, Part II. Bacteriastrum *through* Coscinodiscus, p. 494–1086. Lehre, J. Cramer.

——. 1969. *Catalogue of the fossil and recent genera . . .*, Part III. Coscinosphaera *through* Fibula, p. 1087–756. Lehre, J. Cramer.

——. 1971. *Catalogue of the fossil and recent genera . . .*, Part IV. Fragilaria *through* Naumena, p. 1757–2385. Lehre, J. Cramer.

——. 1975. *Catalogue of the fossil and recent genera . . .*, Part V. Navicula, p. 2386–963. Vaduz, J. Cramer.

——. 1978. *Catalogue of the fossil and recent genera . . .*, Part VI. Neidium *through* Rhoicosigma, p. 2964–3605. Vaduz, J. Cramer.

6.5 Principles of taxonomic nomenclature

F. J. R. Taylor

The purpose here is to recommend to the phytoplanktologist not familiar with formal taxonomic procedure a course to follow when referring to the organisms under study, and to explain some of the commonest features of nomenclature which may seem confusing to a non-taxonomist.

NAMING ORGANISMS

The names given to species (and their greater and lesser hierarchical levels, all collectively referred to as 'taxa') are the method for recognizing the principal units of biological study. It is therefore appropriate that their use be rigorously governed to avoid misrepresentation and confusion. The ultimate authority for this rests with the Codes of Nomenclature, consisting of published rules and recommendations which aim at the recognition of only one name for each kind of organism within a specified classification. Jeffrey (1973) has provided a very useful introduction to the codes which will assist biologists in using them, and Loeblich and Tappan (1966), Sournia et al. (1975) and Taylor (1976d) have referred specifically to their application to members of the phytoplankton.

There are several codes governing different groups of organisms (plants by the International Code of Botanical Nomenclature, or ICBN, animals by the International Code of Zoological Nomenclature or ICZN, and there are codes for bacteria and for cultivated plants). While these are essentially similar, there are some procedural differences (those between the ICBN and ICZN summarized by Sournia et al., 1975), and for present purposes only the ICBN provisions will be briefly outlined, to give an indication of the matters which are important in dealing critically with biological names.

The *correct name* for a phytoplankton organism, in accordance with the ICBN (Stafleu, 1972, or more recent edition when available), is the name which fulfils several stringent requirements:

1. The following must be made available in printed form to the public or institutions: a *description* of the essential characters (or named illustration only, if published before 1 January 1908) plus an *illustration* (obligatory after 1 January 1958), and a *Latin diagnosis* (short, formal description, Latin being obligatory after 1 January 1958). There must also be a designation of the material, or illustration, to serve as the *type* by which later identifications can be judged (essential after 1 January 1958). If used to typify a genus or higher grouping, the type may simply be a name. Only names in, and subsequent to, Linnaeus'

Species Plantarum, 1753, can be considered. Whenever possible, an actual specimen (or specimens) should be deposited in a museum or herbarium to serve as a reference type.

2. The name must be *correctly formulated* for its rank, being in a Latinized form, even if the root is from another language, the species and genus names agreeing in gender. An appropriate standardized ending must be used if for a higher rank than genus (endings for each rank are given in the code and in Jeffrey, 1973). A very valuable aid for the correct usage of botanical Latin has been provided by Stearn (1973).

3. The name must not have been used previously in that rank for a member of the plant kingdom (if it had it would be referred to as a *homonym*) and it must have *priority*, being published before any other name applied within the same rank to the same kind of organism (others being referred to as *synonyms*). The actual date when the publication became available is taken into account when two names appear within the same year relating to the same organism, or when the date printed on the publication is not the year in which the publication effectively took place.

4. The name must be applied at the appropriate rank (e.g. species, variety, form) and does not have priority outside of the rank in which it was originally used.

When new combinations result from transferring a species from one genus to another, the species name should be retained, unless it has already been used with the genus to which the species is being brought, in which case a new species name is provided. After 1 January 1953, the *basionym* (the combination under which the species first appeared) must be cited, with publication details (author, year, page number).

The names of organisms are conventionally italicized (or underlined, when mentioned in manuscripts) except for the abbreviated term which indicates a variety, form or subspecies level, if one is used. The infraspecific name should also be italicized or underlined. In formal usage the name of the author of the taxon is given immediately following the Latin and is not italicized. If well known it may be abbreviated. Citation of the author's name safeguards against confusion in cases where the same name (homonym) has been created by different authors. It also serves as a guide as to where to start looking for references if one is not familiar with the taxon. Two authors are indicated by the Latin *et* or an ampersand (&) between their names. An initial may be given if there have been two or more authors with the same family name publishing within the group (e.g. A. Schmidt or J. Schmidt in diatoms, D. L. Taylor or F. J. R. Taylor in dinoflagellates). An author's name in parentheses means that he or she originally described the taxon but its name has been changed (the name of the changing author is always given after the parentheses in botanical literature and zoologists are also beginning to follow this convention). The initial letter of the generic name is always capitalized; those of species or infraspecific names are not usually capitalized although if the name refers to a person's name it may be (recently this practice seems to be declining). If an author has introduced a new name or nomenclatural revision, he does not usually put his name at the end, but instead indicates the kind of novelty which he has proposed (sp. nov., comb. nov., stat. nov., emend., etc.). Later authors referring to the taxon he has revised or introduced cite his name in place of the revision type designation if they accept the change. The following examples may serve to illustrate some of these points:

Coscinodiscus argus Ehrenberg [or: Ehr.]

Pyrodinium monilatum (Howell) F. J. R. Taylor;

Ceratium symmetricum Pav. var. *coarctatum* (Pav.) Graham et Bron.

In the last example the citation of author of the species (Pavillard) would usually be omitted except in the most formal usage, but would be cited in connection with the varietal name (Pavillard used it first for another species of *Ceratium* but it was shifted to the level of a variety by Graham and Bronikovsky).

SOME RECOMMENDATIONS

The above is only a highly condensed summary of the much more carefully and extensively worded articles of the ICBN. Anyone contemplating introducing a new name or revision must consult the ICBN, or an authority familiar with it, before publishing the new name. If in doubt, avoid introducing new names into publications, and avoid using new names under circumstances in which the requirements listed above are not fulfilled (e.g. in circulated manuscripts or before a full description is provided).

The organisms found in a region can be listed in a systematic arrangement following the classification of the source used (specify it), or they can be listed alphabetically. The latter is not only safer if one is not familiar with the systematics of the group, but makes it easier for others to use the list.

The use of a name implies its critical determination. Usually the reader has no means of checking the identification and it must be taken on faith. Because of the tediousness of community analyses, it is unlikely that the samples will be re-examined by another analyst and this places great responsibility for caution and accuracy on the analyst. Samples should be retained for as long as possible to permit re-examination in cases of special interest. If the publisher will permit it, the illustration of important or taxonomically difficult taxa serves as a useful guide to the reader as to the degree of trust to place in the identifications. At the very least, the taxonomic guides used for identification should be cited so that knowledgeable readers can apply corrections if appropriate. If the identifications have been made by others, this should be specified; and cultures studied should be identified by strain numbers if they are available so that observations based on misidentified cultures can be reassigned and not wasted.

In cases of doubt, name the organism to the lowest hierarchical level of certainty. If possible, indicate what feature(s) of the organism render its identification doubtful. Unfortunately, data related only to the generic level are usually of very limited application (genera such as *Chaetoceros*, or *Ceratium*, for example, containing ecologically and physiologically dissimilar species), but a misidentification that is not detectable merely throws confusion into an already difficult field of study.

Taxonomists that recognize many taxa based on very small differences are colloquially referred to as 'splitters', whereas those who gather taxa together into a few large assemblages are 'lumpers'. Analysts who are beginning a floristic study may wonder which to follow. The work of 'lumpers' may appear highly attractive to one wrestling with subtle differences in the material. From the ecological standpoint there seems to be little doubt that the field analyst should be a splitter as far as possible. It is a relatively easy matter to combine data for two taxa which turn out subsequently to be the same, but it is an impossible task to discriminate data for two taxa that were not discriminated previously—all the data relating to them becoming suspect. Furthermore, although some of the supposed varieties and forms may turn out to be taxonomically trivial, these infraspecific expressions may contain valuable information about life-cycle changes with time or environmental condition (such as summer and winter 'forms'), or nutrient starvation, etc.

There are differences of opinion as to the meaning of formal infraspecific categories. To zoologists the term 'subspecies' generally carries with it a notion of geographic isolation (and consequent lack of interbreeding), and it is difficult to see how this can be applied to marine phytoplankton, although this is the only infraspecific category recognized by the ICZN. Botanists usually use the categories of *variety* and *form*, although the meaning of these is not agreed upon. Hustedt (1930) and Taylor (1976*d*) conceive of varieties as small differences in genotype, with forms representing responses of organisms with the same genotype to differing ambient conditions, but there are others who have used these terms in the opposite way. Consequently, when creating a new infraspecific taxon, the type of variation that the taxon is suspected to represent should be indicated.

Finally, a new species should not be proposed for a single novel specimen. At least two specimens, and preferably more, should be seen, to indicate that it is not an aberrational freak and to obtain some idea of variability included within the taxon. However, it would be good practice to publish an illustration of the suspected new taxon when first seen. This establishes the fact of first discovery (remember that species have often been named after their discoverer by a later author) and draws attention to the possibility of a new taxon.

This section can only offer the briefest summary of these matters. The safest course to follow is to consult with others more experienced whenever in doubt and to check procedure even when not!

163

7
Estimating cell numbers

7.1   General principles

7.1.1   What to count?

Theodore J. Smayda

Historically, phytoplankton abundance has primarily been measured and expressed as cell numbers based on enumeration of the phytoplankton in an aliquot of the sample. Such measurements have established four basic characteristics of phytoplankton dynamics: abundance usually varies with depth, season and region, and a species succession occurs. The use of numerical abundance as a measure of the phytoplankton standing stock is not without problems, however, partly attributable to the considerable interspecific and intraspecific differences in cell size characteristics of the phytoplankton. The cell volume of the diatoms collected in the Gulf of Panama varied from about 50 μm^3 (*Nitzschia delicatissima*) to $12 . 10^6$ μm^3 (*Rhizosolenia acuminata*), i.e. 240,000 *Nitzschia* cells equalled one *Rhizosolenia* cell in volume (Smayda, 1965). The significance of such size variability is that species which are relatively unimportant numerically may be very important in terms of biomass. Hence, abundance based on a numerical census tends to overestimate small cells and underestimate the contribution of large cells (see also Paasche, 1960). Moreover, cell number *per se* has limited value as an expression of phytoplankton abundance in certain food-chain and nutrient-budget analyses where a measure of the carbon, nitrogen or some other constituent is required. In such instances, a measure of biomass is needed rather than a cell census. Biomass can be measured directly from proximate analyses (chlorophyll, ATP, carbon, nitrogen, etc.), or indirectly from the cell-volume characteristics of the enumerated population (see Chapter **8.5**). Biomass measurements do not estimate phytoplankton concentration *sensu stricto*, but, rather, provide a measurement of some constituent common to the entire population which reflects numerical abundance. Thus, biomass measurements may supplement, but cannot substitute for a numerical census.

The advantages of microscopic counting over other methods include (Lund and Talling, 1957): (a) the algae are seen, permitting the detection and evaluation of changes in appearance, including size, colony formation, spore formation and attack by parasites; (b) estimations can be made of populations whose abundance is too low for detection with equal accuracy by proximate analyses; (c) species identifications can be made and, hence, the taxonomic structure of the population can be determined. Moreover, an insight into species succession and calculation of species diversity and other statistical indices are possible.

Ideally, a phytoplankton census will detect, enumerate and identify to species all cells living at the time of collection, even if preserved samples are counted. Natural communities always include, in addition to living cells, a variable fraction of apochlorotic or empty cells representing many taxa. In the Gulf of Panama, for example, the mean incidence of moribund specimens in the individual samples was 10 to 15 per cent of the total phytoplankton community (Smayda, 1966). Dead cells and detritus are ordinarily not enumerated, since such particles, obviously incapable of growth, will not lead to subsequent community alterations. Also, where conversions to biomass or attempts to relate physiological processes to population density are made, inclusion of dead cells, if particularly abundant in such analyses, may lead to spurious conclusions. The distinction between living and dead cells is sometimes difficult; the investigator may subjectively evaluate whether chlorophyll is present, the intensity of pigmentation, whether the cell is intact, and the general condition of the cell. Where applicable, staining (see Crippen and Perrier, 1974 and Chapter **4** above) may permit distinction between living and dead cells.

Some species of phytoplankton are heterotrophic, notably dinoflagellates. If it is desirable to estimate their abundance, such as when certain phytoplankton–zooplankton relationships are focused on, care must be taken not to ignore them as dead cells.

Most pelagic species of diatoms form chains of attached cells. However, since the cell is the basic replicating unit, each cell in a colony must be counted during a census. Counting of colonies is also required in some studies, such as size-class distribution and other statistics. The trichomes of blue–green algae, such as *Oscillatoria* (= *Trichodesmium*), often collect into bundles. For such species, in which enumeration of the individual trichomes is difficult, the number of colonies is enumerated.

Unidentifiable cells should also be recorded. For this purpose, one may use either such broad designations as 'monads' and 'flagellates' (see Chapter **6.1**), or some appropriate description that may eventually permit taxonomic assignment.

7.1.2   How many cells to count?

Elizabeth L. Venrick

The problem of how many cells to count to give satisfactory precision has received considerable attention, although the criterion of satisfaction has varied, as has the index of precision employed. In general, precision refers to the variability of repeated estimates of mean abundance. Many studies have considered only precision of the estimate of the mean in the sample from which the enumerated material was drawn. However, the counting of cells represents the ultimate level of subsampling and should not be considered out of context of the entire design (see Chapter 5.1). The precision of a count may not represent the overall precision of a multilevel sampling programme. Indeed, it is easy to become preoccupied with the precision of the count, even though it need be no greater than is justified by the precision of replicate field samples and this, because of the heterogeneity of the population in the field, is often rather poor. It is often preferable to make rather imprecise estimates on several replicate samples than to make a precise count of a single sample.

In the following discussion, two situations must be distinguished. In the first, the final subsample is completely enumerated, as is often true of material concentrated in a settling chamber for examination on an inverted microscope. In this case, the number of cells must be regulated by the factor of concentration or dilution. In the second situation, only a fraction of the final material is counted, as is usually the case with the Sedgwick–Rafter counting chamber in which the number of cells counted may be regulated by the number of fields examined within the S–R cell. In the enumeration of natural populations, which are composed of many species of different sizes and abundances, a combination of both methods is often employed. The various procedures discussed below apply equally to each situation but the relevant precisions are different. In the one case we have replicate fields and their variability within a single chamber, and in the other case we have replicate counting chambers and their variability within a single sample. The statistical aspects of partial counts have been discussed in Chapter 5.1. If it is desirable to obtain an expression for the precision between replicate chambers, each estimated from a partial count, reference should be made to the formulae developed in that section. In addition, the formulae for optimal allocation of resources (cf. Chapter 5.1) may be applied to determine the most efficient numbers of chambers and fields within chambers (Uehlinger, 1964; McAlice, 1971; Woelkerling et al., 1976).

In either situation, some method of randomization is necessary. This is generally assumed to be accomplished by physically mixing (randomizing) the sample before removal of the subsample to be counted. When the final subsample is further divided into fields for counting, the simplest way to accomplish

167

this is to divide the sample chamber into a grid of non-overlapping fields from which those to be counted are selected at random. The grid may be a physical grid on the top or bottom plate of the chamber or a series of coordinates on the vernier scale of the microscope stage. Some workers partition the sample into strips, although Uehlinger (1964) and Woelkerling et al. (1976) found this method more time-consuming and less precise than the use of random fields. To avoid biasing the count, cells more than half out of the field should not be counted. In practice it is easier to exclude any cell which overlaps the left and distal boundaries (or right and proximal boundaries) of the field than to attempt to judge the median axis of a cell. Further aspects of this subject are discussed by Lund and Talling (1957) and Cassie (1971).

The problem of how many cells to count has been approached in many ways. In the classical theoretical treatment of the distribution of yeast cells in a haemacytometer, Student (1907) concluded that the relative precision depended only on the number of cells counted and recommended that the material be diluted to the point where counting can be done most rapidly and the count based on as many cells as time allows. Following Student's general approach, Lund et al. (1958) recommended counting 100 cells to give a 95 per cent confidence interval of the estimate within $\pm 20\%$ $\bar{x}$ and 400 cells for a precision of $\pm 10\%$ $\bar{x}$. Using a very different approach, Frontier (1972) concluded that a count of 100 individuals would give a relative precision of 31 per cent. Utermöhl (1958) also recommended a count of 100. Allen (1921) determined empirically that a count of 50 individuals was sufficient to keep the deviation from their mean within ± 25 per cent. Holmes and Widrig (1956) recommended counting at least 15 to 30 individuals while Gillbricht (1962) recommended no more than 15 to 35 in each of 5 to 10 subsamples. Koltsova et al. (1971) used two measures of diversity as indicators of the number of individuals which must be counted to obtain an adequate representation of the population. They concluded that the minimum number was a function of the population density and was approximately the number of individuals contained in 100 ml.

Several studies have been directed specifically to counting with a Sedgwick–Rafter cell. Littleford et al. (1940) demonstrated empirically that 40 fields of an S–R cell were adequate when 1,000 or more cells were present but a larger number of fields was necessary as the abundance decreased. Kutkuhn (1958) determined that 10 fields in each of 4 cells gave most efficient results. The procedure of optimum allocation of resources (cf. Chapter **5.1**) led McAlice (1971) to recommend 20 to 30 fields in each of 3 cells, and Woelkerling et al. (1976) to conclude that the most efficient design employed a few fields within each of many S–R cells. Serfling (1949) recommended adjusting the total number of cells counted by regulating the area of the fields rather than their number, which he maintained at 10, and adjusting the area so that the true mean number of cells exceeded 5 per field, permitting the use of parametric statistical procedures.

The disparity of recommendations in the literature reflects different laboratory methods as well as different theoretical concepts and assumptions. It emphasizes both the need to examine directly the error components of a particular procedure and the need for more realistic theoretical models.

PARAMETRIC PROCEDURES

The following discussion presents several formulae for determining the number of cells to count to obtain a satisfactory estimate of the abundance in the sample. These formulae are based upon the assumption of a non-aggregated distribution of individuals which can be described by a Poisson, binomial or normal distribution. This may or may not be a valid assumption. The work on subsampling variability has been summarized in Chapter 5.1; the distribution of individuals into subsamples sometimes is random and sometimes is not. Additional work has been done on the distribution of organisms within a counting chamber (e.g. Student, 1907; Serfling, 1949; Kutkuhn, 1958; Utermöhl, 1958; Uehlinger, 1964; McAlice, 1971; Rassoulzadegan and Gostan, 1976). The results are best expressed by Uehlinger who concluded that the presence of a random distribution in round chambers (or, apparently, other chambers) is not common, is never known in advance, and cannot be obtained voluntarily. As at every stage of the sampling programme, the researcher must make his own evaluation of his particular techniques and the resulting distributions of the biological material under consideration.

With proper handling it does appear that aggregations within the counting chamber can be minimized and the theoretical formulae may give results which are 'not too bad'. At the very least, they provide estimates of minimum sample size and these may provide baselines from which to work. An alternative is to employ a transformation to normalize the data (Chapter 8.2). Under these conditions, parametric formulae for precision will give the appropriate sample size only if the subsequent analyses are based on similarly transformed data.

Counting for precision

Under the assumption of Poisson distribution, the number of cells to be counted in a subsample for any desired precision at that level may be obtained from tables or graphs of the fiducial limits to the Poisson (e.g. Ricker, 1937; Holmes and Widrig, 1956; Uehlinger, 1964; Fisher and Yates, 1970). Figure 29 presents some confidence bands about means of less than 50. For more than 50 cells per sample, the normal approximation has been used (Ricker, 1937). This gives the probability interval for the true mean based on a count of x individuals as $x \pm z_{\alpha}\sqrt{x}$ (the symbols have been defined in Table 4, Chapter 5.1). The relative precision is then given by $z_{\alpha}(100\%)/\sqrt{x}$.

The coefficient of variation of the mean, $CV_{\bar{x}} = s_{\bar{x}}(100\%)/\bar{x}$ has also been used as a measure of precision (Gilbert, 1942; Cassie, 1971; Woelkerling *et al.*, 1976). Under the assumption of a Poisson, this simplifies to $100\%/\sqrt{(n\bar{x})} =$

169

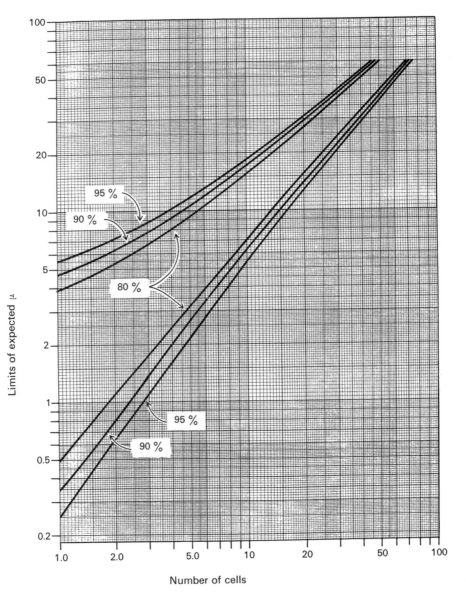

Figure 29
Limits of expectation of population means based on single estimates of abundance from a
Poisson distribution, at three levels of significance: 95, 90 and 80 per cent ($\alpha = 0.05, 0.10$ and 0.20
respectively).

170

$100\%/\sqrt{(\Sigma x)}$ where n is the number of samples counted and Σx is the total number of cells. Gilbert (1942) found $130\%/\Sigma x$ to be a useful estimate of precision for cells which are joined in chains or colonies. Under these assumptions, the precision of the estimate is independent of the number of samples counted.

When more than one subsample is counted, the Studentized normal variate $t_{(\alpha)(n-1)}$ rather than the standard normal variate z_α may be used in the calculation of the number or size of samples from the expression $\bar{x} \pm t_{(\alpha)(n-1)}\sqrt{(s^2/n)}$

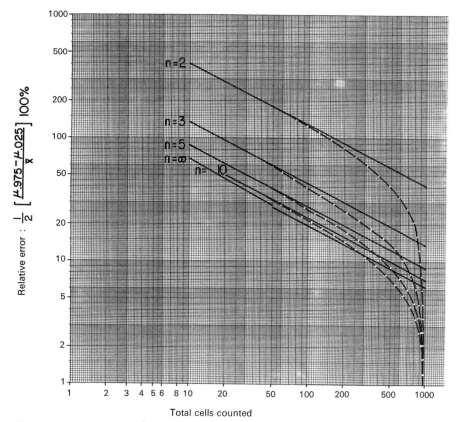

Figure 30
Relative error of the estimate of the population mean, μ, at $\alpha = 0.05$ for various sample sizes. Relative error of finite sample sizes calculated from:

$$\frac{t_{(0.05)(n-1)}\sqrt{\bar{x}/n}}{\bar{x}}(100\%).$$

The innermost interval (df $= \infty$) is derived from fiducial limits of the Poisson for $n \leqslant 50$ (Ricker, 1937) and normal approximation for $n \geqslant 50$,

$$\frac{z_{(0.05)}\sqrt{\bar{x}}}{\bar{x}}(100\%).$$

Dotted lines indicate relative error when finite population correction is applied to samples from a population of 1,000 individuals.

171

which under the assumption of a Poisson becomes $\bar{x} \pm t_{(\alpha)(n-1)}\sqrt{(\bar{x}/n)}$. The use of this parametric formula is appropriate as long as the mean number of cells exceeds 5 (Serfling, 1949). Since $t_{(\alpha)(n-1)}$ is a function of the number of samples, n, this expression depends both upon number and size of samples. For the same total number of cells, a more efficient estimate will be obtained by counting a larger number of smaller samples. The increase is particularly great when n is small, as demonstrated in Figure 30.

The use of the Student's t statistic decreases the apparent precision of a sampling design relative to that calculated using the z variate. It may, however, be a more realistic estimate. The choice of the standard normal variate over the Student t depends upon the validity of the assumption of a Poisson distribution. Only if it is valid is it appropriate to use the theoretical expression for the variance (using the observed mean rather than the observed variance as an approximation for σ^2) and to use the standard normal deviate. If there is any doubt about the validity of the Poisson (and there generally is), use of the observed variance and the Student's t is preferred. When only a single sample has been taken, this option is not available.

In any of the above procedures, a finite population correction (FPC) may be appropriate (cf. Chapter 5.1) and this will reduce the number of cells to be counted. If the entire sample contains a total N possible subsamples (or a counting chamber contains N fields), of which n are counted, application of FPC$-1 = [(N - n)/N]$ gives the expression for the confidence interval of the mean as:

$$\bar{x} \pm t_{(\alpha)(n-1)}\sqrt{s_{\bar{x}}^2[(N - n)/N]} \text{ or } \bar{x} \pm t_{(\alpha)(n-1)}\sqrt{\bar{x}(N - n)/nN}$$

Figure 30 shows the effect of the finite population correction when the true population contains a total of 1,000 cells; the effect is significant only when n exceeds $N/10$ or $N/20$ (Cassie, 1971), but, as an increasing portion of the sample is counted, the relative error decreases more and more rapidly, approaching zero when the entire sample is counted.

Counting for discrimination

A second useful formula for calculating the number or size of samples is based upon the desired discrimination ability. The expression

$$n \geqslant 2(\sigma^2/\delta^2)\{t_{(\alpha)(n-1)} + t_{(2\beta)(n-1)}\}^2$$

may be used to calculate the number or size of samples necessary to have a known certainty (1-β) of successfully distinguishing two populations with variance σ^2 when their true difference is greater than some specified amount, δ (Winer, 1962; Sokal and Rohlf, 1969). Alpha is the accepted probability of finding a difference when there is, in fact, none. Some estimate of population variance must be available. If a Poisson is assumed, $\bar{x}$ may be substituted for σ^2 and the formula solved for number of cells, or number of samples.

172

Example

How many samples must be counted to be 80 per cent certain of successfully distinguishing two populations at a significance level of 0·05 if the larger is as large or larger than 125 per cent of the smaller, and if the mean of the smaller one is approximately 100 cells per sample?

As an estimate of the variance, we shall assume a Poisson and use the variance of the larger population, $\sigma^2 = 125$.

$$\alpha = 0·05; \beta = (1 - 0·80) = 0·20; \delta = 125 - 100 = 25.$$

Since $t_{(\alpha)(n-1)}$ is a function of n, this formula must be solved by reiteration. As a first approximation, assume $(n - 1) = 20$, so that $t_{(0·05)(20)} = 2·086; t_{(0·40)(20)} = 0·860$.

$$n \approx 2(125/25^2)(2·086 + 0·860)^2 = 3·5.$$

Trying a smaller value of n, say $n = 6$, the expression becomes

$$n \approx 2(125/25^2)(2·571 + 0·920)^2 = 4·9.$$

But using $n = 5$ gives

$$n \approx 2(125/25^2)(2·776 + 0·941)^2 = 5·5.$$

So the best estimate of n is 6, which will give a slightly greater discriminatory ability than specified if the distribution to be sampled is truly normal.

SHORT-CUT COUNTING METHODS

The frequent benefits of accepting lower precision in order to reduce counting time have led to the development of short-cut methods for estimating abundance. Legendre and Watt (1972) have reviewed a procedure of estimating abundance from the frequency of occurrence of species in a set of random fields (or sub-samples); recording only presence or absence in each field results in considerable savings of time. This procedure is based upon the assumption of a Poisson or binomial distribution of organisms in the sample. The density of material, or the size of a field, should be regulated so that the most abundant species is present about 80 per cent of the time; it is usually sufficient to count about 30 fields.

Other researchers have used frequency classes rather than point estimates of abundance (Colebrook, 1960; Frontier, 1969, 1973; Frontier and Ibanez, 1974; Ibanez, 1974, 1976). The greater error tolerances allow considerable

reduction in counting time without substantially weakening the final interpretation of the data. Colebrook (1960) has discussed the effect of the category
method on the estimation of population means and standard deviations.

DISTRIBUTION-INDEPENDENT METHODS

If a number of replicate samples are available, it may be possible to abandon the
assumptions of Poisson or normal distributions and determine directly the
relationship between the subsample mean and variance. The sampling characteristics of this function may provide the desired information about sample
size. Cassie (1971) used the relationship $\sigma^2 = \mu + c\mu^2$ to describe overdispersed
distributions and the coefficient of variation, $s/\bar{x}$, as a measure of precision. There
is a limiting value of the CV and a sample size beyond which further increase
produces very little reduction in the coefficient (Cassie, 1971). The maximum
precision and the largest useful sample size may be a function of the amount of
overdispersion of the population. Forty to fifty individuals in a single sample
are recommended by Cassie for random to moderately overdispersed organisms.
Beyond this, it may be more efficient to count two or more small samples than
one large one.

Using a different approach, Frontier (1972) found a transformation ($N^{1/3}$)
which stabilized the variance of replicate zooplankton counts, making the
relative precision a function of the number of individuals counted. A count
of 50 individuals gave a relative precision of 40 per cent, 100 individuals a
relative precision of 31 per cent, 200 individuals a relative precision of 25 per
cent and it was necessary to count 2,500 individuals to obtain a relative precision
of 10 per cent.

Finally, optimal sample size may be determined empirically as was done
in the exhaustive study by Uehlinger (1964) who examined many aspects of
plankton counting. By comparing the true mean (determined by counting an
entire chamber) with the mean estimated from an increasing number of fields
within the sample, she determined that 30 to 40 random fields (out of a total of
489), representing 75 to 100 individuals, gave an estimate within $\pm 2\,s$ of the
true mean.

In the case of non-random distributions, these distribution-independent
procedures offer obvious advantages. Their major disadvantage is the need
for a more or less extensive preliminary study which may not be expedient
except in the case of major sampling programmes. Unfortunately, at the present
time, the generality of individual results is not known.

COUNTING FOR RANK ORDER OF ABUNDANCE

If the data are to be analysed using non-parametric statistics based on rank
order rather than absolute abundance, a smaller sample size may be sufficient.

The author knows of no previous attempt to determine the minimum sample size necessary to assign ranks with a predetermined accuracy. The following is one possible approach; it is not necessarily the only one.

Two criteria, both based upon the binomial distribution, have been employed. These are combined in Figure 31. The diagonal line gives the minimum number of the most abundant species (n_a) for any total number of two species $(n_a + n_b)$ which will give a probability of α or less that they are in the incorrect rank order regardless of their true ratio. The vertical lines give the minimum sample size necessary to assign correctly ranks (with a probability of misranking $\leqslant \alpha$) if the true ratio of the species abundances is equal to or greater than some predetermined value, $C = N_a/N_b$ where $N_a > N_b$.

The phytoplankton should be counted in randomly selected fields and the total abundance of each species cumulated. After each field, or group of fields, the count of the most abundant species (n_a) is compared with that of the next most abundant (n_b) and the point $n_a + n_b$, n_a is located on the figure appropriate to the desired level of α. If it falls above the diagonal line or to the right of the selected value of C, species a is assigned rank 1 and counting of that species is discontinued. Its absolute abundance may be estimated from n_a and the fraction of the sample counted. Species b may then be compared with species c and if successfully ranked, c may be compared with d, etc. Elimination must procede in order, from rank 1 down. Counting is then continued with a reduced species list and the procedure repeated until the entire sample is enumerated or until ranks have been assigned to the desired number of species.

Either criterion (i.e. the vertical line or the diagonal line) may be used exclusively; however, a combination is likely to be more efficient. For instance, the diagonal line may not distinguish between two species of similar abundances until the entire sample has been counted. On the one hand, if the sample contains a large number of cells, this may be undesirably time-consuming. On the other hand, determining a minimum total to be counted in order to rank correctly two species with some preselected ratio, C, will result in more cells than necessary being counted for those species whose true ratio is greater than C.

Derivation of Figure 31

From the binomial distribution in which $p(a) = p(b)$ (i.e. $N_a = N_b$) there is some pair of observations n_a, n_b such that this pair and all less equitable pairs are likely to occur out of a sample size $n = n_a + n_b$ with a probability of α or less. The critical value of n_a (where $n_a > n_b$) is represented by the diagonal line and may be calculated from the expansion of the binomial series:

$$\alpha/2 \geqslant p(n_a, n_b/n) = \sum_{x=0}^{n_b} [n!/x!(n-x)!](\tfrac{1}{2})^n.$$

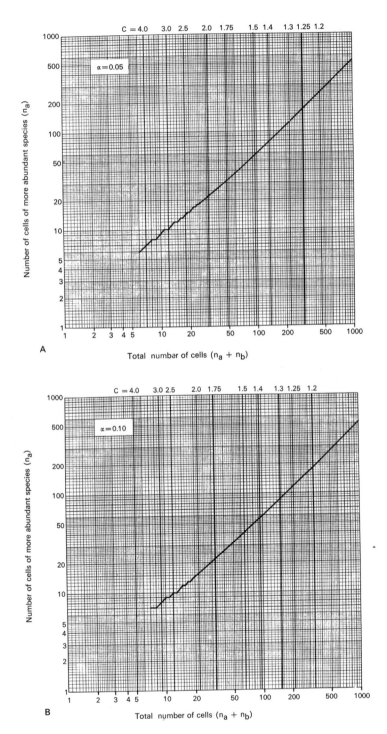

A

B

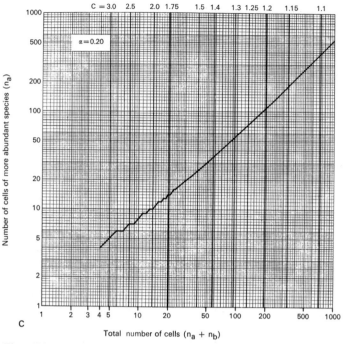

Figure 31

Minimum sample size needed to rank accurately two species at three probability levels: (A) 0·05; (B) 0·10; and (C) 0·20. Diagonal line indicates minimum number of more abundant species, n_a, which must occur out of total number of two species, $n_a + n_b$, to assure correct ranking. Vertical lines indicate minimum total, $n_a + n_b$, to assure correct ranking if the true population ratio, $C = N_a/N_b$, is greater than some preassigned value.

Values may also be obtained from tabulations of the binomial distribution, available in many statistical texts and tables, or from a table of critical values for the sign test (e.g. Dixon and Massey, 1957; Tate and Clelland, 1957; Conover, 1971). Critical values of n_a for $n > 50$ may be approximated from

$$n_a = (n + 1 + z_{(\alpha)}\sqrt{n})/2$$

(Tate and Clelland, 1957).

The vertical lines are calculated on the basis of the acceptance of a region of potential misclassification,

$$\frac{1}{2} \leqslant \frac{N_a}{N_b} \leqslant C.$$

The minimum sample size necessary to correctly rank species outside this interval 95 per cent of the time may be calculated from the normal approximation of the binomial:

177

G

$$\frac{(np - \frac{1}{2}n)}{\sqrt{npq}} \leqslant Z_\alpha$$

where $p = p(a) = N_a(N_a + N_b) = C/(C + 1)$

$q = p(b) = N_b/(N_a + N_b) = 1/(C + 1)$

Solving for n gives:

$$n_a + n_b = n = 4z_{(\alpha)}^2 C/(C - 1)^2.$$

The above formula may be solved and graphical references prepared for any desired values of α or C.

Example

One wishes to count sufficient numbers of species so that the probability of misranking each pair is 0·05 or less if their true ratio is greater than 1:3. After counting several fields within the sample, the following number of species a, b, c, and d are found: $a = 115$, $b = 110$, $c = 50$, $d = 37$.

1. The total of species a and b is 225. This is greater than the critical sample size (220; Fig. 31) needed to distinguish two species if their true population ratio, C, is greater than 1:3. Therefore, species a may be assigned rank 1 and eliminated from the count.
2. Comparison of species b and c shows that the point $n_b + n_c$, n_b (160, 110) falls above the diagonal line of Figure 31A, so that species b may be assigned rank 2 and eliminated from the count. However, similar comparison of species c and d indicates that the point $n_c + n_d$, n_c does not meet either criterion and further individuals must be counted before ranks can be assigned.

Selection of an appropriate value of C depends upon the desired accuracy of the ranking and the amount of material it is feasible to count. Accurate ranking of two species with very similar abundances (small C) will demand counting a very large number of cells. However, any species with true ratios smaller than C will have a probability of being misranked which is larger than α. Selection of α is also arbitrary. For convenience, the same value of α has been used in the calculation of the diagonal and vertical lines. This need not be so.

The α used here, the probability that a species pair will be incorrectly ranked, must be distinguished from α^*, the probability that the entire data set will contain one or more misrankings. The α^* is approximately $(m - 1)\alpha$, where m is the number of species in the set and $m - 1$ is the number of comparisons (this formula is derived in Chapter **8.2**). Thus, if we have 21 species and make each comparison at the 0·05 level, we expect to get 0·05(20) = 1 pair misranked by chance alone. To maintain the overall α^* at 0·05 we would have to use a significance level of 0·05/20 = 0·0025 for each single decision.

COUNTING FOR PRESENCE AND ABSENCE

Counting for presence and absence is the fastest method of counting because of the rapidity with which the presence of the common species is established, allowing them to be eliminated from consideration. Precise determination of absence, however, usually necessitates scanning a large fraction of the material. In order to establish the absence of a species from a sample at some significant level (α) it is necessary to count $(1 - \alpha)$ per cent of the sample. Thus, if one wanted to know at a significance level of 0·05 that a species was absent in a sample of 1,000 individuals, it would be necessary to examine 950 of them.

Derivation

Derivation of $(1 - \alpha)$ per cent is based upon the hypergeometric distribution. Given a population of N individuals of which one belongs to the species of interest, the probability of not seeing it when the first individual is examined is $(N - 1)/N$, and the probability when the second individual is examined is $(N - 2)/(N - 1)$. Thus, the probability of not seeing this individual when a total of n individuals is examined is given by:

$$\frac{(N - 1)}{N} \cdot \frac{(N - 2)}{(N - 1)} \cdot \frac{(N - 3)}{(N - 2)} \cdots \frac{(N - n)}{(N - n + 1)} = \frac{(N - n)}{N}$$

Since α is the accepted probability of not seeing a species which is in fact present, we have

$$\frac{N - n}{N} = \alpha, \quad n = (1 - \alpha)N.$$

Table 7 summarizes the preceding discussion, which at first glance may appear a bit overwhelming. Two general principles are buried therein: greater information (in the sense of abundance estimates rather than ranks or presence/absence) requires more counting, and greater precision requires more counting. In view of the enormous time which is often involved in phytoplankton counts, neither maximum information nor maximum precision should be an end in itself. Instead, consideration should be given to the minimum values consistent with the goals of a particular programme. In many cases, accepting lower precision at the counting stage may allow time for additional field samples, which ultimately increase the knowledge gained about the population in the field.

TABLE 7. Determining how many cells to count[1]

I. *If the final data are estimates of mean abundance*
    A. If individuals are distributed according to a Poisson distribution:
        1. Small mean values ($\mu < 50$): fiducial limits may be obtained directly from tables or graphs (e.g. Fig. 29).
        2. Large mean values ($\mu > 50$): normal approximation:
           (a) Single chamber counted: $x \pm z_\alpha \sqrt{x}$.
           (b) Several chambers or fields within a chamber counted: $\bar{x} \pm t_{\alpha(n-1)} \sqrt{(\bar{x}/n)}$.

    B. If individuals are distributed approximately according to a normal distribution or have been normalized by a transformation:
        1. For a specified precision: $\bar{x} \pm t_{\alpha(n-1)} \sqrt{(\sigma^2/n)}$.
        2. For a specified discrimination ability: $n \pm 2(\sigma^2/\delta^2)\{t_{(\alpha)(n-1)} + t_{(2\beta)(n-1)}\}^2$. (Solve by reiteration.)

    C. If the distribution is aggregated, or unknown:
        1. Stabilize the variance (Frontier, 1972).
        2. Fit observed distribution to a theoretical function (Cassie, 1971).
        3. Determine sample size empirically (Uehlinger, 1964).

II. *If the final data are rank orders of abundance*
    Count according to Figure 31.

III. *If the final data are presences and absences*
    The proportion of the sample necessary to establish an absence $= (1 - \alpha)$.

---

1. The formulae may be solved for number of samples (n) or number of cells (x) to give a precision of the estimate at a given probability level (α); details are given in the text.

7.2 Using the standard microscope

7.2.1 Treatment of an aliquot sample

H. J. Semina

Phytoplankton cells can be counted with a standard (non-inverted) light microscope furnished with a counting stage. The volume to be sampled depends on the phytoplankton abundance. Before counting, the sample is agitated thoroughly by continuous shaking. Water with phytoplankton is taken with an eye pipette, or with a special Stempel pipette (0·05 ml), and placed into a 0·05 ml chamber. If the Stempel pipette is used, water may be transferred on to a glass slide. The chamber, or the drop on the glass slide, is covered by a cover-glass and placed on the microscope counting stage. Cells are counted under a × 10 or × 20 objective; very small forms have to be counted with a magnification of × 40. Depending on the cell concentration, either the whole sample or a part of it is counted (see Chapter **7.1.2**). If the sample was taken with a bottle, the cell number per litre is calculated after making an adjustment for the sample having been concentrated. For example, suppose the initial sample volume (V_1) was 1 litre. Then the sample was concentrated to 10 ml (V_2), from which a subsample of 0·05 ml (V_3) was taken. Assuming that no cells were lost during concentration, the number of cells per litre is obtained by multiplying the result of the count by $N = V_2/V_3 = 200$.

Poor samples are condensed to 1 ml, and cells are counted in this whole volume; for this purpose, all the concentrate is gradually transferred to the chamber, or placed drop by drop on to the glass slide. No recalculation is necessary in this case, if one assumes again that no cells were lost.

In order to count rare (usually larger) forms, one can first count all the cells in a 0·05 ml aliquot and store the sample for several days; then the water is carefully siphoned down to 1 ml and a special 1 ml chamber is used for counting the larger forms only. As this chamber is rather thick, only a low-power objective (× 10) can be used.

Here is an example of a calculation of cell number per cubic metre. The sample was taken with a net that had a mouth 0·1 m² large and a 0 to 100 m column was hauled; thus 10 m³ were sampled (V_1). The volume of the sample collected in the bucket was 100 ml (V_2) from which an aliquot of 1 ml was subsampled (V_3) for counting. The conversion coefficient for cell numbers per cubic metre is $N = (1/V_1)(V_2/V_3) = 10$. One should bear in mind, however, that the efficiency of the net is never 100 per cent (see Chapter **3.3**).

Cells whose position is inconvenient for taxonomic determination can be examined more closely after the count, the coverslip being slightly touched with a preparation needle to make such cells turn round.

181

7.2.2 Counting slides

Robert R. L. Guillard

Two techniques are available. In the usual one, a concentrated living or preserved sample is examined in one or more counting slides having chambers appropriate to the dimensions and abundances of the algae in the sample. In the second method, a preserved sample is concentrated by sedimentation in the separable ('combined') plankton chamber designed for use with the inverted microscope (see Chapter **5.2.1**). The whole bottom of the shallow ('plate') chamber is first scanned with low-power objectives of the standard microscope. The central portion of the chamber is then examined for abundant small species using a water-immersion ('dipping') lens after removing the coverglass. (This technique was recommended by W. Rodhe, pers. comm., *c.* 1955). The chief merit of this second method is the excellent optical resolution permitted without use of the inverted microscope.

Lund *et al.* (1958) considered the fundamental problems of all counting techniques, illustrating relevant elementary statistical methods. Guillard (1973) described in detail the use of counting slides for measuring cell concentrations in algal cultures. Points raised in these papers also pertain to the use of slides for enumerating phytoplankton in natural samples.

Data in Table 8 (see below) will aid in selecting counting slides appropriate to the size and abundance of the phytoplankton species of concern. Irregularity of cell distribution in the slide chambers can be gauged by the *chi*-square test (Lund *et al.*, 1958; see also Chapter **7.1.2**).

Only those chambers which are in current use in hydrobiological or planktological work will be dealt with here. However, it occurred to the author and to the editor that there are other types of chambers used in biology and medicine that may well be worth applying to plankton research after appropriate testing. These are known under the names Thoma, Malassez and Nageotte, among others, and can be purchased, at least in Europe, from manufacturers such as Brand, Prolabo and Schreck.

ACCESSORIES FOR THE STANDARD MICROSCOPE

1. A condenser of long working distance to permit Köhler illumination with thick slides is desirable. Phase-contrast illumination is frequently useful and often nearly necessary for observation, and, as relief from bright-field illumination, aids in preventing fatigue.
2. The mechanical stage should traverse left-to-right (horizontally) without unintentional vertical movement. For use with the dipping lens and plate

chamber it may be necessary to make an adapter to secure the plate chamber in the slide holder of the mechanical stage.

3. Objectives should include achromatics of $\times 20$ to 25, numerical aperture (N.A.) $\geqslant 0.5$, and $\times 40$ to 45, N.A. $\geqslant 0.6$, both with the longest working distances available (heights of counting slide chambers are given later). Correct combinations of objectives and oculars permit magnification of at least 500 with all slides. Wide-field $\times 12.5$ to 15 oculars are commonly used.

4. Water immersion (dipping) objectives of $\times 40$ to 55 magnification and N.A. 0.7 to 0.85 are recommended. Some of these are small enough in diameter at the tip so that most of the area of the separable plate chamber is accessible. Note that the dipping objective must be one especially constructed for use in seawater.

5. No. $1\frac{1}{2}$ coverglasses (about 0.18 mm thick) of appropriate size should be used for all slides to improve resolution and reduce the overall height of the various chambers.

6. Most counting slides have no marking grids, hence a grid or other field-limiting device (Lund et al., 1958) is needed in the ocular. The commercially available Whipple disc (e.g. Bausch & Lomb, Scient. Optical) provides a square field divided into 100 squares; one of the four central squares is further divided into 25 smaller squares which are useful for estimating cell sizes. Typically, at a magnification of $\times 125$, the side of a Whipple square subtends c. 700 μm in the image plane; the smallest squares measure c. 14 μm. The exact area subtended by the Whipple square must be measured using a stage micrometer; in the example given, it is about 0.5 mm^2 at $\times 125$.

7. Ethanol (90 per cent), commercial 'denatured' alcohol, isopropanol or acetone can be used for the necessary scrupulous final cleaning of counting chambers and coverglasses, following removal of each sample by distilled water. Clean surgical gauze (cheesecloth) cut into c. 10-cm squares is used for wiping.

SAMPLE PREPARATION

Samples of natural phytoplankton can differ in cell concentration by a factor of at least 10^4, even considering just samples collected from the euphotic zone. Single-species concentrations as low as a few cells per litre occur and are considered significant. Merely to detect such species requires scanning material derived from a few hundred millilitres of the original sample, which means that for accurate counting, the original sample material must be concentrated many hundredfold by appropriate techniques (see Chapter 5 above). The most difficult task is to enumerate species that are both small and scarce, in water containing detrital material that interferes with observation. In all cases, correct choice of counting device and concentration factor is critical (the concentration factor is the ratio 'volume of original sample/volume of final concentrate').

To prevent evaporation from altering cell distributions while cells are settling in a counting slide, put the slide into a petri dish or other closed container on a glass triangle or other support to hold it off the bottom and add a few drops of water to maintain a saturated atmosphere.

THE SEDGWICK-RAFTER SLIDE

See McAlice (1971) for detailed treatment of this counting device and its literature. To fill the slide, set a coverglass diagonally across the chamber leaving a space open at opposite corners. Deliver 1·0 ml of concentrate to the chamber through one opening with a large-bore pipette, then slide the cover-glass into position to seal the chamber. Before counting, examine the chamber under low power to detect obviously unsatisfactory distributions of the algae. If the coverglass interferes with dispersal of large species, it may be best not to use it, though visibility, especially near the edges of the chamber, is lowered in quality, and evaporation alters the distribution of specimens (McAlice, 1971).

When counting, tally individual organisms or colonies, as appropriate (see **7.1.1** or Lund *et al.*, 1958). Count specimens in one or more entire chambers, or in a given number of horizontal strips (of known area relative to the 1,000 mm^2 of the whole chamber), or in a given number of randomly selected Whipple fields, of which the area subtended, at the magnification used, is known. Count specimens until the number required for the desired level of statistical reliability is attained (see Chapter **7.1.2**). Convert counts to cell concentrations in the original sample using the relationships between areas just mentioned and the concentration factor.

THE PALMER–MALONEY SLIDE (PALMER AND MALONEY, 1954)

Cover the chamber with a round or square 22-mm No. $1\frac{1}{2}$ coverslip, making secure contact with the glass or plastic chamber ring. Tilt the slide slightly (to right or left) and add 0·1 ml of suspension via the lower of the two entry channels with a large-bore Pasteur pipette. A small piece of paraffin paper placed over each loading channel will prevent evaporation and does not inter-fere with observation. Examine the slide for satisfactory distribution of cells and count as just described for the Sedgwick–Rafter cell. Note that a half chamber can be counted relatively easily using the centre of the loading channels as a marker. (The channel width can be measured.) A few horizontal or vertical strips can be searched if they pass near the centre of the chamber; these strips can be taken to have the length of the diameter of the cell (17·9 mm) and the width of the Whipple field.

TABLE 8. Properties and uses of counting slides

Sedgwick–Rafter (Clay Adams) Figure 32A	The chamber is without rulings and is rectangular (50×20 mm), 1 mm deep, of area 1,000 mm^2 and volume 1·0 ml. With No. $1\frac{1}{2}$ coverslips, most $\times 20$ N.A. 0·5 objectives can be used, permitting magnification to $\times 500$. The largest phytoplankters can be held in this slide, and many species as small as 10 µm can be recognized in it. It is best suited to large and relatively scarce organisms, which can be detected in the concentrate at just a few per ml. A concentration of 10^4 cells/ml in the concentrate yields *c.* 5 cells/ Whipple field at $\times 125$ magnification.
Palmer–Maloney (Arthur H. Thomas Co.) Figure 32B	The chamber is without rulings and is circular, of diameter 17·9 mm, depth 400 µm. area 250 mm^2, and volume 0·1 ml. It has a loading channel (slot) on each side. High-dry objectives can be used. Cells (of some species) as large as 150 µm will enter and be reasonably well distributed in this chamber. Even the smallest phytoplankters can be detected in it. Species in the concentrate at over 10/ml should at least be detected; 10^4 cells/ml yield *c.* 2 cells/Whipple field at $\times 125$ magnification.
Haemacytometer, 0·2 mm deep (Many manufacturers, e.g. Hausser Scientific) Figure 32C	Generally available with Fuchs–Rosenthal ruling (Fig. 33), which consists of sixteen squares each 1 mm on a side (and each further divided into sixteen squares 250 µm on a side). Slides with two such rulings (thirty-two 1-mm squares) thus hold a total of 0·0064 ml of plankton concentrate in the ruled areas. High-dry objectives can be used if the coverglass is thin; wide-field oculars are recommended. Species larger than *c.* 75 µm will seldom distribute themselves well, and long thin species, or those forming long colonies, will usually accumulate near the entry slit or at the chamber edges. Cell densities as low as 10^3/ml in the concentrate are detectable; 5×10^3 cells/ml yield an average of one cell per 1-mm square. (Note that some manufacturers supply slides 0·2 mm deep but with the Neubauer ruling, described next. For these, 5×10^3 cells/ml also yield one cell per 1-mm square.)
Haemacytometer, 0·1 mm deep (Many manufacturers, e.g. American Optical Corp.)	Available with Neubauer ruling or Improved Neubauer ruling (Fig. 34). Either ruling has nine 1-mm squares variously subdivided; most slides have two such rulings and hence hold 0·0018 ml of plankton concentrate in the ruled areas. The slide is useful only for high cell densities because 10^4 cells/ml yield only one cell per 1-mm square. Remarks apply as for haemacytometers 0·2 mm deep, except that the maximum size of cells that can be be counted adequately is roughly half, *c.* 30 µm.
Petroff–Hausser bacteria-counting slide, depth 0·02 mm (Hausser Scientific) Figure 32D	This slide has but one chamber, 0·02 mm deep, with Improved Neubauer ruling (see above). The total volume in the ruled area is 0·00018 ml. High-dry and oil-immersion objectives can be used. This slide is useful only when dealing with dense populations of cells of bacterial dimensions, e.g. *Synechococcus* sp. or *Micromonas* sp. A concentrate density of 5×10^4 cells/ml yields an average of one cell per 1-mm square.

HAEMACYTOMETERS (SEE TABLE 8)

Filling the chambers by immersing the slide as suggested by Lund *et al.* (1958) is not routinely practical. To fill a slide, seat the coverglass firmly on the glass supports. Use a smooth-tipped pipette of adequate-bore diameter for the species concerned. Hold the pipette at an angle of *c.* 45° if the bore is small, but lower the pipette towards the horizontal if the bore is large—this controls the flow rate. Place the pipette tip next to the entry slit of the chamber, release the liquid

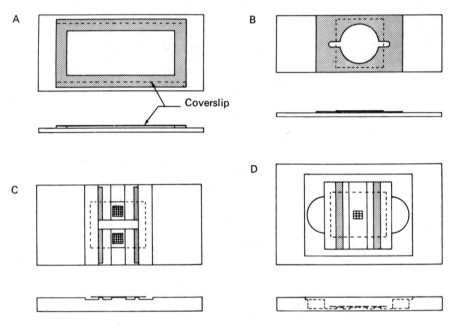

Figure 32
Counting chambers: (A) Sedgwick–Rafter chamber; (B) Palmer–Maloney chamber; (C) one type of haemacytometer (blood-counting chamber); (D) Petroff–Hausser bacteria-counting chamber.

flow and almost simultaneously remove the pipette from contact, so that liquid flows quickly and evenly into the chamber, exactly filling it. Fill the pipette anew for each chamber. The time taken from filling the pipette to flooding the chamber should be short, to minimize settling of cells in the pipette.

Examine the slide for satisfactory distribution of cells, then tally organisms in as many squares or fractions of squares as is indicated by the desired level of statistical reliability. Spread the total count over as much area as possible, e.g. counts from four quarter-squares are preferable to a count of one whole square in one place. Sample each chamber on slides having more than one chamber.

For all haemacytometers the fundamental measurement is the average

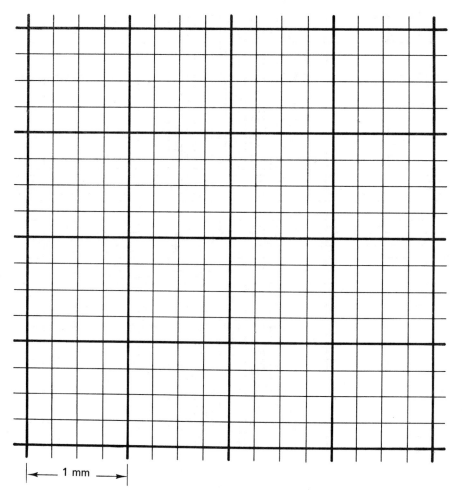

Figure 33
Fuchs–Rosenthal ruling.

number of specimens per 1 mm square, from which the cell density is readily computed (Table 8; also Guillard, 1973).

THE SEPARABLE SEDIMENTATION CHAMBER
AND WATER-IMMERSION LENS (ZEISS; WILD)

Combined chambers are described in Chapter **5.2.1**; their volumes range from 10 to 100 ml. The preserved plankton sample as collected may be put into the combined chamber for settling, or the sample may be concentrated by sedimentation first, in which case the concentration factor must be taken into account.

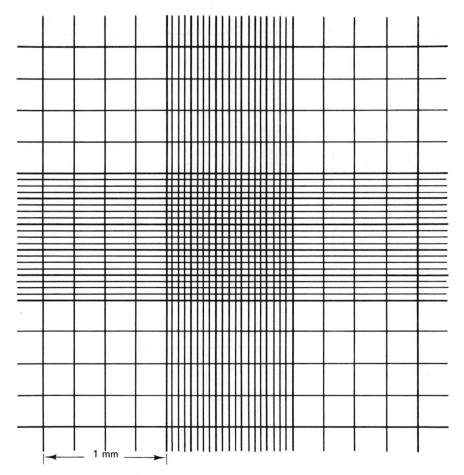

Figure 34
Improved Neubauer ruling. Division of the ruling into nine 1-mm squares is by double or triple lines, not shown in the figure.

After the sample or concentrate has sedimented into the plate chamber and the separable cylinder has been removed, the plankton is scanned with the low-power objective. The depth of the chamber (4 mm) is too great for most objectives of × 20 or greater.

The plate chamber has a round opening 26 mm in diameter with area c. 530 mm^2. There is no ruling, thus the whole chamber or representative diameters or fields must be counted. If, as before (Table 8), the Whipple square covers 0·5 mm^2 at × 125, an average of 1 cell/square means there are $532/0·5 = 1,064$ specimens in the plate chamber; these specimens represent the entire population of the concentrate (or sample) put into the combined chamber. By way of comparison with the other slides, note that if a cell suspension has

188

10^4 cells/ml, the settling of 10 ml into a separable chamber yields c. 100 cells/ Whipple field at $\times 125$.

Following the low-power survey, the coverglass is gently slid aside and randomly chosen Whipple fields (of known area) are scanned with the dipping objective of high magnification. Survey as much of the area as can be reached with the objective (as close to the edge as possible).

Convection currents, evaporation and shifting of the objective lens will ultimately disturb the specimens, which limits the number of fields that can be counted with acceptable reliability. The technique can be checked using preserved cultures of known density or a plankton sample previously counted by a reliable method.

7.2.3 Membrane filters for estimating cell numbers

Robert O. Fournier

Enumeration of phytoplankton on membrane filters is basically no different from any other technique in which the concentrated cells rest on a flat surface and are viewed microscopically. However, several problems exist which are not as important in other approaches, e.g. inability to manipulate cells, difficulty in recognizing cells because of filtration-induced distortions and finally the fact that the sample will be biased towards the most robust organisms (Chapter 5.4.2).

Enumeration on membrane filters does have one problem in common with all other techniques, i.e. that of whether or not the cells are randomly distributed. This question of random distribution of cells on filters has been addressed by McNabb (1960) and Holmes (1962). The former author felt that they were distributed randomly while the latter disagreed. On the basis of personal experience, I would agree with Holmes that visual inspection frequently shows clumping at the filter periphery. Perhaps this is the result of circulation patterns set up by uneven flow through the filter. Since uneven cell distributions are a common occurrence the only certain method of estimating cell numbers is to enumerate the entire filter. Counting cells in random fields would certainly bias the results.

Counting an entire filter can be long and arduous if too much water has been filtered and too many cells are present. The sample volume should be adjusted empirically so as to yield between 200 and 400 counts (Lund et al., 1958). Counting should be done in the same way as one would approach the inverted-microscope method. Enumerate at the lowest magnification which will permit recognition of the organisms in question (magnifications higher than necessary reduce the visible field without offering any noticeable advantage). Move the filter from side to side until the entire surface has been examined.

7.3 Using the inverted microscope

Grethe Rytter Hasle

Chapter **5.2.1** dealt with the treatment of samples prior to microscopical examination, including description of chambers for the combined sedimentation and counting. While these chambers may be used with some success on a standard compound microscope (see Chapter **7.2.2**), the inverted microscope can hardly be used for estimating cell numbers in water samples without using chambers of the type described.

MICROSCOPICAL EQUIPMENT

The *microscope* is inverted in the sense that the light source and condenser illuminate the chamber from above and the objectives view the specimens from below through a thin bottom plate of the chamber (Fig. 35).

Microscope lamp. A powerful light source is needed (e.g. 6 V/20 W), particularly for phase contrast and photomicrography. The lamp may be powered by a step transformer or a regulating transformer. A field diaphragm is necessary to set up Koehler illumination.

Condenser. A long-working condenser must be used in connection with chambers; phase-contrast illumination is advantageous for examination of most planktonic diatoms, flagellates and smaller dinoflagellates; bright-field illumination may be better for examination of coccolithophorids and larger dinoflagellates. A condenser with a rotating-phase diaphragm charger and iris diaphragm is thus preferable.

Mechanical stage. The microscope should be equipped with a good-quality movable mechanical stage with vernier scales. The option of restricting movement to a certain range by limiting screws (e.g. to 1 cm^2 by using both screws or to a band 1 cm wide) is offered for some inverted microscopes (e.g. Wild and Zeiss). The specimen holders provided by Wild and Zeiss accept the commercial combined plate chamber (Fig. 36). The Throndsen chamber is designed to fit into the same holders. Coaxial controls for the stage suspended by a shaft (Fig. 35) lessen the strain on the right arm of the investigator and leave the left hand free for the focusing controls.

Objectives. For counting, × 6·3 or 10, × 16 or 20 phase objectives should be available; a × 40 phase and a × 60, 90 or 100 phase water or oil immersion objective should be present in the nosepiece for more critical examination. All objectives should be of such a quality that they can be used for bright field, although the use of highly corrected optical systems is not urgent in most routine phytoplankton investigations. For photomicrography and identi-

191

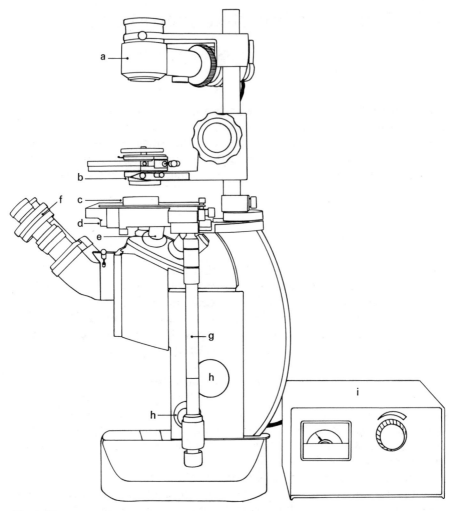

Figure 35
Inverted microscope: (a) lamp; (b) condenser; (c) counting chamber; (d) mechanical stage; (e) objectives; (f) eyepieces; (g) shaft for coaxial controls; (h) focus controls; (i) transformer (in more recent models the transformer is built-in rather than separate).

fication at high magnification (×40 to 100 objectives) a quality better than achromat may be of importance.

Tube. A binocular or a trinocular (for camera attachment) tube should be used.

Eyepieces. One eyepiece should be equipped for counting by two parallel threads intersected by a third one (home-made: Lund *et al.*, 1958) or by parallel threads which are movable in some commercial 'counting eyepieces'. In the latter case the width of the field (the stripe) examined at the time can be adjusted to density of material present, the parallel threads being fairly close when sediment is dense, and further apart when sediment is sparse. The counting eyepiece with movable threads is longer than an ordinary eyepiece. The second eyepiece therefore is adjusted to the length of the counting eyepiece. The magnification of the counting eyepiece is usually × 10 although eyepieces of higher magnification may be used. When computing the total magnification of the

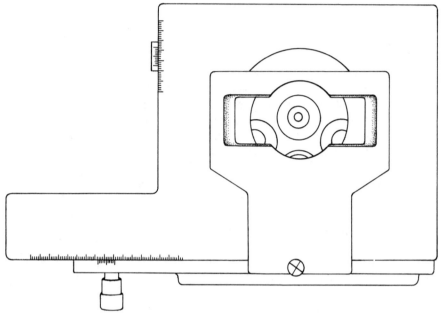

Figure 36
Mechanical stage of an inverted microscope (after Wild Ml, Catalogue 140c XII.69).

system the magnification factor of the optics carrier must be taken into account; this is usually of the order of × 1·25 or 1·5. Thus, the total magnification of the system is calculated as the product of the magnification of the objective, the magnification of the eyepiece and the magnification of the optics carrier. Since measurements are made frequently during counting, it is convenient to have an ocular micrometer inserted into one of the eyepieces. The micrometer should be as far away from the centre of the eyepiece as possible so that it does not interfere with the image of the organisms.

ANCILLARY DEVICES

Laboratory *counters* with five or more keys are useful for counting predominant species.

A *tape recorder* may be useful when counting very diverse samples.

Micropipettes for identification purposes. To get specimens, particularly of dinoflagellates and pennate diatoms, in a position convenient for identification a special device such as the pipette described by Throndsen (1969a) is useful.

COUNTING PROCEDURE

The procedure described here is a version of the Utermöhl method (Utermöhl, 1931, 1958) used in our laboratory where the interest is mainly focused on botanical, hydrographical and ecological problems. Other laboratories practise other modifications, and readers are referred to additional papers by Lund *et al.* (1958), Hobro and Willén (1975) and Willén (1976), all based on freshwater plankton, by Gillbricht (1959) based on marine plankton, and by Margalef (1969c) based on freshwater as well as marine plankton. Although the method has usually been practised in laboratories on land, it was used with success on board a ship by Travers and Travers (1971).

Depending on the purpose of the investigation all organisms encountered are identified to species, group of species, genus or algal group (see Chapter **6.1**). Either the cell numbers of each taxon are recorded if an estimation of the total population is wanted, or only selected species or groups are enumerated, e.g. the most numerous ones. The organisms in question, lying between the parallel threads of the counting eyepiece, are counted as they pass the third, crossing thread through the movement of the mechanical stage. (Some investigators prefer to move the mechanical stage vertically, others horizontally.) Care should be taken that those organisms not completely inside the field limited by the two parallel threads are counted only once.

A phytoplankton sample must usually be examined under more than one magnification. For larger forms a magnification of $\times 80$ to 200 (objective $\times 6 \cdot 3$ or 10 combined with eyepieces $\times 10$ to 15 and a magnification factor of the optics carrier $\times 1 \cdot 25$ to $1 \cdot 5$) is adequate. The smaller forms demand a magnification of $\times 200$ to 450 (objective $\times 16$ or 20) or sometimes $\times 600$ to 750 (objective $\times 40$) as for counting monads in a dense sample. Fortunately, the smaller species are usually, but not always, the predominant ones, and reliable counts may be obtained from a smaller subsample than is used for the larger species. The two subsamples may be two separate chambers or different fractions (one smaller and one larger) of the bottom of the same chamber (see below).

Medium-sized phytoplankton concentrations

A 2-ml chamber (e.g. Throndsen chamber) is useful for counting species attaining cell numbers (chain numbers in case of chain-forming species) of 15 to 20 and more in 2 ml. All individuals observed are counted, larger forms as well as smaller ones. This examination is done under a total magnification of $\times 200$

to 450. Afterwards a larger chamber (10 or 50 ml) under a total magnification × 80 to 200 is used for counting species rarer than 15 to 20 cells or chains in 2 ml. Higher magnification (objectives of × 40 or more) is used frequently during the counting to check identification of organisms. If use of immersion objectives is necessary, it is an advantage to read the position of the specimen in question by means of scales and verniers of the mechanical stage so that the identification can be done after the counting is finished.

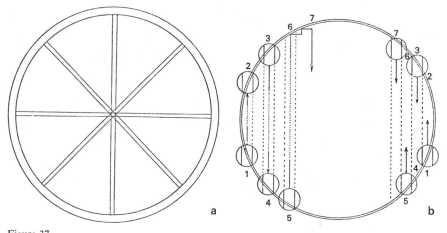

Figure 37
(a) Counting of diameter transects (after Utermöhl, 1958). (b) Counting of: right, every stripe; left, every second stripe (after Utermöhl, 1958).

Instead of examining two separate chambers, a two-step examination of one chamber may be carried out. A part of the bottom area of a 10- or 50-ml chamber is examined under high magnification as described above. This part may consist of one or more crossed-diameter transects (Fig. 37a), or every second, third, fourth or more stripes (Fig. 37b), or simply the first half of the bottom (cell counts from the two halves are rarely similar, however; see Hasle, 1969). Afterwards the whole bottom area is examined under lower magnification as is done when two separate chambers are examined. It should be emphasized, however, that many planktologists prefer to scan the whole bottom area under low magnification before examination of two crossed-diameter transects by a higher-power objective (Margalef, 1969c).

Examination of two crossed-diameter transects can readily be done by moving the mechanical stage horizontally to give one transect and then vertically to give the second one. Counts of more diameter transects can easily be obtained in chambers without a rectangular bottom plate (e.g. the chambers designed by Utermöhl, 1931), and the mechanical stage of most inverted microscopes in use these days (e.g. Wild M40) also permits rotation of chambers with a rectangular (or in older models, square) perspex bottom plate.

195

The ratio of the whole bottom area to that of one diameter transect is $\pi n/4$ where n is the ratio between diameter and width of diameter transect. The total number of cells in the whole chamber is found by multiplying the cell numbers found in the diameter transect by $\pi n/4$.

Dense phytoplankton concentrations

In this case the entire count is often based on the sample in the 2-ml chamber, even though larger and/or rarer species will usually be excluded from observation, leading to a reduction in the number of species recorded (Hasle, 1959). If concentrations of predominant species exceed several hundred cells per 2 ml, a two-step examination of the 2-ml chamber is recommended: one or more diameter transects (the 2-ml Throndsen chamber can be rotated) should be examined under a total magnification of × 200 to 450 and the whole bottom area under a total magnification of × 80 to 200. Small cells, e.g. flagellates, 'monads' and small centric single-celled diatoms are easily overlooked in dense sediments. To include them in the examination one or more narrow-diameter transects may be scanned under a still higher magnification than the one obtained by a × 16 or 20 objective. A three-step examination will thus be performed. (For 'bloom' investigations other types of chambers and microscopes than those described here may be more useful.)

Sparse phytoplankton concentrations

The sediment of 50 ml, 100 ml or more is examined directly in the plate chamber of a 50-ml combined chamber or transferred to a 2-ml chamber (see Chapter 5.2.1). Examination of the whole bottom area with a × 16 or 20 objective is recommended since an important part of sparse phytoplankton populations often consists of small forms.

RELIABILITY

Few attempts to test the reliability of the procedure described have been published. It should be noticed that the expected error in the estimates calculated by Willén (1976) was based, with reference to Lund et al. (1958), on the assumption of random distribution. Holmes and Widrig (1956) and Hasle (1969) found that individuals of chain-forming species were not selected at random during subsampling. Moreover, their results for two species of dinoflagellates were contradictory and, therefore, inconclusive. This apparently means that not only individual phytoplankton species but also composition of the population and possibly also chemical and physical properties of the water sample are decisive for distribution. If the investigations by Holmes and Widrig (1956) and Hasle (1969) are representative for marine material in general, the actual samples should be tested before expected error of estimates is calculated.

196

7.4   Using the fluorescence microscope

Gabriel A. Vargo

A quantum of light that strikes a pigment molecule and thereby sets the molecule into an excited state may be converted to chemical energy, lost by conversion to thermal energy, transferred to other pigment molecules or re-emitted as fluorescence (Govindjee *et al.*, 1967).

The autofluorescence of chlorophyll upon excitation by blue-violet light ($\sim$450 nm) and its subsequent emission of red light at 685 nm (Govindjee *et al.*, 1967) was used by Tchan (1953) for the study of soil microalgae. Wood (1955) and Wood and Oppenheimer (1962) adapted the method for the enumeration of phytoplankton. Chlorophyll-containing cells, which fluoresce brick-red against a dark-blue background, are counted in a concentrated, unstained sample. Total cells, with and without chlorophyll, can be counted after staining with a DNA-binding fluorochrome, acridine orange (Eastman Kodak Co.). Stained cells, both living and dead, fluoresce green; however, an orange or red fluorescence may occur if the DNA (or RNA) is heat-denatured (Daley and Hobbie, 1975). Loveland (1970) also notes that nuclear DNA will fluoresce green and cytoplasmic RNA will fluoresce red when stained with acridine orange in 'low' concentrations at pH 6.

Quantitative and qualitative assessments of both pigmented and non-pigmented organisms in water samples or attached to particulate material may be made rapidly and with relatively unsophisticated equipment that is very amenable to shipboard use, even in rough seas (Wood, 1965).

EQUIPMENT

Any standard microscope fitted with an Abbe or aplanatic condenser with a numerical aperture of 1·4 may be used. A light source of 30 watts or greater is, however, required. If a tungsten light source is used, such as the Unitron Model LKR (Unitron Instruments Inc.), a monocular microscope head is required since light is lost through the prisms of a binocular head. High-intensity light sources with greater production of ultraviolet (UV) wavelengths, such as a high-pressure mercury-vapour lamp, are required if the binocular head is used on the microscope. It is advisable, however, when using high-intensity light sources, to add a UV-absorbing filter between the stage and ocular to preclude any inadvertent passage of UV light to the ocular. A silver mirror should be replaced by one of polished aluminium if reflected light is used, as silver has a low reflectance of UV (Loveland, 1970).

A blue excitation filter (BG 12, Schott Inc.) which passes light at approximately 450 nm is used between the lamp and sample with a secondary yellow

197

or orange filter (OG 1 or OG 5, Schott Inc.) on the ocular. The condenser is used at full aperture to allow for maximum light transmission, and a drop of non-fluorescing oil (Wood, 1955, 1962, used cedar oil or paraffin oil) is placed between the condenser and the slide to minimize light scattering. A thin counting chamber, such as the Petroff–Hausser, should be used to minimize light loss (although a Palmer–Maloney chamber is also satisfactory). A Whipple disc may also be inserted in the ocular (see Chapter **7.2.2**).

The basic equipment, microscope with monocular tube, light source and filters, can be mounted on a board (25 × 75 cm) and easily secured to a bench or table for shipboard use.

Commercially available fluorescent microscopes include essentially the same filter combination and condensers as indicated above and are normally equipped with mercury vapour or xenon illuminators. The high-intensity illuminators allow for the use of oil-immersion objectives of higher magnification and thus great definition of the image, which aids in distinguishing phytoplankton from other fluorescing particles.

PROCEDURE

Samples are collected using any standard sampling device, although the volume required will vary according to the biomass present. Wood (1962) and Vargo (1968) used a 3·75 litre sample throughout the Florida Straits and Caribbean.

Samples must be concentrated before counting, especially when small volumes are counted (i.e. the Petroff–Hausser chamber). Kimball and Wood (1964) used a continuous-flow centrifuge for this purpose, which was devised by modifying a commercial blender. Continuous centrifuges have now become available commercially. These are reported on elsewhere (Chapter **5.3**) together with cup-type centrifuges.

The centrifugate is removed from the cup by rinsing (with 0·45 μm filtered sample water and a rubber policeman) into a test-tube or centrifuge tube marked at a known volume. Several rinses may be required to remove all material from the cup, after which the centrifugate is diluted to volume (normally, 10 ml). Concentrated samples should be stored at temperatures approximating those from which they were collected if they are not counted or scanned immediately.

Fields, paths or Whipple units may be counted although each must be calibrated (see Jackson and Williams (1962) for calibration of counting chambers). Basically, the volume of a single field is calculated and expressed in millilitres (3×10^{-5} ml using the Petroff–Hausser chamber on a Wild M20 microscope with a × 10 objective and × 10 ocular; Wild Heerbrugg Ltd); the counts are normalized per field, and the cell concentration in the diluted centrifugate expressed as cells per millilitre or cells per litre. Dividing by the concentration factor yields the cell concentration in the original sample (e.g. if 1 litre is concentrated to 10 ml, the concentration factor would be 100).

Wood (1962) recommended counting fields since the grid of the Petroff–Hausser chamber is not visible when low-intensity light sources (tungsten lamps) are used; he counted 10 fields if there were less than 5 cells per field and 4 fields if there were more than 5 cells per field. For small forms, the use of fields rather than paths is recommended (Anon., 1974) since in scanning a slide some cells may be missed if low magnification is used.

An initial count of an unstained sample is made to estimate the number of red fluorescing cells (i.e. containing chlorophyll). A second count is then made for total organisms after staining with a distilled water solution of acridine orange. A final dilution of 1:5,000 of acridine orange to sample must be used to achieve the green cytoplasm fluorescence since, at higher concentration, cytoplasmic RNA may fluoresce red (Loveland, 1970). Subtracting the unstained count from the stained count indicates total non-chlorophyll-bearing cells. All counts should be made in a darkened room after allowing for dark adaptation of the human eye.

Ordinary light may be used to evaluate taxonomic or cytological detail or to measure cells by removing the filters. The position of cells which warrant further observation should, however, be noted (by stage coordinates) and returned to after the fluorescent count is made, since dark adaptation may be lost by switching between light types.

EVALUATION

Wood and Oppenheimer (1962) report that heat-killed phytoplankton continue to fluoresce for up to 24 hours after death while fluorescence in cells treated with formalin, iodine or copper acetate decreased by 90 per cent after two hours. Derenbach (1970) also reports changes both in the intensity and colour of chlorophyll fluoroescence in both heat-killed and preserved (formalin) phytoplankton. The intensity of fluorescence was dependent on the medium in which cells were resuspended, the storage temperature and the light intensity during storage. Similarly, both live and dead cells will fluoresce when stained with acridine orange. Since cells stained with this fluorochrome may also fluoresce red under certain conditions, as noted above, the differences between stained and unstained counts and live or dead cells is difficult to interpret. Use of high-intensity light sources with a greater proportion of UV light than tungsten lamps may also cause degradation or quenching of chlorophyll fluorescence if the cell is continuously exposed for more than 3 to 5 minutes, the brick-red colour changing to bright yellow (Wood and Oppenheimer, 1962). Counts or qualitative observations must therefore be made relatively rapidly.

Identification of small cells is extremely difficult using the fluorescence method, even if high-intensity lamps and greater magnification are used. A single chloroplast from a ruptured cell will fluoresce brightly and be counted as a cell since it is difficult to distinguish between it and a small naked flagellate or a small coccolithophorid.

199

The fluorescence counting method, as described, also incorporates the use of centrifugation as a means of concentrating samples to obtain sufficient material for counting in small-volume chambers. Since retention capacity varies with community composition (see Chapter **5.3**), this part of the method must be evaluated both seasonally and locally for the type of centrifuge used and for reliable enumeration. Use of the Petroff–Hauser counting chamber may also discriminate against larger diatoms and dinoflagellates because of cell-size to chamber-volume relationships or poor distributional patterns within the chamber. A suggested approach for increasing enumerative reliability might be to incorporate other concentration techniques (see Chapter **5**) and a larger-volume counting chamber.

Transmission fluorescence microscopy is eminently satisfactory as a method for the qualitative evaluation of chlorophyll containing organisms which are either included in faecal pellets (Gerber and Marshall, 1974), associated with sand grains or detritus, or involved in symbiotic relationships (Wood and Oppenheimer, 1962). Additionally, fluorescence may be used as a means of differentiating cytological and physiological aspects of phytoplankton cells through the use of enzyme or organ-specific fluorochromes (Derenbach, 1970).

EPIFLUORESCENCE MICROSCOPY

Reflected rather than transmitted light is used for epifluorescence microscopy. The method has been used primarily for the enumeration of bacteria attached to particles (Rodina, 1961, 1967; Munro and Brock, 1968) or concentrated on filters (Francisco et al., 1973; Zimmermann and Meyer-Reil, 1974; Jones, 1974).

Daley and Hobbie (1975) offer a complete review and evaluation of the methodology, including recommendations for light sources, excitation and emission filter combinations, types of membrane filters for concentrating cells, and an evaluation of a variety of fluorochromes for staining (acridine orange, fluorescein isothiocyanate and euchrysine).

Briefly, the method involves staining of a sample with a fluorochrome and filtering it on to either a black membrane filter (Francisco et al., 1973; Daley and Hobbie, 1975) or polycarbonate (Nuclepore) filter (Zimmermann and Meyer-Reil, 1974). The filter is mounted in a low-fluorescing oil while moist, and a second drop of oil and the coverglass added, and examined using a microscope equipped with a vertical illuminator and excitation turret. A Whipple grid, inserted in the ocular, aids in counting. Most cells, live or formalin-preserved, fluoresce green when stained with acridine orange, fluorescein isothiocyanate or euchrysine 2GNX. Red fluorescence may also occur which Francisco et al. (1973) suggest may be related to physiological changes within the cell.

Daley and Hobbie (1975) recommend using a 200-watt ultrahigh-pressure mercury lamp, combined with a 2-mm KG 1 (Schott Inc.) heat filter, a 4-mm

BG 38 (Schott Inc.) red-suspension filter, a K480 (Carl Zeiss Inc.) sharp-cut filter and two KP490 (Carl Zeiss Inc.) excitation filters. The K480 and KP490 filters contribute a reduction in background fluorescence and fading as well as a reduction in the autofluorescence of chlorophyll. Samples fixed with formalin at a final concentration of 2 per cent (v/v) formaldehyde, stained for three minutes with acridine orange (final concentration 5 mg l^{-1}), filtered on to either 0·45 μm or 0·22 μm black Sartorius filters, rinsed with distilled water (recommended for both freshwater and estuarine samples), and mounted while moist in a low-fluorescing oil, yielded the best results. The method is applicable to both freshwater and marine samples with some simplification in technique (no K480 filter, no distilled-water rinse and dry filter storage) possible in seawater (Daley and Hobbie, 1975).

The method may have application for phytoplankton counting, either on filters or settled into small-volume chambers (Coulon and Alexander, 1972); however, the problems of cell rupturing and poor phytoplankton distributions on filters must be kept in mind.

7.5   Electronic counting

7.5.1   Sensing-zone counters in the laboratory

R. W. Sheldon

In this section we deal with automated counting and sizing of phytoplankton and other particles by 'sensing zone' counters. The instruments described are laboratory instruments. They can be operated either on board ship or in a laboratory on shore but water must first be collected (e.g. by bottle sampler) before a sample can be counted. However, sensing-zone counters are not limited to the laboratory. In principle it is just as logical to put the sensing zone into the water as it is to bring the water to the sensing zone, and this in fact has been done (see Chapters **7.5.2** and **7.5.3**). Either living or preserved samples may be counted but because precipitates can form in preserved samples it is preferable to count living samples whenever possible.

Sensing-zone counters measure and count particles suspended in a fluid. In general, all particles within a given size range are counted and there is no discrimination between phytoplankton and other particles (except as described below). However, by means of other measurements, a particle count can usually be related to a phytoplankton count without undue difficulty. In contrast to the situation a decade ago when only one electronic counter suitable for use with phytoplankton was produced commercially (Sheldon and Parsons, 1967a), there are now several different instruments available. No single instrument will perform all phytoplankton counting measurements equally well, but any one instrument will be excellent for a specific task. In the past an approach to a research problem often had to be modified because of the limitations of the counter, but we can now select the instrument which will most readily produce the data we need. However, all sensing-zone instruments work on similar principles and it is necessary to understand their advantages and limitations if studies are to be planned effectively.

COINCIDENCE

In all sensing-zone counters a sample is made to flow in a restricted zone. When a particle passes through the zone, a sensor detects and measures a property that can be related to its size. Different instruments use different methods to pass a sample through the sensing zone and different particle properties are measured, but if an accurate count is to be made, the particles must pass the sensing zone one at a time. If this condition is not satisfied two (or more) particles will occupy the sensing zone at the same time and will be counted as one.

The number of coincident counts will depend on the size of the sensing zone and on the particle concentration in the sample. At low concentrations, coincidence is small and can be ignored, but as the concentration increases coincidence rises rapidly. It is important to know the relationship between

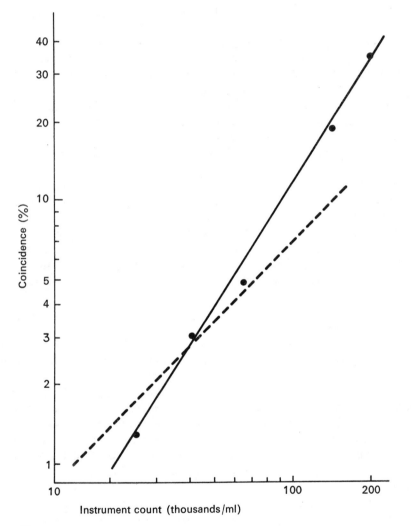

Figure 38
The relationship brtween coincidence and particle concentration for a 100-μm Coulter aperture.
●———● Experimental data (line drawn by eye)
－－－－ Theoretical relationship according to a Poisson distribution

coincidence and particle concentration so that limits to the count accuracy can be set. These can be calculated or measured directly.

Because the water has to flow through a restricted zone the sample can be visualized as a long thread with particles dispersed randomly along it. In effect, the sample is simply a series of sensing zones laid end to end. The probability of two particles falling within one sensing zone should therefore follow a Poisson distribution, but in practice we find more coincident passages at high particle concentrations than theory would predict (Fig. 38).

The magnitude of the discrepancy probably depends on the shape of the sensing zone and the flow pattern through it, and this will vary from instrument to instrument. It is better therefore not to rely on calculations but measure coincidence directly. To do this a count is made at a high particle concentration. The sample is then diluted with particle-free water and a further count is made. This process is repeated until the concentration is so low that coincidence can be neglected. By multiplying the final concentration by the dilution factors the true counts at each dilution can be calculated. Comparing these with the counts actually obtained gives the coincidence.

It is somewhat surprising to find that when the particle concentration is such that 10 per cent coincidence occurs, there will be, on average, only one particle for each eight zone lengths of the sample. This means that at least 80 per cent of the counting time is spent in looking at clear water and only 20 per cent or less in counting particles. This is a fundamental limitation to sensing-zone counting techniques and it depends on the nature of random dispersions. It cannot be overcome either by instrument design or by sample manipulation. It is essential therefore that in all methods of sensing-zone counting the samples are organized so that the particles enter the sensing zone one at a time.

SPURIOUS COUNTS

It was noted at the beginning of this section that sensing-zone counters count all particles in suspension. Unfortunately, under certain conditions they may even register counts that are not caused by particles. We should bear in mind the fact that sensing-zone counters count voltage pulses caused by discontinuities in the sample fluid, and we arrange our procedures so that particles are the only discontinuities in the sample. However, the instruments will count anything that will register as a discontinuity and they will also count anything that will produce electrical pulses similar to those caused by particles, even when these pulses do not originate from the sample. All counts caused by something not in the sample—often electrical line interference or radiation—are referred to as 'noise'. With good instrument design and careful operation noise should never be an insurmountable problem. It can either be eliminated or allowed for, depending on the circumstances.

Spurious counts caused by discontinuities in the sample fluid are more difficult to detect as these produce voltage pulses which are not easily distin-guished from those caused by particles. Bubbles are the most common source of counting error in this respect. According to Medwin's data on natural bubble distributions in the sea (Medwin, 1970), the concentration of bubbles can, on occasion, be roughly the same as that of the phytoplankton. Fortunately, it is easy to remove bubbles from a sample before a count is made.

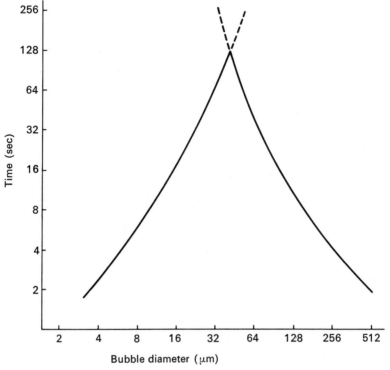

Figure 39
The relationship between residence time and size for bubbles in a 10-cm-deep container. (Data from Blanchard and Woodcock, 1957.) (For explanation see text.)

Air bubbles form when water is poured from one container to another. These will rise to the surface, because of their buoyancy, and will also shrink because the air in the bubbles dissolves in the water. Surface tension increases the air pressure in the bubble and it will dissolve even when the surrounding water is supersaturated (Blanchard and Woodcock, 1957). The lifetime of a bubble depends on its size and on its position in the water; a large bubble will rise quickly but a small bubble will shrink quickly. The relationship between residence time and size for bubbles in a 10-cm deep beaker is shown in Figure 39. The two processes of shrinking and rising are treated separately, even though both occur together. On the left-hand side of the diagram is the relationship between the diameter of the bubble and the time required for it to shrink away to nothing. On the right-hand side is the relationship between bubble size and the time taken to rise 10 cm. The longest-lived bubbles have diameters in the range 30 to 40 μm. These are also the most abundant bubbles found at the sea surface (Medwin, 1970).

As many diatoms and dinoflagellates are also in this size range, it is important to make sure that water samples are free of bubbles if plankton are to be counted accurately. It is easy enough to leave a sample to stand before a count is made and it is clear from Figure 39 that the longest-lived and most favourably placed bubbles could not exist in a 10-cm beaker for more than 2 minutes. In this time they must either rise to the surface or disappear by shrinking. If the sample is stirred, this should be done gently because cavitation at the stirrer blades can also cause bubbles to form.

Although a sample will clear itself of bubbles on standing, particles may be formed during the clearing process. Bursting bubbles cause particles to form in seawater (Sutcliffe *et al.*, 1963; Baylor and Sutcliffe, 1963), and when a bubble shrinks away it leaves behind it a particle of organic material (Johnson, 1976). The size of these particles is not known with certainty, although those produced by bubbles from breaking waves are around 5 μm (Sutcliffe *et al.*, 1971). If all bubble-formed particles are similar in size, then the concentration of material derived from the bubbles caused by sample handling will usually be much less than the concentration of other particles in the sample. However, it is desirable to develop sample-handling techniques so that bubble formation is minimized.

ACCURACY

A fairly extensive literature has appeared on the accuracy of sensing-zone counters (mainly Coulter Counters), but very little of this is relevant to the study of phytoplankton. The reason is simply that for many particle-counting applications extreme accuracy is necessary (for instance, for counts of human blood cells or of pharmaceutical powders) and it is important therefore in these cases to know exactly how the sensing zone is 'seeing' a particle. But in phytoplankton studies it is not usually necessary to know cell size to an accuracy of 1 or 2 per cent, and for practical purposes the various factors that give rise to inaccuracies in size determinations (e.g. variation in particle shape, resistivity, refractive index, orientation) can be ignored.

An impression of inaccuracy is sometimes given because sensing-zone counters do not necessarily 'see' a particle in the same way that an observer with a microscope would. For instance, the dinoflagellate *Ceratium tripos* may be 200 μm long but an electrical sensing-zone counter (Coulter-type counter) would see this as a 40 μm sphere. (As different instruments measure different particle properties—volume, area, etc.—the 'diameter' of a particle is always taken as the equivalent spherical diameter, i.e. the diameter of a spherical particle with the same volume, area, etc., as the particle measured.) Even more extreme differences are met with if chain-forming diatoms are counted. Provided that the diatom is small relative to the size of the sensing zone, a Coulter-type counter will measure the total volume of the cells in the chain, but it will 'see' a 250 μm chain of *Skeletonema costatum* as a 16-μm particle.

SOME COMMERCIALLY AVAILABLE
SENSING-ZONE COUNTERS

Many kinds of sensing-zone instruments have been built by individuals for various specialized purposes, but there are now three commercial kinds available which are suitable for use with phytoplankton.

Coulter-type counters[1]

These were first introduced about twenty years ago and they have been used fairly extensively in aquatic research for about the past ten years (Sheldon and Parsons, 1967*a*). In these instruments a sample flows through a small pore across which an electrical potential is applied. This produces an electrically sensitive zone in the vicinity of the pore (Fig. 40A). In effect, the zone around the pore becomes a small electrical resistor. When a particle passes through this zone it causes a change in the electrical resistance, which is proportional to the volume of the particle. As the formation of the sensing zone depends on the application of an electrical potential, the sample fluid must be an electrolyte. Seawater is, of course, an excellent electrolyte and marine phytoplankton can be easily counted. Coulter-type counters will work at quite low salinities ($<5\%$) but freshwater plankton cannot be counted directly. A small amount of concentrated electrolyte has to be added to the water immediately before counting to give a final electrolyte concentration of about 0·5 per cent. This has no immediate effect on the phytoplankton (Mulligan and Kingsbury, 1968).

If the particles are to be sized correctly, their diameters must lie between 2 and 40 per cent of the pore diameter. In principle, particles of any size can be measured but the instruments available commercially will only work well with pores of from 30 to 2,000 μm diameter. With these a particle-size range from 0·6 to 800 μm can be measured. Counts are most easily made over a particle-size range from about 1 to 100 μm, and most of the phytoplankton normally encountered fall within this range. The amount of sample from which counts are taken can be varied from a fraction of a millilitre to more than a litre depending on the particle size and concentration; but as about 20 ml are needed simply to wet the electrodes, this is effectively the lower limit of sample size. For phytoplankton counts, samples of up to 100 ml are most commonly used.

1. These were originally developed and are still manufactured by Coulter Electronics Inc. Instruments working on the same principle can also be obtained from the following manufacturers: Ljungberg & Co.; Particle Data Inc.; Telefunken AEG; Toa Electric Co.; VEB Transformratoren & Rontegemwerk.

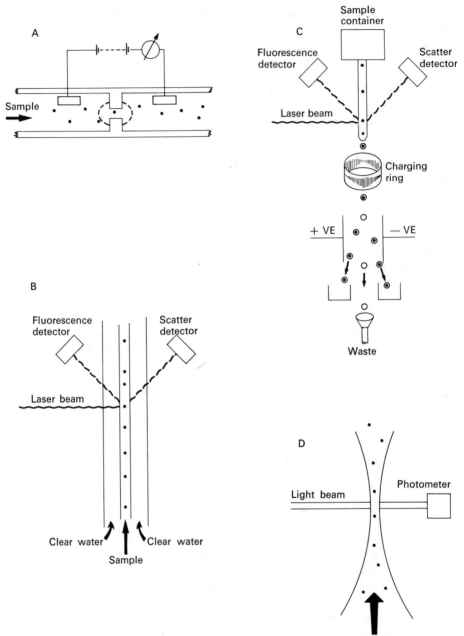

Figure 40
The operating principles of sensing-zone counters: (A) Coulter-type counters; (B) laser-beam counters; (C) laser-beam counters with cell sorting; (D) optical (photometric) counters.

Optical counters using a laser

At present the most commonly used instrument of this type is the Cytofluorograf (Bio/Physics Systems Inc.). In this instrument the sample is injected as a thin (25 μm) thread within a larger (250 μm) column of particle-free water. The sample flow is laminar. The sensing zone is formed by a laser beam with a diameter of 30 μm which shines through the sample thread (Fig. 40B). As the particles pass through the laser beam the light is scattered. The amount of scatter varies with particle area and refractive index but for most practical purposes it is essentially proportional to particle area. Particles from 2 to 100 μm can be measured but the instrument is most easily used with particles in the size range of from 2 to 40 μm. The sample flow rate is low and only small samples can be taken (about 5 ml). In waters with low particle concentration such a sample would not include a statistically adequate number of large particles. For these situations a particle size range of from 2 to about 15 μm could be measured.

A laser beam carries a considerable amount of energy and this will cause certain compounds to fluoresce strongly enough for measurements to be made on single particles. For instance, the blue light of an argon laser (488 nm) will cause the chlorophyll in phytoplankton to fluoresce red, so that in a sample of phytoplankton both the scatter (size) and red fluorescence (chlorophyll content) can be measured cell by cell. In a mixed population containing both phytoplankton and particles with no chlorophyll (organic detritus, inorganic material or zooplankton) all particles will scatter light but only the phytoplankton will fluoresce. It is possible therefore to separate the counts due to phytoplankton from the counts due to other particles. Separate counts for different groups of phytoplankton, or for phytoplankton and microzooplankton, may also be possible if specific fluorescent stains are used. (See Chapter 7.5.3.)

A further application of fluorescence was devised several years ago by Fulwyler (1965). This has been developed by Bonner *et al.* (1972) and Steinkamp *et al.* (1973), and has been well described by Herzenberg *et al.* (1976). Particles with different optical properties are not only identified but are separated into two populations (Fig. 40C). Instruments based on Fulwyler's method have recently been put into commercial production: the TPS-1 Cell Sorter (Coulter Electronics Inc.), and the Fluorescence Activated Cell Sorter (Becton Dickinson Electronics Laboratory). The following description applies to the TPS-1 Cell Sorter, but the two instruments do not differ significantly in principle. The sample thread is surrounded by particle-free fluid. It is passed through a 70-μm pore and ejected into the air as a thin thread. The scatter and fluorescence caused by the interaction of the particles and a laser beam are measured at this point. After passing the laser the thread is broken into droplets. As most of the sample thread will not contain particles most of the droplets will also contain no particles. These fall away to waste, but the droplets containing particles are automatically identified and separated. To illustrate how this is done,

H

consider the example given above of a mixed population of phytoplankton and particles with no chlorophyll. As a phytoplankton cell passes the laser, both fluorescence and scatter are recorded and measured. The instrument responds to the scatter and fluorescence signal by causing the charging ring (see Fig. 40c) to become charged either + or −, so that when the droplet reaches the charging ring it also becomes charged. If a non-fluorescing particle passes the laser, only scatter is recorded and the instrument responds to this by causing the charging ring to become charged in the opposite sense. When droplets with no particles pass the charging ring, neither ring nor droplet acquire a charge. The droplets continue to fall and pass between permanently charged plates. The droplets with phytoplankton are deflected to one side: the droplets with non-fluorescing particles are deflected to the other side, and droplets with no particles continue to fall free.

In addition to separating particles simply on the basis of fluorescence or non-fluorescence, it is also possible to separate populations of particles with differing levels of fluorescence. A Coulter aperture can also be put into the system to measure particle volume. These instruments would seem to have great potential in studies of algal physiology.

The flow rate is limited by the diameter of the sample thread and it is also limited by the fact that it has to pass through a pore in order to be broken into standard-sized droplets. Particles from about 1 to 40 μm can be measured and sample size can be varied from about 2 to 20 ml.

Optical counters using a light beam

Photometric particle counters of various kinds have been produced for many years but their use has tended to be restricted to biomedical applications—mainly the analysis of blood cells. However, instruments have recently been developed for industrial use which have the capacity to take large samples and to measure particles with a wide range of size. They can be applied directly to phytoplankton counting. Perhaps the most widely used instrument of this type is the HIAC Automatic Particle Size Analyzer (Pacific Scientific Co, HIAC Division). In this instrument the sample is pumped (or forced) through an optical cuvette (Fig. 40D) which is intersected by a light beam. The sensing zone is formed where the light beam strikes the sample, and a photosensing device is placed on the opposite side of the cuvette from the light source. As the particles pass through the light beam the light is blocked and this attenuation is measured. The attenuation varies with particle area. The sample flow is turbulent and consequently the particles tumble as they pass through the sensing zone. In this way the maximum projected area is measured.

A single sensor will measure particles whose diameters vary by a factor of 60. Depending on the sensor, particles from 1 μm to 9 mm can be measured. The sample size can be varied from a few millilitres to many litres depending on the particle size and concentration in the sample. The sensing zones of these

210

instruments are relatively larger than those of the other instruments described and consequently their sample flow rates are somewhat greater. This could make them particularly suitable for situations where large samples with moderate particle concentrations are likely to be taken (e.g. oligotrophic waters). Continuous flow systems can be set up, using one or more sensors, for shipboard operation with continuously pumped samples.

THE DETERMINATION OF PARTICLE STANDING STOCKS

A major problem faced when using counting techniques to measure particle standing stocks is how best to express the measurements. As particles of many different sizes are counted it is fairly obvious that the data can conveniently be presented as a histogram; but it is not immediately apparent which kind of histogram would be best. In many cases the easiest way to collect the data is to count the particles occurring in arithmetically equal size classes, but this is not very informative (Sheldon and Parsons, 1967b). A better way is to measure the mass or volume of particulate material which occurs in logarithmically equal-size intervals (Sheldon and Parsons, 1967a, 1967b; Sheldon et al., 1972).

A logarithmic scale is a natural scale to use, for three reasons. First, many natural particle distributions are log normal (Aitchison and Brown, 1957, Chap. 10). Second, the tremendous range of size of phytoplankton can only be conveniently expressed on a log scale.[1] Third, when considering food webs, it is the ratio of the size of organisms that is important, not their absolute dimensions (Ursin, 1973; Kerr, 1974).

It is of small importance whether the size scale is of particle volume, cross-sectional area or equivalent spherical diameter, provided that it is logarithmic. The form of the scale is also not critical but there are certain advantages to be gained by adopting a standard scale (Krumbein and Pettijohn, 1938). A scale to base 2 is probably the best one to use (Sheldon and Parsons, 1967a, 1967b; Strickland and Parsons, 1972; Sheldon et al., 1972; Platt and Denman, 1978).

The size distributions of particles occurring in near-surface water during a diatom bloom are shown in Figure 41. Some samples were prepared for microscopic observation and the diatoms were seen to be mainly Chaetoceros chains. A few simple measurements were sufficient to show that the equivalent spherical diameter of these chains was around 30 μm. This diatom population can be clearly seen in the frequency distributions on the left of Figure 41. In these distributions the concentration of particulate material per size interval is plotted against the logarithm of particle size. It can also be seen that there were probably two populations of diatoms present; one with an average diameter of about 20 μm (equivalent sphere), which occurred mainly at the surface water,

---

1. It is not generally appreciated that the relative size difference between a small flagellate and a large diatom is similar to that between a mouse and an elephant.

211

and a second population with an average diameter of around 50 μm which occurred throughout. The average size of the larger diatoms varied with depth. In contrast, using the same data, the distributions on the right of Figure 41 (particle number versus log size) indicate only that the particle concentration varied with depth. This is simply because in any sample of natural water there

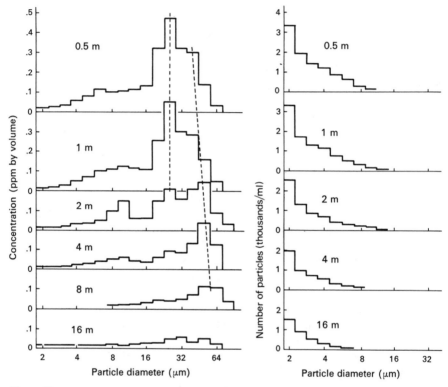

Figure 41
Particle distributions in near-surface water off Halifax Harbour, Nova Scotia, Canada, during a diatom bloom. Left: Distribution of particle concentration. Right: Distribution of particle number. Note that the left and right histograms represent the same samples. The only difference is the method of setting out the data. Sample depth is shown on each histogram.

are always greater numbers of small particles than large ones. The diatom bloom is not apparent in the number distributions because the number of cells, when compared to the smaller particles, is negligible; yet the biomass of the diatoms was greater than the mass of the small particles.

The visual effect of the large numbers of small cells can be reduced and the numbers of the larger cells enhanced by putting the count on a log scale. Mathematically this is a somewhat dubious procedure but, provided it is used only for illustrative convenience, there is no harm in it. This has been done in Figure 42.

212

It can be seen that the distributions of particle number are still not easy to interpret. A visual check with a microscope showed that there was a diatom bloom in the near-surface water; at a depth of 50 m the suspended particles were mainly organic detritus with an admixture of living organisms of various kinds and at a depth of 150 m (9 m above the bottom) the particles in suspension were mainly flocculated fine-sediment grains. The differences between the

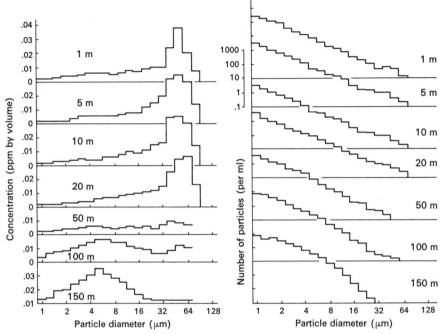

Figure 42.
Particle distributions in a vertical profile from near the surface to near the bottom, Chedabucto Bay, Nova Scotia, Canada. Left: Distributions of particle concentration. Right: Distributions of the log of particle number. Sample depth is shown on each histogram.

distributions, reflecting the differences in particle occurrences as indicated by the microscope observations, are clearly evident in the histograms on the left of the figure (concentration versus log diameter) but the differences are by no means as apparent in the histograms on the right (log number versus log diameter).

The overwhelming advantage of using concentration versus the log of diameter for the histograms is that it aids clarity of observation and suggests where further observations could profitably be made. It also helps in the interpretation of food-web patterns. It is the concentration of material relative to its size that one must consider when investigating how material moves through a food web.

There are, however, certain disadvantages to using concentration as the ordinate of a size-frequency distribution. For instance, the form of the curve is sensitive to changes in the grade scale of size employed, but if a standard scale is used then all curves can be compared. This has been the solution adopted by civil engineers and others for the measurement of grain-size distributions (Krumbein and Pettijohn, 1938). A histogram of concentration also cannot be mathematically integrated; to find the total concentration between any two sizes one must add the ordinates for each of the grades.

Finally, it must be emphasized that, although particle-size analysis is a powerful tool for investigating the distribution and interaction of plankton, one must never forget that all the particles in suspension are measured. Therefore, in order to confirm that any changes in (or difference between) size distributions are due to phytoplankton or other organisms it is necessary to check every sample or group of samples by another method. If this is not done, quite fundamental mistakes can be made. For instance, a slow process of precipitation or a rapid growth of micro-organisms look very similar when presented as a series of size-frequency distributions, and it would be tragically easy to confuse the two if only counting data were considered. A glance through a microscope or a quick chemical check (e.g. ATP or chlorophyll) can assist in showing what the true situation is.

214

7.5.2   Perspectives of *in situ* counting

Carl M. Boyd

Counting of phytoplankton cells *in situ* has not yet been undertaken, although the techniques of counting and determining sizes of zooplankton organisms *in situ* as employed by Maddux and Kanwisher (1965) and by Boyd and Johnson (in press) can be rather easily applied to counting phytoplankton cells. The electrodes used in the zooplankton counting techniques can be made of almost any size, and Boyd (unpubl.) has made electrodes to count 20-μm phytoplankton cells for laboratory studies. These electrodes may be rather easily adapted to *in situ* studies by addition of a pumping unit and a prefilter mesh to retain large particles that would plug the orifice. The procedure offers the interesting feature of enabling the effective counting of particles in a minute core of water from surface to depths. Such information on the variability of phytoplankton abundance on the scale of centimetres should give some insight into the availability of particles to copepods at scales that are of interest to both the ecologist and the copepod.

The on-axis holographic technique considered by Knox (1966) was incorporated in a prototype instrument developed at the Scripps Institution of Oceanography (United States) to provide *in situ* holograms of planktonic organisms in the size range of large phytoplankton to microzooplankton. The use of on-axis holography (see Chapter **5.5.5**) held promise for allowing the *in situ* observation of organisms in relatively large volumes (e.g. tens of litres) at depth without drastically disturbing their spatial arrangement. However, the variable, but often large, amount of small particulate material in natural seawater, even at considerable depth in the ocean, caused the diffraction of the reference beam of the laser. The result was that an adequate ratio of reference beam to scene beam could not be achieved, and good resolution was limited to those organisms a few centimetres into the scene (J. R. Beers, pers. comm.).

7.5.3 Perspectives for automated identification and counting

Carl M. Boyd

In order to identify cells by automated techniques one must take advantage of natural or induced differences in the cells. The most obvious difference—that of 'shape' or morphology—is among the most difficult to incorporate into an automated technique. Wied *et al.* (1970) discuss an extended research project that had as its goal the automated identification of mammalian cells for biomedical applications. The limited success of that project, which employed a microspectrophotometer to produce images to be processed by a large and carefully programmed computer, demonstrates the complexity of the problems of signal processing and image recognition.

Certain studies, however, do not require that algal cells be divided into more than a few categories, and existing automated apparatus might be quite adequate in these cases. Instruments such as the Omnicon (Bausch & Lomb, Analytical Systems Div.) and the Quantimet (Imanco; Cambridge Instruments) employ a television camera attached to a microscope; the image of objects on the microscope stage is displayed on a television screen. Electronic signal-processing modules allow the items on the screen to be counted, and individual items can be processed to yield measurements of surface area, length of major and minor axes, perimeter, optical density, and estimated volume (Wilkinson *et al.*, 1974; Gibbard *et al.*, 1972). Fawell (1976) was able to separate zooplankton samples into three categories (copepods, euphausiids and chaetognaths) with the Quantimet 720, and where enumeration into categories as broad as these is relevant, the technique offers many advantages. Certainly, division of this resolution is beyond the present capabilities of resistive-type counters.

A very attractive technique of pattern recognition is based on laser holography as discussed by Almeida and Eu (1976) and Cairns *et al.* (1972). A complex optical spatial filter representing the Fourier transform of the image of a known species of alga (diatoms in these studies) is prepared by photographic processes, and this filter is then matched against a similar optical transform of an unknown cell. Correlation of the two transforms allows one to identify the unknown species. The potential of this technique for automated identification of cells is remarkable and will bear watching in the future; however the procedure must be regarded as a developing technology at this point.

Devices that take advantage of the natural fluorescence of chlorophyll have been used by several scientists to count phytoplankton cells. The Cytofluorograf (Bio/Physics Systems Inc.), as an example, employs optical techniques and light-sensing electronics such that individual particles in a stream are illuminated. All particles are counted by their light-scattering properties, and the magnitude of the scattering gives one an indication of the size of the particle. Those particles that fluoresce are detected separately to give the operator an indication of the number and intensity of fluorescence of individual particles. Additional information on this technique may be found in Chapter **7.5.1**. These devices are useful in counting cells but are still far from offering a means of identification of a population of mixed algal cells.

A fruitful approach to identification of phytoplankton cells would be based on discriminate labelling and subsequent detection and separation by techniques peculiar to that label. The familiar technology of radioisotope labelling would not seem well-suited since most isotopes are taken up in a rather similar manner by different phytoplankters, and the system to detect isotope levels in individual cells would be slow and cumbersome. The techniques of selective fluorescent labelling seem to offer the most promise. The process of cell identification using this technology requires an instrument such as the Cytofluorograf; individual cells are subjected to an intense monochromatic light, and the intensity of the fluoresced light is measured. As used by Herzenberg *et al.* (1976) and Steinkamp *et al.* (1973) on mammalian cells, however, the cells to be identified are first treated with a non-toxic fluorescent stain that is specific for certain components in certain cells, and the cells are categorized by virtue of their relative fluorescence. A wide variety of fluorescent stains exists and their specificity at least in mammalian cell systems is well documented. Conceivably, stains exist that would label specific groups of phytoplankton cells in a mixed population, and that would, through the additional use of antigens, enable sites to be created on phytoplankton cells so that classification may be refined to those cells having, for example, certain enzyme systems. This technique seems particularly well suited to biochemical or physiological studies employing phytoplankton cultures, though the procedure might also be used to enumerate certain species in natural populations (species for which specific fluorescent stains would have been established).

Instruments currently exist that will accomplish separation of algal cells into rather coarse categories and will do this with the speed (and inherent complexity) of specialized electronic instrumentation. Clever scientists can surely use these devices to obtain results otherwise unobtainable, but the task of species identification familiar to most algologists is still in the domain of the microscopist. The fundamental advances in the technology of automated species identification will probably not be made by biologists but will come from basic studies in image recognition, holographic processing and solid-state physics. However, biologists should attempt to be cognizant of developments, and will profit by employing the advances as they occur to increase the power of their experiments.

H*

7.6 The dilution-culture method

Jahn Throndsen

The main purpose of the serial dilution-culture method is to provide material for identification and enumeration of the non-preservable fraction of the phytoplankton. In addition, unialgal cultures for physiological and chemical, as well as morphological and microanatomical studies, may be obtained. Compared with other methods for studying living material it has the advantage that the number of specimens of each species is increased by cultivation. The method has been used infrequently in phytoplankton studies (e.g. Knight-Jones, 1951; Bernhard et al., 1967; Throndsen, 1969b) and it has also been applied to studies on heterotrophic flagellates (Lighthart, 1969).

OUTLINE OF THE METHOD

The two basic assumptions for use of the serial dilution-culture method are that (a) cells viable in the sea will also grow under the culture conditions offered, and (b) the cells are evenly distributed in the sample. From the presence or absence of a species in different amounts of water the most probable number (MPN) per unit volume can be determined statistically. Presence of a species will become apparent and a culture of that species (often, however, in mixture with other species) will be available for identification.

Many nanoplankton flagellates apparently grow well in culture and hence they may be recorded and enumerated by the serial dilution-culture method. Some species with special requirements will regularly grow up in dilution cultures though they will not survive subculturing (e.g. *Hemiselmis virescens* Droop). A modified 'Erd-Schreiber' medium (Throndsen 1969b; and Table 9 below) appears to meet the requirements of a large variety of nanoplankton flagellates. The number of species encountered can be increased if the sample is cultured in media based on water collected from the same locality in which the sample is taken.

MATERIAL AND EQUIPMENT

The water sample to be used for inoculation and preparation of the culture medium should be collected by a non-toxic water sampler (see Chapter 3.1) or directly in a pyrex-quality glass bottle. Samples to be used as inocula should be placed in an insulated container immediately after collection to avoid thermal and light shocks.

Each series of dilution cultures will require the following: 29 pyrex test-tubes (culture-clean: treated with hot diluted NaOH, KOH or NaCO$_3$, followed by hot diluted HCl, and rinsed thoroughly in tap water followed by glass-distilled water), 25 with approximately 9 ml of medium, 4 with 12 to 15 ml of medium. One sterile disposable (graduated) 10-ml syringe with eccentric opening (the toxicity of the brand has to be tested in advance by a sensitive flagellate culture). Test-tube racks. Sterile cotton-wool plugs (dentist's tampons are

TABLE 9. Modified seawater 'Erd-Schreiber' medium for serial dilution cultures (slightly modified from Throndsen, 1969b)

Medium	Quantity (ml)
Seawater (preferably from the same sample as the inoculum)	250
Soil extract (1 kg garden soil boiled with 1,000 ml glass-distilled water to give 500 ml extract)	3
NaNO$_3$ (5 g in 100 ml glass-distilled water)	0·25
Na$_2$HPO$_4$.12H$_2$O (1 g in 100 ml glass-distilled water)	0·25
NaFeEDTA (120 mg in 100 ml glass-distilled water)	0·15
Vitamin mixture (Thiamine (B$_1$) 200 mg, Biotin 1 mg, Cyanocobalamine 1 mg, in 1,000 ml glass-distilled water)	0·50

The medium is pasteurized at 80°C for 15 minutes, then cooled down to incubation temperature in running water. If possible, add the vitamins to the medium after it is cooled.

suitable). Culture room or cabinet with fluorescent light. Sterile glass tubes (diameter about 3 mm) for subsampling the culture tubes. Compound microscope with phase contrast, high-power objectives and a heat-absorbing filter before the lamp.

PROCEDURE

Inoculation

It has proved suitable to use five dilution steps (the inocula corresponding to 1, 10^{-1}, 10^{-2}, 10^{-3} and 10^{-4} ml) with five parallels of each. When the simplified method to be described here is followed, 25 test-tubes are each filled with approximately 9 ml of 'Erd-Schreiber' medium. Four more test-tubes are filled with 12 to 15 ml of medium.
1. The sample to be surveyed is subsampled by use of the 10-ml disposable syringe (with eccentric opening), and one ml is given to each of five test-tubes (with approximately 9 ml medium) (Fig. 43A–B). The sample should be gently, but thoroughly, mixed before the subsample is drawn.
2. The contents of the syringe, except for the last 1 ml, are ejected (c), and 9 ml of medium are drawn into the syringe from one of the test-tubes containing 12 to 15 ml (D) to accomplish a 1:10 dilution of the original sample.

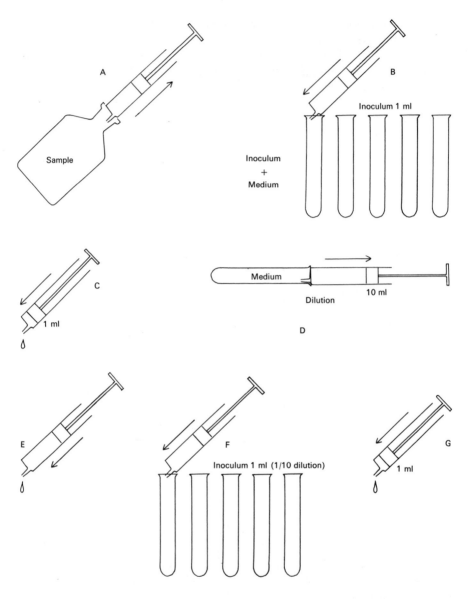

Figure 43
The main steps in the simplified serial dilution-culture method: (A) the sample bottle is subsampled by means of a 10-ml disposable syringe (of non-toxic material); (B) 1 ml of the subsample is given to each of five test-tubes containing approximately 9 ml of growth medium; (C) the rest of the subsample, except for the last 1 ml, is expelled from the syringe; (D) 9 ml of medium is added from a test-tube (held mouth-to-mouth with the syringe in a horizontal position) to produce a 1:10 dilution; (E) 1 to 2 ml of the diluted (sub)sample are expelled; (F) five test-tubes (with approximately 9 ml of medium) are inoculated with 1 ml each of the 1:10 dilution; (G) all but the last 1 ml is expelled as in C, and the procedures D to F are repeated to achieve the dilution step 1:100.

Note that the homogeneity of the phytoplankton suspension produced in this way depends on the eccentric opening of the syringe and the speed with which the medium is drawn in.

3. One or two ml of the 1 : 10 dilution is expelled from the syringe (E) and another series of 5 test-tubes (with 9 ml medium) is inoculated with 1 ml in each tube (F).

4. Proceed as 2 and 3 to produce the next dilution and inoculation, and repeat until all 25 tubes have been inoculated.

Incubation

The serial-dilution cultures should preferably be incubated in a room or cabinet with controlled temperature (2° to 20° C depending on the sea temperature), and under controlled light conditions, e.g. fluorescent tubes giving 1,000 to 2,000 lux or 8 to 16 . 10^{14} quanta $cm^{-2} s^{-1}$ at the level of the culture tubes. When test-tubes with a rim are used, a simple incubation rack can be made from a plate with holes of suitable diameter.

After two to six weeks (depending on the culture conditions, especially temperature and light) the cultures are examined microscopically and the presence of each species in the tubes noted. Subsamples for the examination are to be drawn with sterile narrow tubes if continued growth in the culture tubes is desirable.

Calculation

When the growth pattern (presence or absence) of each species through the culture series has been determined, the MPN (most probable number) can be estimated from Table 10. This table covers a range of three dilution steps, and a set of three dilutions has to be chosen out of the five cultured. The choice depends on the original concentration of the species in question, and how well it survives in competition with the other species present. The growth of sensitive species is often inhibited in the less-diluted part of the series, whereas that of less-sensitive species (e.g. *Emiliania* = *Gephyrocapsa huxleyi*) appears to be good throughout the series. In the latter case the MPN value is independent of the choice of dilution steps to be used for the estimation.

For further information on MPN estimates and their statistical reliability, Swaroop (1956), Woodward (1957) and Del Vecchio and D'Arca Simonetti (1959) may be consulted.

REMARKS

The accuracy of the dilution procedure described above is inferior to that accomplished by the use of graduated pipettes (e.g. Knight-Jones, 1951), but other factors are probably more important for the reliability of the results.

221

TABLE 10. Most probable number (MPN) in 10 ml of sample, based on the presence and absence of growth in tubes from three subsequent steps of a dilution-culture series: 5 tubes with 1 ml inoculum each, 5 tubes with 0·1 ml and 5 tubes with 0·01 ml (modified from Anon., 1955)

Growth in			MPN	Growth in			MPN	Growth in			MPN
1·0	0·1	0·01 ml		1·0	0·1	0·01 ml		1·0	0·1	0·01 ml	
0	0	0	0	1	0	0	2	2	0	0	4·5
0	0	1	1·8	1	0	1	4	2	0	1	6·8
0	0	2	3·6	1	0	2	6	2	0	2	9·1
0	0	3	5·4	1	0	3	8	2	0	3	12
0	0	4	7·2	1	0	4	10	2	0	4	14
0	0	5	9	1	0	5	12	2	0	5	16
0	1	0	1·8	1	1	0	4	2	1	0	6·8
0	1	1	3·6	1	1	1	6·1	2	1	1	9·2
0	1	2	5·5	1	1	2	8·1	2	1	2	12
0	1	3	7·3	1	1	3	10	2	1	3	14
0	1	4	9·1	1	1	4	12	2	1	4	17
0	1	5	11	1	1	5	14	2	1	5	19
0	2	0	3·7	1	2	0	6·1	2	2	0	9·3
0	2	1	5·5	1	2	1	8·2	2	2	1	12
0	2	2	7·4	1	2	2	10	2	2	2	14
0	2	3	9·2	1	2	3	12	2	2	3	17
0	2	4	11	1	2	4	15	2	2	4	19
0	2	5	13	1	2	5	17	2	2	5	22
0	3	0	5·6	1	3	0	8·3	2	3	0	12
0	3	1	7·4	1	3	1	10	2	3	1	14
0	3	2	9·3	1	3	2	13	2	3	2	17
0	3	3	11	1	3	3	15	2	3	3	20
0	3	4	13	1	3	4	17	2	3	4	22
0	3	5	15	1	3	5	19	2	3	5	25
0	4	0	7·5	1	4	0	11	2	4	0	15
0	4	1	9·4	1	4	1	13	2	4	1	17
0	4	2	11	1	4	2	15	2	4	2	20
0	4	3	13	1	4	3	17	2	4	3	23
0	4	4	15	1	4	4	19	2	4	4	25
0	4	5	17	1	4	5	22	2	4	5	28
0	5	0	9·4	1	5	0	13	2	5	0	17
0	5	1	11	1	5	1	15	2	5	1	20
0	5	2	13	1	5	2	17	2	5	2	23
0	5	3	15	1	5	3	19	2	5	3	26
0	5	4	17	1	5	4	22	2	5	4	29
0	5	5	19	1	5	5	24	2	5	5	32

Growth in			MPN	Growth in			MPN	Growth in			MPN
1·0	0·1	0·01 ml		1·0	0·1	0·01 ml		1·0	0·1	0·01 ml	
3	0	0	7·8	4	0	0	13	5	0	0	23
3	0	1	11	4	0	1	17	5	0	1	31
3	0	2	13	4	0	2	21	5	0	2	43
3	0	3	16	4	0	3	25	5	0	3	58
3	0	4	20	4	0	4	30	5	0	4	76
3	0	5	23	4	0	5	36	5	0	5	95
3	1	0	11	4	1	0	17	5	1	0	33
3	1	1	14	4	1	1	21	5	1	1	46
3	1	2	17	4	1	2	26	5	1	2	64
3	1	3	20	4	1	3	31	5	1	3	84
3	1	4	23	4	1	4	36	5	1	4	110
3	1	5	27	4	1	5	42	5	1	5	130
3	2	0	14	4	2	0	22	5	2	0	49
3	2	1	17	4	2	1	26	5	2	1	70
3	2	2	20	4	2	2	32	5	2	2	95
3	2	3	24	4	2	3	38	5	2	3	120
3	2	4	27	4	2	4	44	5	2	4	150
3	2	5	31	4	2	5	50	5	2	5	180
3	3	0	17	4	3	0	27	5	3	0	79
3	3	1	21	4	3	1	33	5	3	1	110
3	3	2	24	4	3	2	39	5	3	2	140
3	3	3	28	4	3	3	45	5	3	3	180
3	3	4	31	4	3	4	52	5	3	4	210
3	3	5	35	4	3	5	59	5	3	5	250
3	4	0	21	4	4	0	34	5	4	0	130
3	4	1	24	4	4	1	40	5	4	1	170
3	4	2	28	4	4	2	47	5	4	2	220
3	4	3	32	4	4	3	54	5	4	3	280
3	4	4	36	4	4	4	62	5	4	4	350
3	4	5	40	4	4	5	69	5	4	5	430
3	5	0	25	4	5	0	41	5	5	0	240
3	5	1	29	4	5	1	48	5	5	1	350
3	5	2	32	4	5	2	56	5	5	2	540
3	5	3	37	4	5	3	64	5	5	3	920
3	5	4	41	4	5	4	72	5	5	4	1,600
3	5	5	45	4	5	5	81	5	5	5	2,400 +

It should be stressed that the mere cell numbers which can be read from the presence and absence of growth in the MPN tables may not be equally reliable for all species. Each species has to be considered separately and in relation to the rest of the community as growth of some species is often suppressed by others.

Three main reasons for finding too-low MPN estimates are:

1. Natural variation in viability of the inoculated cells, and/or loss of viability during handling.
2. Growth may be suppressed by other species, or the medium may be inadequate for the species in question.
3. Some species may have a short vegetation period and die off before routine inspection of the series, unless the series is examined at short (e.g. weekly) intervals.

It is generally agreed that quantitative estimates obtained by the serial dilution culture method are too low (Ballantine, 1953; Bernhard *et al.*, 1967), but it is the only practical technique for a quantitative and qualitative survey of the fragile 'naked' flagellates.

8
Interpreting the observations

8.1 Qualitative and autecological aspects

8.1.1 Biogeographical meaning; indicators

Theodore J. Smayda

In certain studies on phytoplankton dynamics, abundance is measured by proximate analysis of some chemical constituent such as chlorophyll or carbon, and species composition and numerical abundance are ignored. This contrasts with other studies dealing with certain aspects of annual cycles, phytoplankton-grazer interactions, regional dynamics, and species succession, in which community composition is described, statistical analysis of species groupings sometimes attempted (see Margalef and González Bernáldez, 1969; McIntire and Overton, 1971) and the fluctuations in species representation and abundance with time and regionally recorded. Unfortunately, field experimentation is usually not carried out concurrently during such studies and, thus, investigators who seek to explain the observed behaviour of individual species must rely heavily on clues to species growth requirements suggested by studies on both natural history and autecological experimentation. Since such autecological data are still very limited, biogeographical observations provide the primary source for impressions of species apparent growth requirements despite the obvious limitations of these observations to provide such information (the latter must be based on experimental data). This chapter describes some general biogeographical aspects of the phytoplankton used by ecologists to facilitate data interpretation. While the distributional patterns of only a relatively few species of phytoplankton have been described, the patterns referred to here probably illustrate the principal types characteristic of the phytoplankton.

Despite the geological age of the different groups of planktonic marine algae and a pattern of oceanic circulation seemingly favourable to an exchange of genetic material, a single community of cosmopolitan species has not evolved.

225

Rather, unique and persistent geographical distributions characterize individual species and certain higher taxonomic groups (Smayda, 1958; Hasle, 1976).

The question is, to what extent are such patterns regulated by basic, quantifiable laws rather than stochastic processes? That is, are the major distributional patterns developed and maintained primarily through indigenous species responses to regionally unique and definable combinations of light, nutrients, temperature and other 'direct' factors of growth (see Braarud, 1935), or are they a consequence of regional circulation patterns which influence the dispersal, accumulation and isolation of phytoplankton communities? The latter mechanism approaches a stochastic mechanism not amenable to quantification or modelling, given the vagaries of wind-driven circulation and the associated species responses to unmeasured growth conditions along the transport gradient. Resolution of this question goes beyond this biogeographical consideration; it is relevant to all efforts attempting to quantify environmental regulation of phytoplankton behaviour in the sea.

Indirect evidence supports Gran's (1902) view that such biogeographical patterns are indeed regulated primarily by environmental factors rather than representing periodic invasions and withdrawals of species' assemblages ('plankton types') resulting from water-mass movements (Cleve, 1897). The partially successful application (see Braarud, 1961, 1962) of autecological data to explain certain distributional features of individual species supports this view. This has led to the equally fundamental and related working hypothesis that an individual organism, or an assemblage of phytoplankton species, conveys information about the ecological conditions of its environment, and that this information content can be generally inferred from knowledge of the environmental conditions accompanying previously reported occurrences or absences of a species. Hence, from such impressions of apparent environmental preferences, the probable causes of a species presence or absence and of its observed successional and distributional patterns can be inferred. Margalef's (1961a) tabulation of certain ecological conditions associated with the occurrence of 379 species of phytoplankton illustrates the ecologists' attempt to define specific ecological and geographical preferences of species (Fig. 44).

The primary biogeographical classification of phytoplankton is into regional groups which describe their distributional range in terms of nearshore versus offshore occurrences, as well as latitudinally. A species may thus be described as being neritic, sometimes as brackish, estuarine or coastal (see Smayda, 1958, however) or oceanic, and by a (more or less) latitudinal descriptor such as Arctic, Boreal, Temperate, Subtropical and Tropical. Historically (Gran, 1902), these descriptors are combined to classify species as being, for example, arctic-neritic; 'oceanic with a pronounced northern distribution' (= Chaetoceros boreale, Gran and Braarud, 1935, p. 359) or temperate-atlantic-neritic plankton element (Gran, 1902, p. 80), etc. Cosmopolitan species also occur. Representative distributional patterns are illustrated in Smayda (1958) and Hasle (1976), who also give other examples of biogeographical descriptors.

The ecologist's interest in such biogeographical classification is in the associated description of environmental parameters. The latitudinal descriptor suggests something about the species tolerance (preference) to temperature, while the onshore versus offshore classifier suggests something about its possible requirements, or tolerance, for nutrients and its osmotic needs. Coastal

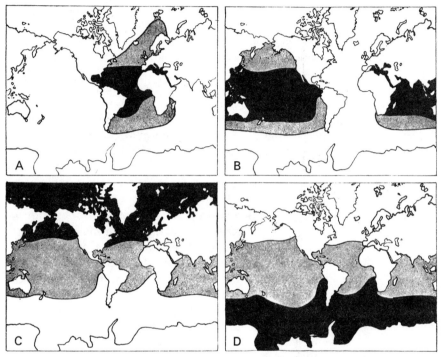

Figure 44
An attempt to delimit geographical types of phytoplankton distribution. From a compilation of the records available for 379 species of planktonic algae, the following twelve types were proposed:

(I)   Cosmopolitan, 'euryoic' (i.e. tolerant to a large set of environmental variables): combination of C plus D.

(II) Cosmopolitan, temperate and warm waters: A + B.

(III) Cosmopolitan, warm waters: A + B, black areas only.

(IV) Cosmopolitan, cold waters: C + D, black areas only.

(V)   Atlantic, euryoic: A.

(VI)   Atlantic, warm waters: A, black area only,

(VII) Indo-Pacific, euryoic: B.

(VIII) Indo-Pacific, warm waters; B, black area only.

(IX)   Boreal, euryoic: C.

(X)   Boreal, cold waters: C, black area only.

(XI)   Austral, euryoic: D.

(XII) Austral, cold waters: D, black area only.

(From Margalef, 1961*a*: rearranged here by editor after consulting the author.)

species may be more tolerant to eutrophic water than oceanic species; this view has led to efforts to apply nutrient-uptake kinetics data to account for species distribution and succession (e.g. Dugdale, 1968; Eppley *et al.*, 1969; Smayda, 1973). Thus, an arctic-neritic designation suggests a preference for cold, nutrient-rich waters and, depending on time of occurrence, for low or high light intensities and for short or long day length. A tropical-oceanic designation suggests a preference for warm, oligotrophic waters and, depending on depth of occurrence, for either high or low daily light intensity. These few examples illustrate the general approach and the many combinations of apparent environmental preferences suggested by field and autecological data, including the extent to which a species may be widely (eury-) or narrowly (steno-) tolerant of a given factor.

This approach requires an investigator to have a fairly good insight into the regional and seasonal occurrences and abundances of phytoplankton generally and, especially, of the species in question, as well as knowledge of the appropriate experimental autecological studies. Since the approach is also qualitative at best, and given the importance of factor interactions (rarely evident in biogeographical observations) in regulating species and community dynamics, extreme caution should be used in ascribing biogeographical meaning or regulation of species occurrence based on such observations alone.

Clearly, attempts to arrive at biogeographical meaning are simply efforts to use the presence or absence of one or more species, or stages in their life cycles and, sometimes, their morphological appearance as indicators of certain environmental conditions, including hydrographic events, eutrophication or warming trends, or long-term changes symptomatic of environmental disturbances. The extent to which this is successful depends, *inter alia*, on the environmental condition being monitored and whether the appropriate indicator species is known, if it occurs at all. The information sought from the indicator species sometimes requires of the investigator minimal autecological insight, such as using the organism to indicate certain water-mass incursions. At the other extreme, considerable autecological knowledge is required to find and use an indicator organism that would be symptomatic of a very specific condition, for example a certain pollutant. In the latter case the organism is used similar to a miner's canary, i.e. to bioassay a unique environmental disturbance.

The diatom *Planktoniella sol* may be an indicator species of the Gulf Stream near its entrance into the Norwegian Sea (Smayda, 1958). Such attempts to use certain species as indicators of given water-masses have been replaced by more complicated attempts (see Thorrington-Smith, 1971; Venrick, 1971) to relate specific phytoplankton populations to their associated water-mass, sometimes designated as phytohydrographic regions. Such statistical approaches also allow species associations to be established and potential functional groups to be identified. That is, a knowledge of the autecology of one member of species pairings might be extrapolated to certain co-occurring species on the reasonable assumption that their inevitable co-occurrence reflects similar physiological and ecological characters (see Chapter **8.4**). The use of statistical procedures in the analyses of numerical census data, and in attempts to relate species occurrences and community structure to environmental conditions,

228

represents a more objective and rigorous approach to evaluating the biogeographical and indicator value of individual species or assemblages. Such methods are recommended over the more traditional, non-statistical treatment of numerical census data sets.

Coccolithophorid species have been successfully used as indicator species of palaeo- and recent temperatures (McIntyre and Bé, 1967; McIntyre et al., 1970). However, similar attempts to use indicator species for eutrophication and other anthropogenic disturbances have been marginally successful (see Dunstan, 1975). The wide range of responses which characterize the species-rich and physiologically opportunistic planktonic marine algae reduces the potential indicator value of the more easily sampled and identifiable species. Braarud's (1945) early comments on attempts to classify and use marine phytoplankton as universal indicator organisms for polluted waters remain valid, and are very relevant to the general problem of their selection and use as indicator species. Based on a thorough analysis of phytoplankton communities in both polluted and unpolluted sections of Oslo Fjord (60° N), he concluded (p. 71): '... as a general rule the concentration of [nutrient] salts or other chemical factors seems to be of minor importance as compared with other milieu factors for the occurrence of the various species . . .' and (p. 104), '. . . a general ecological classification based upon relationship to the complex nutrition factor, pollution, would seem to be of little use. It might even be misleading as to the ecological character of the various species.'

8.1.2 Eco- and morphotypes

Theodore J. Smayda

Phytoplankton generally have evolved an opportunist stratagem to ensure growth and survival in nutritionally dilute seawater characterized by great and frequent changes in overall growth conditions. The inducible enzyme alkaline phosphatase, for example, is formed to allow utilization of phosphorus present in organic compounds when preferred inorganic sources become limiting (Kuenzler and Perras, 1965).

Morphotypic and ecotypic responses also occur. Phytoplankton species characteristically exhibit a variable cell morphology (habitus) which is easily and routinely recognized during microscopic examination of natural and cultured populations. The factors which induce changes in morphology and, hence, the particular significance of such expressions are usually unknown. Morphotypic expressions which are phenotypic, sometimes teratological, probably occur within several cell divisions (days) after initial environmental triggering. They are thus short-term manifestations, both in the time required for their induction and in their persistence.

Ecotypes, in contrast, are generally detectable only through experimentation; they usually lack morphological manifestations based on available observations for *Thalassiosira pseudonana* (see below), and probably develop in response to environmental pressures which are more or less continuous over several million cell divisions (i.e. millennia) at a minimum. Ecotypes thus gradually evolve through long-term physiological modification which then persists, possibly as a genotypic expression.

Some species, notably within the diatom genus *Rhizosolenia* and the dinoflagellate genus *Ceratium*, are di- or polymorphic, i.e. characterized by two or more forms. *Rhizosolenia hebetata* occurs as forma *semispina* and f. *hiemalis*. Such polymorphism may be seasonal (cyclomorphosis), as suggested by Proshkina-Lavrenko's (1955) separation of the diatom *Chaetoceros sociale* into forma *autumnale* and f. *vernale*. Experimental evidence is still lacking to evaluate the popular view that changes in temperature and/or silica levels induce seasonally occurring diatom morphs, although cell-wall structural patterns used taxonomically are influenced by silicon concentrations (see Paasche, 1973a). The species *Thalassiosira rotula* and *T. gravida* are possibly morphotypes of the same species, whose particular expression may be dependent on temperature (Syvertsen, 1977; see Fig. 45 herein). Salinity is another variable which can induce morphotypes (Paasche *et al.*, 1975; and Fig. 46 herein).

Many species of planktonic diatoms form colonies which break up during unfavourable growth conditions, as during nutrient limitation (see Smayda and Boleyn, 1965, 1966). Individual cells separated from these colonies are some-

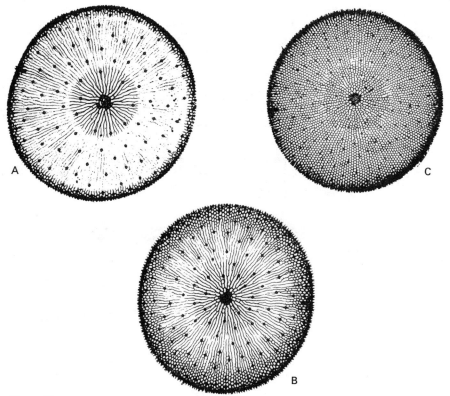

Figure 45
Thalassiosira rotula Meunier, clone 'Lvb', grown at different temperatures (*c.* × 1,500):
(A) 17° C, *T. rotula* type; (B) 10° C, transition type; (C) 3° C, *T. gravida* type.
(From Syvertsen, 1977; slightly modified here with respect to page-setting.)

times misidentified; a *Thalassiosira* specimen may be erroneously assigned to
the genus *Coscinodiscus*, for example. Different morphological stages are sometimes produced during the life cycle or cell division of a species which can be
assigned to different taxa using current taxonomic criteria. Thus, three distinct
morphological stages (= 'species') occurred during the life cycle of *Coscinodiscus concinnus* (Holmes and Reimann, 1966), and dividing cells of *Nitzschia
alba* produced cells assignable to the genera *Nitzschia* and *Hantzschia* (Lauritis
et al., 1967). Cytomorphological manifestations of environmental conditions
also occur (Holmes, 1966), but their detection by microscopy is more difficult.

Evidence is increasing that physiological clones (races) of the same species
occur, since Braarud's (1961) initial experimentation and consideration of this
possibility. The more recent evidence comes primarily from experiments carried
out with several clones of the diatom *Thalassiosira pseudonana* isolated from
nearshore, continental shelf, and oligotrophic, tropical oceanic waters. These
clones differ significantly in their growth rates and responses to temperature

231

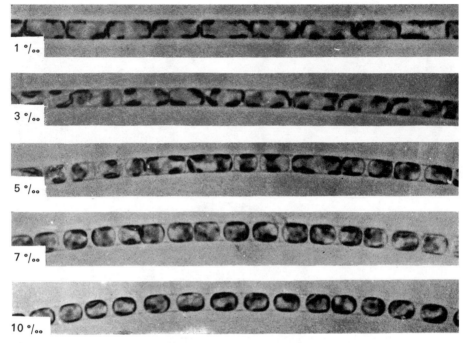

Figure 46
Skeletonema subsalsum (A. Cleve) Bethge grown in increasing salinities (× 800). (From Paasche *et al.*, 1975; stages corresponding to 2, 4, 6 and 8‰ salinity have been omitted here.)

and salinity (Guillard and Ryther, 1962), in their rates of respiration and the effect of temperature on these (Ryther and Guillard, 1962), and in uptake kinetics of nitrate (Carpenter and Guillard, 1971) and silicon (Guillard *et al.*, 1973). Electrophoretic banding patterns of enzymes also suggest that these clones represent separate ecological races (Murphy and Guillard, 1976).

Morphotypic variations clearly present more serious problems to taxonomists than to field ecologists. Such variability is undoubtedly responsible for some misidentifications which affect calculations of species diversity index and certain phytosociological parameters. However, such errors probably have an insignificant effect on data interpretation. The occurrence of ecotypes, however, presents a more serious problem. Should ecotypes similar to those found in *Thalassiosira pseudonana* generally characterize geographically or environmentally isolated populations of the same phytoplankton species, then the routine extrapolation of experimental autecological data to explain species behaviour *in situ* may sometimes be inappropriate. The physiological differences between clones may be so great, for example, that applying growth or physiological constants from one clone to account for the *in situ* behaviour of a geographically isolated population may potentially lead to erroneous conclusions.

232

8.1.3   The size of cells

H. J. Semina

PRECAUTIONARY NOTES

The diameter of phytoplankton cells varies from 1 to 2,000 µm. The cell size partially depends on the sampling equipment and on the processing method. A bottle sample usually contains smaller cells than a net sample. Cell size in net samples depends on the gauze mesh opening and on the water volume being hauled. Species forming large colonies with individual small cells can be caught both with a sampling bottle and with a net. A net hauled through a large water volume collects colonial forms and larger cells which ordinarily are scarce and less likely to occur in a bottle sample. For example, a Juday net with a 37-cm or 80-cm mouth opening and 180-µm mesh easily catches the species of the genera *Chaetoceros, Climacodium, Ethmodiscus, Hemiaulus,* several *Nitzschia,* many species of the genera *Ceratium, Pyrocystis,* etc. (for more detail, see Semina, 1962).

The technique of sample fixation also affects the resulting size composition of phytoplankton. Unpreserved samples contain a great number of smaller flagellates which are destroyed by fixation and missed in the subsequent determinations. The destruction may be so heavy that one can no longer tell an organism from detritus.

Measurement of smaller cells (from less than 5 to 20 µm) requires an objective of × 20 or × 40. Large cells can be counted using a × 10 lens. Depending on the purpose of the investigation, cell sizes are determined for a single species or for total phytoplankton. In either case, the necessary number of cells is taken and measured; for rare species, one has to be satisfied with a few measurements or even a single one.

When estimating the cell size, different authors may speak of different size groups. So far, there is no universally accepted classification of phytoplankton size groups. Division into ultra-, nanno- and microplankton is made by many authors (see references in Kisselev, 1969). Sournia (1968), following the limnological classification of Dussart (1965, 1966), placed the upper limit of ultranannoplankton at 2 µm and that of nannoplankton at 20 µm, larger cells being regarded as microplankton. An author concerned with phytoplankton size should give a clear definition of his size groups. He ought not to forget that what he describes as a 'small' cell may be a 'large' cell to another author or vice versa, which may lead to confusion.

CELL MEASUREMENTS

The data to be discussed here are either obtained from direct linear measurements of cell diameter, or calculated from cell-volume estimations if such data have already been made available by electronic counting (Chapter **7.5.1**) or microscopical study (Chapter **8.5**).

1. The average diameter $\bar{D}$ directly measured in a sample is:

$$\bar{D} = \frac{\Sigma \, d_i \cdot n_i}{\Sigma \, n_i},$$

where: d_i = cell diameter of each species in the sample; and n_i = cell number of each species. When estimating $\bar{D}$, one should bear in mind that statistically this figure is more reliable in a rich sample than in a poor one (see also Chapter **7.1.2**). With poor samples, results for several samples may be averaged, for example, by aggregating data on samples from different depths into an overall average for the entire sampled layer (the sizes are calculated as average weighted values).

Direct measurement of cell diameter is convenient, because it spares much calculation effort. The diameter can be measured in any kind of sample. A dinoflagellate's diameter is measured at the transverse groove (for cells of the *Dinophysis* type in lateral view, and for such cells as *Peridinium* or *Ceratium* in ventral or dorsal view). For diatoms of a round cylindrical shape—genera such as *Coscinodiscus* or *Thalassiosira*—it suffices to measure the cell diameter. The same is true of the diatoms having elongated cells of cylindrical cross-section. For diatoms such as *Nitzschia* or *Thalassiothrix*, their width is measured along the transapical axis. All these measurements produce figures closely approximating to the average diameter (see below). In some phytoplankton species and genera, e.g. in *Chaetoceros*, it is difficult to measure the diameter, since the observer sees a *Chaetoceros* chain in girdle view; it should be measured along the apical axis, where the cell size is more constant than along the pervalvar axis.

2. If cell volumes are already available, then the average cell volume $\bar{V}$ can be evaluated. For any given sample,

$$\bar{V} = \frac{V}{\bar{N}},$$

where V = cumulative volume of all cells; and $\bar{N}$ = number of cells in the sample.

3. If cell volumes are too cumbersome for handling, one can use the calculated average diameter $\bar{d}$. It can be calculated like $\bar{D}$ (see above) from calculated cell diameters of individual species d. Alternatively, as an approximate value, $\bar{d} = \sqrt[3]{v}$, where v is the cell volume. (More precisely, $d = \sqrt[3]{(6v/\pi)}$, but even the first formula shows good agreement with d_i above.)

234

INTERPRETATION OF RESULTS

Cell-size studies are commonly applied to taxonomic studies and to biomass estimations. We shall deal here with some other aspects.

Cell size has long been considered as an important feature of the phytoplankton life cycle, for instance, by investigators of auxosporulation in diatoms (see references in Semina, 1972 or Round, 1973, etc.). Individual cell-size variations were considered in connection with environmental factors, such as temperature and salinity (see Wimpenny, 1936, 1946, 1956, 1966a, and others). Estimation of phytoplankton cell size is essential for plant physiology, and for research on feeding of zooplankton and fish. Literature in these fields is abundant (see, e.g. Jørgensen, 1966; Mullin *et al.*, 1966; Parsons *et al.*, 1967).

Less investigated is the influence of phytoplankton size on the optical properties of water (see, e.g. Hart, 1966; Frackowiak and Januszczyk, 1975). Generally speaking, small cells exert a greater influence than larger ones on the optical properties of water, but the cell shape and the formation of colonies should also be taken into account.

The variation of cell size is a major feature in describing a plankton community. The description of cell size and the interpretation of the possible governing factors is an interesting task for the biologist. Communities can be compared by their cell sizes in various ways. For example, phytoplankton can be divided into size groups (Parsons, 1969), one predominant size group can be identified, and these data can be correlated with the features of the habitat. In particular, a predominance of smaller forms at the divergences of currents has been observed (Margalef, 1963; Kozlova, 1964; Beklemishev, 1969); the cell size of the predominant group varies from season to season, and this can be taken to characterize the more mature and the less mature stage of a community (Margalef, 1963, and others).

A community also can be characterized by the proportion of various size groups. The proportion should be estimated both by cell number and by biomass. If either characteristic is used alone, an erroneous or biased image of the community may result (an estimation based on cell number may exaggerate the significance of smaller forms which often are more numerous, while an estimation based on biomass may exaggerate the significance of larger forms). It should be borne in mind that smaller forms have a more intensive metabolism and divide faster; they should be regarded as the main producers when there is a high concentration of nutrients, and as potential producers when there is a low concentration of nutrients. Larger forms may be less abundant than smaller ones, but their biomass may be higher; this is particularly important in poor areas, where metabolism of small forms is depressed, and where large forms serve, as it were, as reservoirs of organic matter (Semina, 1976).

The size characteristic of a community can be based on the average phytoplankton cell size. This has proved a convenient and methodologically simple approach to the estimation of phytoplankton size distribution. The distribution of the average cell size in the ocean shows certain regular patterns. We have established large-scale patterns (for hundreds and thousands of miles), and medium-scale patterns (tens of miles or less), in the average cell-size distribution (Semina, 1968, 1969, 1971, 1972, 1974a, 1974b; Semina and Tarkhova, 1972; Semina et al., 1976). Seasonal variations of average cell size also have been established (Semina, 1972; Vinogradova, 1973; Lopez Baluja, 1976). Spatial and temporal variations of average cell size are observed simultaneously in different taxonomic divisions of algae, provided that each is represented by enough species and cells. Simultaneous variation of average cell size has been established for diatoms and dinoflagellates on the Cuban shelf (Lopez Baluja, 1976) and for diatoms, dinoflagellates and coccolithophorids in the Pacific (Semina, 1974b).

The next essential step in analysing phytoplankton size pattern is to establish its correlation with the environmental conditions. If physical data are available, one can try to correlate any size characteristic with such environmental factors as temperature, salinity, illumination, nutrients, vertical water movements, and the depth of the upper mixed layer. The main factor affecting the cell size can sometimes be identified by a correlative method.

For example, correlation with the vertical movements has enabled us to establish for the tropical zone a curvilinear dependence of the average cell size on the vertical velocity (Semina, 1972, 1974b). Water velocity can be measured in various ways. We made use of the data of Burkov (1972), calculated from the vortex of tangential wind stress. For correlation with phytoplankton, average velocity for a period of time, say for a season, should be taken (cell size should not be correlated with a vertical velocity computed during a short time, for example, during a single survey). Vertical movements of water are highly variable. Phytoplankton responds to this variation with a certain lag and, compared to these movements, phytoplankton size is rather conservative.

Apart from estimating the average cell size, one should pay attention to the kind of morphotypes (Lebensformen) that determine it. In so doing one should remember that morphotypes do not depend on cell size only. For instance, near the western coast of Africa, where the average cell size is rather small, colonial forms of Chaetoceros, Nitzschia, Thalassiosira and others prevail, whereas some way from the shore in the open sea we observed areas of a large average cell size, typically with single spherical or cylindrical cells (Semina et al., 1976).

Interpretation of the dependence of cell size on the environmental conditions, considered as a separate problem, has been recently carried out by mathematical modelling. Some authors give primary importance to such factors as illumination and rate of nutrient supply to cells (Parsons and Takahashi, 1973a). These authors computed the growth rates of two species, differing in cell size, under different illumination and nutrient conditions, and hence made a hypothesis as to which of the two species would prevail in a community under given environmental conditions. They assumed that the sinking of cells was compensated for by upwelling of a sufficient velocity. Other

236

authors attach more importance to metabolic features of larger and smaller cells, namely their difference in respiratory loss of organic matter (Laws, 1975). Unlike Parsons and Takahashi, Laws emphasizes that the difference of growth rate may prove less important for competitiveness of the cells differing in size than the species resistance to losses through respiration and other causes (including sinking). The influence of environmental factors on cell size is now being debated. Parsons and Takahashi (1973a) disagreed with this interpretation (Semina, 1972), but they in turn were criticized by Hecky and Kilham (1974).

The average size of cells in a community ($\bar{d}$) may depend on: (a) intraspecific variation of cell size (Δd); and (b) number of species (N) of a given average size. (We believe that $\bar{d}$ depends more on the last factor than on intraspecific variation.) A community maintains its productivity under varying environmental conditions owing to a prevalence of alternative species of different sizes. Recently, Lewis (1976) suggested that conservation of surface-to-volume ratios is likely to be widely observed in nature and will affect the balance of phytoplankton species under selective pressure. The issue under discussion is how to determine the factors that are responsible for the prevalence of a given size group.

8.2 Statistical considerations

Elizabeth L. Venrick

The purpose of data analysis is to extract information. To this end, statistics is a tool which helps one make 'wise decisions in the face of uncertainty' (Wallis and Roberts, 1956). In addition to facilitating the summarization and examination of large quantities of data, statistical methods have the advantage of being insensitive to the personal biases of the investigator. However, information may be extracted from data by non-statistical methods, and the degree to which a scientist relies on statistics is partly a matter of personal philosophy as well as of familiarity with statistical procedures. The more elaborate statistical procedures generally make rigid assumptions about the characteristics of the data to be analysed. Manipulating the data to conform to these assumptions may be a laborious process. Indeed, the calculations required by many statistical procedures themselves are time and/or money-consuming activities. A scientist should not become so enamoured with statistics that the time and effort spent in analysis is out of proportion to the information gained. Nor should he hesitate to reject a statistical procedure when the trade is unfavourable. A statement by Colebrook (1969) helps to put statistics in a realistic perspective:

... there are two kinds of workers; those who use statistics and those who do not. And, it must be admitted that, so far, workers who do not use statistics have made a considerably greater contribution to our knowledge of the plankton than those who do.

PARAMETRIC AND NON-PARAMETRIC PROCEDURES

Data may be classified into three types: (a) measurements; (b) ranks, in which n data points are ordered from one extreme to the other and assigned ranks $1, 2, 3, \ldots n$ according to their position in the series; and (b) scores, which represent a classification into categories, usually dichotomous, such as presence/absence, black/white, above the mean/below the mean, etc. While there is a progressive loss of information from measurements to ranks to scores, there is usually a corresponding increase in the ease with which the data are obtained.

Statistical procedures in which a probability is assigned to a hypothesis on the basis of data are derived from some theoretical frequency distribution from which it is assumed the samples were drawn. These so-called parametric statistics of measurements are generally based upon the normal distribution, and include the most highly developed techniques. Their major disadvantage, especially for plankton work, is the necessity of transforming biological data into a normal distribution.

238

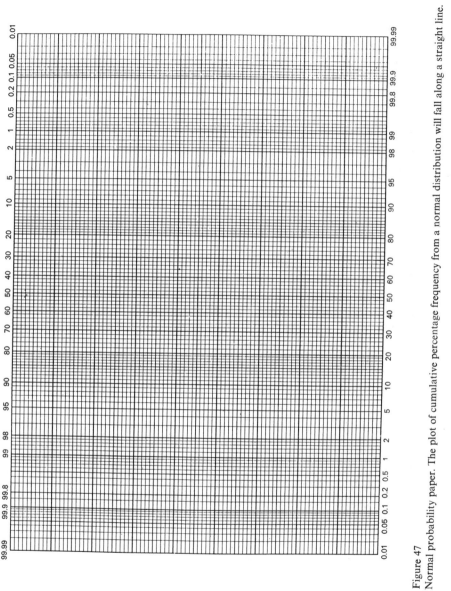

Figure 47
Normal probability paper. The plot of cumulative percentage frequency from a normal distribution will fall along a straight line.

Ranked data fit a rectangular distribution in which all values have an equal probability of occurring. Dichotomous scores can be treated with the binomial distribution. The statistics of these types of data are referred to as non-parametric, or distribution-free statistics, because the success of the transformation into a rectangular or a binomial distribution is independent of the original frequency distribution of the data.

Since measurements are readily translated into ranks or scores (but not the reverse) both types of statistics are potentially available for the analysis of abundance measurements. The choice is largely a matter of personal preference. This author prefers the non-parametric techniques because of their relative simplicity, and because of their freedom from assumptions which are unlikely to be fulfilled, such as normality and stability of the variance. Other workers, including the late R. A. Fisher (1966) discount non-parametric statistics, largely because the loss of information when the data are translated to ranks reduces the sensitivity of the tests.

Testing for normality

If the maximum sensitivity is to be obtained from parametric statistics, the data must be distributed according to a normal distribution with stable variance. Normality may be tested in several ways. The simplest is to plot the cumulative frequency distribution of the data on normal probability paper (Cassie, 1962; Sokal and Rohlf, 1969), a sheet of which is reproduced in Figure 47. Normal data will fall along a straight line. Unfortunately, there are no objective criteria for determining directly how straight a line must be in order to be adequate. Other tests of normality include the chi-square goodness of fit test (e.g. Cassie, 1972); the Kolmogorov–Smirnov test (Tate and Clelland, 1957; Conover, 1971); the Lilliefors test (Conover, 1971); and the use of rankits (Sokal and Rohlf, 1969).

Example

The following is the frequency distribution of pelagic ciliates counted in random fields of a settling chamber by Rassoùlzadegan and Gostan (1976; table IV, sample 2.VII.73). According to the index of dispersion, calculated by the authors, the individuals are significantly aggregated:

(1) x_i (cells/field)	(2) Observed frequency	(3) Cumulative frequency (%)	(4) Expected frequency
0	3	1·4	7·6
1	5	3·8	4·6
2	4	5·7	6·7
3	12	11·4	9·7
4	20	11·8	20·9
5	24	32·3	14·5
6	16	39·9	16·8

240

(1) x_i (cells/field)	(2) Observed frequency	(3) Cumulative frequency (%)	(4) Expected frequency
7	23	50·9	18·7
8	30	65·2	18·9
9	16	72·8	19·3
10	15	79·9	17·6
11	10	84·7	16·0
12	7	88·0	13·4
13	6	90·9	10·8
14	4	92·8	8·0
15	3 ⎤	94·2	5·7 ⎤
16	4 ⎦	96·1	4·0 ⎦
17	1 ⎤	96·6	2·7 ⎤
18	3	98·0	1·5
19	2	99·0	0·8
20	0	—	
21	0	—	
22	1	99·5	
23	1 ⎦	100·0	0·8 ⎦

(Brackets indicate pooling of frequencies for *chi*-square test.)

$n = 210, \overline{X} = 7\cdot77, s^2 = 18\cdot68$

$$x^2_{14} = \frac{(3 - 7\cdot6)^2}{7\cdot6} + \frac{(5 - 4\cdot6)^2}{4\cdot6} \cdots \frac{(3 - 9\cdot7)^2}{9\cdot7} + \frac{(8 - 5\cdot8)^2}{5\cdot8}$$

$$= 35\cdot89.$$

These data may be examined directly for normality. Figure 48 gives a plot of x_i (col. 1) against cumulative percentage frequency (col. 3) on normal probability paper. The straight line represents the normal distribution of the same mean and variance. Also shown is the cumulative frequency of the data transformed according to $\log(x_i + 1)$ as recommended by the authors. In this case, the transformation improves the fit to a normal distribution only at the higher values.

A *chi*-square goodness of fit test may be applied to assess more objectively the deviation from normality. The frequencies expected from a normal distribution (col. 4) may be calculated using tables of the cumulative frequency of the standardized normal variate $(x_i - \mu)/\sigma$, or they may be read from the graph. Frequencies in the higher tail were pooled (indicated by brackets) to keep 80 per cent of the expected frequencies greater than 5, a rule of thumb recommended by many statisticians. Of the $(n - 1) = 16$ degrees of freedom, two are lost because the mean and variance of the normal distribution are estimated from the sample. Thus, the *chi*-square value has 14 degrees of freedom. The calculated value is highly significant ($p < 0\cdot005$) indicating significant departure from normality.

I

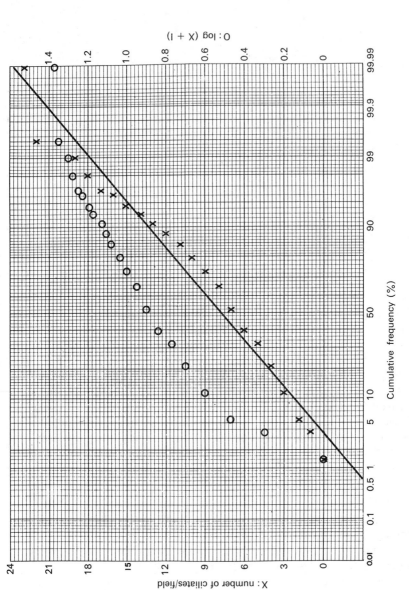

Figure 48

Normal probability paper used to examine the distribution of ciliates within a settling chamber. The raw data (x's) from Rassoulzadegan and Gostan (1976) are compared with the straight line representing a normal distribution with the same mean and variance. The log-transformed data (circles) seem to give a better approximation to a straight line (not drawn) except at the lower tail of the distribution.

242

The stability of the variance can be examined directly only when a series of replicate samples are available (e.g. Barnes, 1952; Frontier, 1969, 1972; Ibanez, 1971, 1976). The Bartlett procedure is recommended to objectively evaluate homogeneity of variances (Dixon and Massey, 1957; Sokal and Rohlf, 1969) although, to my knowledge, it has not been applied to this situation. Generally, stability of the variance is judged by instinct.

Transforming data

When the initial data are not normal, they may often be successfully normalized by use of an appropriate transforming function (Bartlett, 1947; Barnes, 1952). The transform to be used depends upon the distribution of the raw data. If they are periodic, a sine transform may be successful (King and Hida, 1954). An arcsin-square root transform is appropriate for percentages (Cassie, 1971) and a square root transform for data which fit a Poisson distribution (Kutkuhn, 1958; McAlice, 1971). The transformation most frequently used in plankton work is of the form $Y = \log(x + c)$ where c may be zero or some positive value. This is a satisfactory transform for the negative binomial, the log-normal and the Poisson-log-normal distributions (Cassie, 1962, 1963). Woelkerling *et al.*, (1976) found the optimum value of c ranged from 0·1 to 2·0 for different phytoplankton species within the same series of samples. Other workers have used the simplified expression $Y = \log(x)$ (e.g. Barnes and Hasle, 1957). Table 11 summarizes some of the applications of transformations to plankton data. The variety of results indicates the non-generality of a single transform function and emphasizes the need to test the success of the transformation on the data under consideration.

Violation of assumptions

Statisticians have done considerable work on the errors incurred in using parametric procedures when one or more of the underlying assumptions is not satisfied (Cochran, 1947; Ratcliffe, 1968). In general, violation of the assumptions results in a loss of efficiency and introduces errors into the tests of significance, in the direction of producing too many significant results. Cochran (1947) ventured a tentative opinion that these errors are not often great, at least for moderate deviations from the assumptions. However, extreme skewness and marked instability of the variance (the two deviations most often recognized in plankton data) are among the factors most likely to cause serious disturbances. It is unlikely that biological data ever completely satisfy the assumptions underlying parametric statistics. The more that stringent requirements are relaxed, the more the results lose their probabilistic nature and become only qualitative indices. The precise spot on this continuum at which a statistic is no longer informative depends not only on the nature of the data but also on the goals of the programmes and the philosophy of the scientist.

TABLE 11. A review of transformations applied to plankton data. Only studies which examined the success of the transformation are included. Transformations judged satisfactory are noted with an asterisk (*)

Reference	Material	Transformation		Criterion of success
Barnes (1952)	Zooplankton: 25 subsamples from a single sample	$\sqrt{x}$	*	Stability of variance
		$\sqrt{x + 0.05}$	*	
		$\sqrt{x + (3/8)}$	*	
	Zooplankton from modified Hardy Plankton Recorder	$\sqrt{x + 0.05}$	*	Stability of variance
		$\sqrt{x + (3/8)}$	*	
	Zooplankton from a series of net tows	$\log x$	*	Stability of variance
		$\sinh^{-1}\sqrt{x}$	*	
		$(1/\beta)\sinh^{-1}\beta\sqrt{x}$	*	
		$\sqrt{x}$		
Kutkuhn (1958)	Microplankton: 20 fields within each of 2 subsamples from each of 2 water samples	$\log(x + k/2)$	*	Untransformed data described by negative binomial distribution for which this transform is appropriate
Colebrook (1960)	*Calanus* from Continuous Plankton Recorder	$\log(x + 1)$	*	Stability of variance; reduced skew of distribution of means
Williamson (1961)	Zooplankton from modified Hardy Plankton Recorder	$\log x$		Failed to normalize data according to the frequency distribution
Frontier (1969)	Zooplankton from net tows	$\log^2 x$	*	Stability of variance
		x		
		$\sqrt{x}$		
		$\log x$		
Frontier (1972)	Zooplankton: 3 subsamples from each of 28 collections (all $\bar{x} \geqslant 10$)	$x^{1/3}$	*	Stability of variance
		$\log^2 x$		
		$\log x$		
		$\sqrt{x}$		
Frontier (1973)	Zooplankton from 8 successive series of 4 net tows	$\log^p(x + 1)$	*	Stability of variance; best value of p varied between 1 and 2 depending on the species
Ibanez (1971, 1976)	Zooplankton and phytoplankton from net tows	$\log(x + 1)$	*	Stability of variance and strength of principal axis extracted by principal component analysis
		$\sinh^{-1}\sqrt{x}$		
		$\log^2(x + 1)$		
		$\sqrt[3]{x}$		
		$\sqrt{x + (3/8)}$		
Woelkerling et al. (1976)	Experimental mix of 5 phytoplankton species	$\ln(x + c)$ $(c = 0.1, 0.5, 1, 2)$	*	Lilliefors' test for normality (best value of c depended on the species)
		$\sqrt{x + c}$ $(c = 0, 0.5, 1)$		

Non-parametric statistics

Although there are only a few textbooks devoted solely to non-parametric statistics (e.g. Tate and Clelland, 1957; Conover, 1971; Hollander and Wolfe, 1973), most statistical texts discuss the more useful techniques. The power of a statistical test (T) is the probability that it will lead to rejection of a false hypothesis, and this is a function of the number of samples. The efficiency of a non-parametric test (T_2) relative to its parametric equivalent (T_1) may be measured by the limiting ratio of n_1/n_2 as n_1 approaches infinity, where n_1 and n_2 are the number of samples required for T_2 to equal T_1 (Conover, 1971). The efficiency of many non-parametric tests is 95 per cent or greater when applied to normally distributed data. This was demonstrated empirically on phytoplankton data by Woelkerling *et al.* (1976). The more general tests, especially those based on scores, have relative efficiencies of the order of 60 per cent. The efficiency of a non-parametric test can be expected to increase considerably as the underlying data deviate from normality.

TYPES OF STATISTICAL PROCEDURES

Statistical procedures may be broadly classified into three major categories (although many procedures may fall into more than one, depending on the usage): summary or descriptive statistics; hypothesis-testing statistics; and predictive statistics. Probably every procedure has been applied to plankton data at some time or another and the following discussion is intended to be only the briefest introduction.

Summary statistics

The simplest and most frequently used statistics are those which condense a large body of information into a few values, such as the mean or median, which represent the central tendency, or the variance or range, which represent dispersion.

The calculation of point estimates is a non-parametric procedure since no assumption about the underlying population distribution is made. However, point estimates of the mean or median value are generally accompanied by some measure of precision of the estimate, often in the form of a confidence interval which is that interval within which the true mean or median is likely to fall. This step deserves some development here.

Calculation of a confidence interval does presuppose a specific frequency distribution. When the measure of central tendency is the mean, the calculation of a confidence interval is usually based on the assumption of a normal distribution. If normality is achieved by transforming the data, the calculated mean will not be the same as the mean calculated from untransformed data (Barnes, 1952; Cassie, 1962; McAlice, 1971), and appropriate adjustments must be made.

245

The percentiles of the distribution, however, do not change. Assuming normality, the appropriate expression for the confidence interval of the population mean is

$$\bar{x} \pm t_{(\alpha)(n_1 - 1)}\sqrt{MS_{between}/n_{total}}$$

where:

$MS_{between}$ is the mean square error between replicate field samples (Chapter **5.1**);

n_1 is the number of field samples; and

n_{total} $(= (n_1) . (n_2) . (n_3)...)$ is the total number of enumerations. This will be greater than n_1 if there is more than one count per sample.

$t_{(\alpha)(n_1 - 1)}$ is the appropriate value of the Student-t for the desired significance level (α) and is obtained from standard statistical tables.

If the population distribution is Poisson, and if the sampling strategy is such that the $MS_{between}$ is expected to be equal to the mean, then confidence intervals can be obtained from single counts using the Poisson distribution or Figure 29, Chapter **7.1.2**. It can also be shown from the Poisson that the maximum mean value expected from an observation of 0 cells is 1·6, 2·3, 3·0 or 4·6 at $\alpha = 0·20$, 0·10, 0·05, 0·01 respectively.

In non-parametric statistics the most frequently used measure of central tendency is the middle value or median, formally defined as that value which neither exceeds nor is exceeded by more than half of the observations. The median is less influenced by extreme values than is the mean, and this is an advantage in summarizing skewed distributions. Unlike the parametric counterpart, the procedure for calculating confidence intervals for the population median does not depend upon an estimate of the population variance, and it does not impose symmetry about the sample median. The procedure is based upon the expansion of the binomial distribution in which the probability that a given observation will fall above (or below) the median value $= p = 1/2$ (Tate and Clelland, 1957; Dixon and Massey, 1957). In a set of n observations, the probability of a given number of observations falling above or below the median is given by the appropriate term of the expansion of $(1/2 + 1/2)^n$, so that the rank of the observation which approximately delimits any specified confidence interval can be obtained by cumulating the terms in the expansion from one end or the other until the specified probability is obtained. Figure 49 gives the approximate 95 per cent confidence interval for a median of 50 or fewer observations. Confidence intervals are more extensively tabulated by Tate and Clelland (1957) and Dixon and Massey (1957).

Summary statistics are not limited to measures of central tendency and dispersion. Population structure may be summarized by any of a number of diversity indices (see Chapter **8.3**). Relationships between two bodies of information may be summarized by indices of dissimilarity (Thorrington-Smith, 1971) or similarity (Miller, 1970; Levandowsky, 1972; Haury, 1976a, 1976b) as well as by the probabilistic coefficients of correlation and these indices, in turn, often form the basis for grouping procedures (see Chapter **8.4**). The more complicated

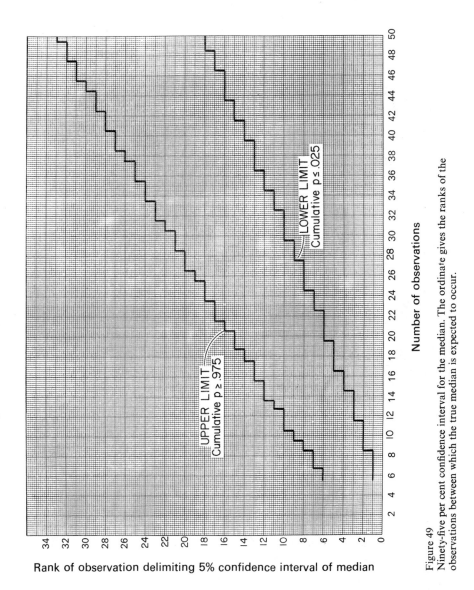

Figure 49
Ninety-five per cent confidence interval for the median. The ordinate gives the ranks of the observations between which the true median is expected to occur.

descriptive procedures include the various ordination procedures (Williamson, 1961, 1963; Colebrook, 1964; Robinson, 1970; Levandowsky, 1972; Ibanez, 1973a, 1976). Although these may be based on probabilistic indices, their initial purpose is generally descriptive.

Hypothesis-testing statistics

The bulk of statistical procedures are those designed to credit or discredit some hypothesis (the null hypothesis) on the basis of a set of data. The so-called significance level (α) is the probability that the null hypothesis may be rejected by chance when it is in fact true. In the vast majority of situations, a hypothesis can only be rejected with a given probability of error; it cannot be proved with a specified probability of error. Thus, two samples with very different means may lead to the rejection of the hypothesis that they came from the same population, but two samples with identical means may come from the same population, or from two different populations with more or less similar means.

The major hypothesis-testing procedures include: the analysis of variance (e.g. Winsor and Clarke, 1940; Barnes, 1951; Anraku, 1956; Barnes and Hasle, 1957; Uehlinger, 1964; McAlice, 1970; Platt et al., 1970) which, for two samples, is equivalent to a t test for differences of means; tests of concordance, correlation and covariance which examine relationships (e.g. Cassie, 1959a, 1959c, 1960; Williamson, 1961; Bernhard and Rampi, 1966; Colebrook, 1966; McAlice, 1970; Haury, 1976a, 1976b); and regression analyses (Cassie, 1959a, 1959c; 1960; Smayda, 1963; Grant and Kerr, 1970; Walsh, 1971; Venrick, 1971). The basic procedures may be performed with either parametric or non-parametric statistics. Parametric procedures, however, tend to be more highly evolved, and the more elaborate methods (such as multilevel nested analysis of variance, or multiple regression) as yet have no non-parametric equivalent.

A potential source of error in hypothesis-testing, which is often overlooked, arises from multiple testing—either applying several tests to the same data set, or applying the same test to several sets of data. When a decision is to be made on the basis of a single significant result, the overall level of significance is no longer the same as the level at which the individual tests are made. For instance, one situation which occasionally arises is a series of observations (for example, replicate phytoplankton samples from 21 different locations) which are to be compared pair-wise by means of t tests. With 21 data sets there will be 20 possible comparisons; if each is performed at a significance level of 0·05, at least one significant test is expected by chance alone, and in fact, the probability of obtaining no significant tests by chance alone is only 36 per cent. A similar situation arises when a matrix of correlation coefficients is scanned for significant correlations. Even performing both tests in a two-way analysis of variance changes the actual probability of finding a significant result by chance alone.

There are several ways to correct for multiple testing, depending upon the particular test being used. In the case of multiple t tests the most sensitive procedure is to use the analysis of variance followed by one of the several *a posteriori* comparisons between individual means (Sokal and Rohlf, 1969, p. 235–46; e.g. McAlice, 1970). Alternatively one can expand the binomial with $p = \alpha$, the significance level of individual tests, to obtain the expected number of 'significant' results, and then scan the complete data set to see if there are too many significant results. This procedure is useful where there are not too many significant results and the null hypothesis is not rejected (e.g. Venrick *et al.*, 1977). It is less useful when the number of significant results is excessive because it does not allow one to select which of the 'significant' results are meaningful. Another approach is to adjust the probability level of the individual tests to some new value of α so that the overall probability of a significant result of the entire study, α^*, remains at the desired level. The approximate formula for this is $\alpha = \alpha^*/(m - 1)$ where m is the number of data sets to be compared and $(m - 1)$ is the number of comparisons.

Derivation

If α is the probability that a single significant result will occur by chance alone, the probability that a single test will give a non-significant result is $(1 - \alpha)$ and $(1 - \alpha^*)$; the probability that there will be no significant results out of the $(m - 1)$ possible comparisons between m data sets is given by $(1 - \alpha)^{m-1}$. Expanding:

$$[1 + (-\alpha)]^{m-1} = 1 + (m - 1)(-\alpha) + \frac{(m - 1)(m - 2)(-\alpha)^2}{2!} + \ldots$$

When α is small, all but the first two terms can be eliminated, leaving

$$1 - \alpha^* = [1 + (-\alpha)^{m-1}] \simeq 1 - (m - 1)(\alpha)$$

$$\alpha^* \simeq (m - 1)(\alpha)$$

$$\alpha \simeq \alpha^*/(m - 1).$$

Predictive statistics

Once a relationship has been tested and accepted as significant it becomes a potential predictive tool. Thus, many correlations and regressions belong in this category as well as in the previous one. For instance, Smayda (1966) found significant correlation between plankton biomass and an expression for up-welling winds, leading to further investigation into the predictive potential of this relationship using linear regressions.

The classic tools of prediction are correlation and regression. Simple predictive relationships are a constant part of most laboratory procedures. The concentration of chlorophyll in an extract may be predicted from the pre-determined linear relationship between chlorophyll and fluorescence, or from solution of the multiple regression equations relating chlorophyll a, b, and c to the absorption observed at three wavelengths.

The use of correlations and regressions in the field for predicting the behaviour of plankton systems has not been extensive. According to Walsh (1971):

Linear regression analysis is appropriate for preliminary insight into a complex system, but is an inadequate description of biological phenomena. Linear relations . . . cannot be expected to fully describe or predict biological relationships which are basically non-linear and consist of thresholds, time lags, and saturation and inhibition effects. Polynomial fits . . . may lead to greater prediction in regression analysis, but the biological meaning of the higher order terms raise a serious problem in the interpretation of the results.

It is for these reasons that the field of biological prediction in the ocean has become the domain of simulation model builders who have largely abandoned the rigours and restrictions of the classical predictive tools in favour of mathematical expressions which more accurately reflect functional relationships.

The history of simulation modelling in biological oceanography has been summarized by Walsh (1972). To date the best-developed models in phytoplankton ecology are probably those pertaining to nutrient kinetics (Dugdale, 1968; Walsh and Dugdale, 1971) and those related to the kinetics of a phytoplankton patch (Kierstead and Slobodkin, 1953; Platt and Denman, 1975; Wroblewski et al., 1975; Wroblewski and O'Brien, 1976; Wroblewski, 1977). In a model, 'the flow of information is cyclic where a simulation model's initial inability to adequately describe experimental and field results is necessary feedback for ongoing work' (Walsh, 1972). At the present time, the value of most biological models is not their predictive ability, but their ability to direct future research. In the ocean, this is likely to be the case for some time.

8.3   Diversity

Ramón Margalef

It seems as if the phytoplankton could function just as well if the number of species were much lower than actually is the case. The multiplicity of species can thus be perceived as a nuisance by the student, but nevertheless it poses a problem worthy of study. Why are there so many species? In dealing with this question and with other more or less related problems it is convenient to have some quantitative expression for global properties of the communities. These treat the number and richness of species and the relative abundances of each of them, properties which are the base of the common qualifiers 'monotonous', 'mixed or polymixed', 'with dominance of one or few species', and so on, and to which reference will be made under the general term *diversity*.

Diversity is used as an abridged expression of how a set (for instance, the community) is distributed into subsets (for instance, the species). The possibilities of sensible applications of such a concept are many. It can be applied to chemical composition, with reference to how an amount of substance is apportioned among different chemical species. It is also relevant to the distribution of the total biomass of an ecosystem in a number of subsets characterized by different turnover rates. In what follows, the concept of diversity will be applied exclusively to express how a set of individuals is distributed into a number of species. Diversity, in this case, is related to dynamics of populations and to history.

The abundance of species can be estimated either by number of individuals, as is more usual, or as biomass, as is sometimes done in phytoplankton studies. Diversity can be computed in both ways, with diverging results, since the size of the individuals is different in different species. Of course, in any given case, if diversity is to have a meaning, it is in relation to the criterion adopted for the partition of the sets into subsets. For example, if the unit is the phytoplankton cell (not the colony), and the species are the subsets, diversity is potentially significant in relation to demography, or dynamics of the populations of cells. A more refined classification going down to the characterization of the individual, and making as many subsets as individuals, would miss the point, because individuals of the same species, even if distinguishable, belong to the same subset in so far as breeding, survival and passing genes to the next generation are concerned. That is, they belong to the same team in the game of life.

DIVERSITY INDICES

A list of species, with attached expression of the abundance, conveys an idea of diversity. How can one represent it by a single number? A suitable index of

251

diversity must be at its lowest if all the present individuals belong to the same species (i.e. if there is only one non-empty subset), and attain an upper limit if each one of the items in the sample belongs to a different species, or to a different subset. The simplest index is the number of species identified in a sample of a suitable number of cells (i.e. 100, 1,000 or 10,000). If species are ranked according to their abundance, the way in which their respective numbers increase or decrease is assumed to be related to diversity. Diversity is considered to be low if only one or a few species are dominant, followed by rapidly decreasing numbers of other species. Taking advantage of the fact that in a collection of replicate samples, number of species (S) is approximately proportional to the logarithm of the number of individuals (N), the ratio of proportionality $(S - 1)/\ln N$ is a convenient index of diversity. This is implicit in the writings of botanists (Gleason, 1922) and is used formally as such by Margalef (1957b). Other proposed indices, such as the *alpha* of Fisher, Corbet and Williams (Williams, 1964), are also based on the assumption of regularity in the distribution of individuals into species. It would result from the play of evolutionary forces and from the internal equilibria achieved in the persistence of any ecosystem.

Other indices of diversity are free from the assumption of a regular distribution of individuals into species. Simpson's index (Simpson, 1949) expresses the probability that two specimens taken at random belong to the same species. If N_i is the number of individuals of species i, and N is the total number of individuals in the sample or in the studied ecosystem, and s is the number of species, such probability B is

$$B = \sum_{i=1}^{s} \frac{N_i(N_i - 1)}{N(N - 1)}.$$

For a large number of species and individuals, and using $p_i = N_i/N$, the following approximation is adequate,

$$B = \sum_{i=1}^{s} p_i^2.$$

The numerical value of this expression is low for a high diversity, and its maximum value is 1, so it may be convenient to use $1 - B$ as a measure of diversity.

The mathematical theory of communication (also named theory of information) provides us with a potential index that is also relevant in other fields of science (physics, study of language) as a measure of the possibilities of choice. It can be expressed in two forms, their respective numerical values converging for large numbers of individuals. The formula adopted by Shannon and Weaver (1949) is:

$$H = - \sum_{i=1}^{s} p_i \log_2 p_i \qquad \text{where } p_i = N_i/N, \text{ as above, and } \sum_{i=1}^{s} p_i = 1.$$

252

The eventual limits of confidence for each one of the probabilities (p_i) may be combined for an estimate of the reliability of the measure H. The units of H are 'bits', with no dimensions attached. The alternative formula is derived from Boltzmann's expression of entropy, and has the form (Brillouin, 1962):

$$H = \frac{1}{N} \log_2 \frac{N!}{\prod\limits_{i=1}^{s} (N_i!)}.$$

The numerical value of this expression refers to a unique combination and has no attached limits of confidence. It depends on the N_i's, but these figures are estimated by counting, with all attendant sampling errors.

Mathematically minded ecologists prefer the Shannon–Weaver expression, which is more elegant, and more practical for the purposes of computation (Table 12). Brillouin's expression is more attractive to the biologist, because it presents diversity as something related to the number of connections that can be established in a system of many elements (relationships such as predator/prey, parasite/host, and so on, most of them only virtual). Moreover, changes of diversity resulting from dynamics of populations can be worked step by step in such a formula. For instance, the replacement of a specimen of species a by one of species b results in a new H:

$$H = \frac{1}{N} \log_2 \frac{N!}{(N_a - 1)! \, (N_b + 1)! \ldots N_s!}$$

$$= \frac{1}{N} \left(\log_2 \frac{N!}{N_a! \, N_b! \ldots N_s!} + \log_2 \frac{N_a}{N_b + 1} \right)$$

giving an idea of the resulting change in the diversity index. Taking out an individual of a rare species and substituting it by another of an already common species decreases diversity. (Incidentally, this is what man is doing throughout nature.) The magnitude of the change, according to the formula, depends on the total number of individuals and on their distribution into species.

The numerical value of diversity reflects both the total number of species and their relative numerical representation. Considering this, several authors have proposed to separate these two components of diversity (Margalef, 1957b; Lloyd and Ghelardi, 1964; Pielou, 1969). The contribution of the unequal representation of the different species to the measure of diversity is referred to as evenness or equitability. This may be expressed as the difference between the actual diversity and the hypothetical diversity if all species were equally frequent, or were distributed according to a given hypothesis.

TABLE 12. Values of $-p_i \log_2 p_i$ as a function of p_i. The first two significant figures of p_i are shown at left, the third one at top (table prepared by Marta Estrada)

	0.000	0.001	0.002	0.003	0.004	0.005	0.006	0.007	0.008	0.009
0.000	0.00000	0.00996	0.01793	0.02514	0.03186	0.03821	0.04428	0.05010	0.05572	0.06116
0.010	0.06643	0.07156	0.07656	0.08144	0.08621	0.09089	0.09545	0.09993	0.10432	0.10863
0.020	0.11287	0.11704	0.12113	0.12517	0.12913	0.13304	0.13689	0.14069	0.14443	0.14812
0.030	0.15176	0.15535	0.15890	0.16240	0.16586	0.16927	0.17265	0.17598	0.17927	0.18253
0.040	0.18575	0.18893	0.19208	0.19519	0.19827	0.20132	0.20434	0.20732	0.21027	0.21320
0.050	0.21609	0.21896	0.22179	0.22460	0.22738	0.23014	0.23287	0.23557	0.23825	0.24090
0.060	0.24353	0.24613	0.24871	0.25127	0.25381	0.25632	0.25881	0.26127	0.26372	0.26615
0.070	0.26855	0.27093	0.27330	0.27564	0.27796	0.28027	0.28255	0.28482	0.28706	0.28929
0.080	0.29150	0.29370	0.29587	0.29803	0.30017	0.30229	0.30439	0.30648	0.30855	0.31061
0.090	0.31265	0.31467	0.31668	0.31867	0.32065	0.32261	0.32455	0.32648	0.32840	0.33030
0.100	0.33219	0.33406	0.33592	0.33776	0.33959	0.34141	0.34321	0.34500	0.34677	0.34853
0.110	0.35028	0.35202	0.35374	0.35545	0.35714	0.35883	0.36050	0.36216	0.36381	0.36544
0.120	0.36706	0.36867	0.37027	0.37186	0.37343	0.37500	0.37655	0.37809	0.37962	0.38113
0.130	0.38264	0.38413	0.38562	0.38709	0.38855	0.39001	0.39145	0.39288	0.39430	0.39571
0.140	0.39711	0.39849	0.39987	0.40124	0.40260	0.40395	0.40529	0.40661	0.40793	0.40924
0.150	0.41054	0.41183	0.41311	0.41438	0.41564	0.41689	0.41813	0.41937	0.42059	0.42181
0.160	0.42301	0.42421	0.42540	0.42658	0.42775	0.42891	0.43006	0.43120	0.43234	0.43346
0.170	0.43458	0.43569	0.43679	0.43788	0.43897	0.44005	0.44111	0.44217	0.44322	0.44427
0.180	0.44530	0.44633	0.44735	0.44836	0.44936	0.45036	0.45135	0.45233	0.45330	0.45426
0.190	0.45522	0.45617	0.45711	0.45805	0.45897	0.45989	0.46081	0.46171	0.46261	0.46350
0.200	0.46438	0.46526	0.46612	0.46699	0.46784	0.46869	0.46953	0.47036	0.47119	0.47201
0.210	0.47282	0.47362	0.47442	0.47521	0.47600	0.47678	0.47755	0.47831	0.47907	0.47982
0.220	0.48057	0.48131	0.48204	0.48276	0.48348	0.48420	0.48490	0.48560	0.48629	0.48698
0.230	0.48766	0.48834	0.48901	0.48967	0.49032	0.49097	0.49162	0.49225	0.49289	0.49351
0.240	0.49413	0.49474	0.49535	0.49595	0.49655	0.49714	0.49772	0.49830	0.49887	0.49943
0.250	0.50000	0.50055	0.50110	0.50164	0.50218	0.50271	0.50324	0.50378	0.50427	0.50478
0.260	0.50528	0.50578	0.50627	0.50676	0.50724	0.50771	0.50819	0.50865	0.50911	0.50957
0.270	0.51002	0.51046	0.51090	0.51133	0.51176	0.51218	0.51260	0.51301	0.51342	0.51382
0.280	0.51422	0.51461	0.51499	0.51537	0.51575	0.51612	0.51649	0.51685	0.51720	0.51755

	0	1	2	3	4	5	6	7	8	9
0.300	0.52108	0.52138	0.52166	0.52195	0.52222	0.52250	0.52276	0.52303	0.52329	0.52354
0.310	0.52379	0.52403	0.52427	0.52451	0.52474	0.52497	0.52519	0.52541	0.52562	0.52583
0.320	0.52603	0.52623	0.52642	0.52661	0.52680	0.52698	0.52716	0.52733	0.52750	0.52766
0.330	0.52782	0.52797	0.52812	0.52827	0.52841	0.52855	0.52868	0.52881	0.52893	0.52905
0.340	0.52917	0.52928	0.52939	0.52949	0.52959	0.52968	0.52977	0.52986	0.52994	0.53002
0.350	0.53011	0.53017	0.53023	0.53029	0.53035	0.53040	0.53045	0.53050	0.53054	0.53058
0.360	0.53061	0.53064	0.53066	0.53068	0.53070	0.53072	0.53073	0.53073	0.53073	0.53073
0.370	0.53072	0.53071	0.53070	0.53068	0.53066	0.53063	0.53060	0.53057	0.53053	0.53049
0.380	0.53045	0.53040	0.53035	0.53029	0.53023	0.53017	0.53010	0.53003	0.52995	0.52987
0.390	0.52979	0.52971	0.52962	0.52952	0.52943	0.52932	0.52922	0.52911	0.52900	0.52889
0.400	0.52877	0.52864	0.52852	0.52839	0.52825	0.52812	0.52798	0.52783	0.52769	0.52753
0.410	0.52738	0.52722	0.52706	0.52689	0.52673	0.52655	0.52638	0.52620	0.52602	0.52583
0.420	0.52564	0.52545	0.52525	0.52505	0.52485	0.52464	0.52443	0.52422	0.52400	0.52378
0.430	0.52356	0.52333	0.52310	0.52287	0.52263	0.52239	0.52215	0.52190	0.52165	0.52140
0.440	0.52114	0.52088	0.52062	0.52035	0.52008	0.51981	0.51953	0.51925	0.51897	0.51869
0.450	0.51840	0.51810	0.51781	0.51751	0.51721	0.51690	0.51659	0.51628	0.51597	0.51565
0.460	0.51533	0.51501	0.51468	0.51435	0.51402	0.51368	0.51334	0.51300	0.51265	0.51230
0.470	0.51195	0.51160	0.51124	0.51088	0.51051	0.51015	0.50978	0.50940	0.50903	0.50865
0.480	0.50826	0.50788	0.50749	0.50710	0.50670	0.50631	0.50591	0.50550	0.50510	0.50469
0.490	0.50428	0.50386	0.50344	0.50302	0.50260	0.50217	0.50174	0.50131	0.50087	0.50044
0.500	0.50000	0.49955	0.49910	0.49865	0.49820	0.49775	0.49729	0.49683	0.49636	0.49589
0.510	0.49542	0.49495	0.49448	0.49400	0.49352	0.49303	0.49255	0.49206	0.49156	0.49107
0.520	0.49057	0.49007	0.48957	0.48906	0.48855	0.48804	0.48753	0.48701	0.48649	0.48597
0.530	0.48544	0.48491	0.48438	0.48385	0.48331	0.48277	0.48223	0.48169	0.48114	0.48059
0.540	0.48004	0.47948	0.47893	0.47836	0.47780	0.47724	0.47667	0.47610	0.47552	0.47495
0.550	0.47437	0.47379	0.47320	0.47262	0.47203	0.47143	0.47084	0.47024	0.46964	0.46904
0.560	0.46844	0.46783	0.46722	0.46661	0.46599	0.46537	0.46475	0.46413	0.46350	0.46288
0.570	0.46225	0.46161	0.46098	0.46034	0.45970	0.45906	0.45841	0.45777	0.45711	0.45646
0.580	0.45580	0.45514	0.45448	0.45382	0.45316	0.45249	0.45182	0.45114	0.45047	0.44979
0.590	0.44911	0.44843	0.44774	0.44706	0.44637	0.44567	0.44498	0.44428	0.44358	0.44288
0.600	0.44217	0.44147	0.44076	0.44005	0.43933	0.43862	0.43790	0.43718	0.43645	0.43573
0.610	0.43500	0.43427	0.43353	0.43280	0.43206	0.43132	0.43058	0.42983	0.42909	0.42834
0.620	0.42758	0.42683	0.42607	0.42531	0.42455	0.42379	0.42302	0.42226	0.42149	0.42071
0.630	0.41994	0.41916	0.41838	0.41760	0.41682	0.41603	0.41524	0.41445	0.41366	0.41286
0.640	0.41206	0.41126	0.41046	0.40966	0.40885	0.40804	0.40723	0.40642	0.40560	0.40478
0.650	0.40396	0.40314	0.40232	0.40149	0.40066	0.39983	0.39900	0.39816	0.39732	0.39648
0.660	0.39564	0.39480	0.39395	0.39310	0.39225	0.39140	0.39054	0.38969	0.38882	0.38796
0.670	0.38710	0.38623	0.38536	0.38449	0.38362	0.38275	0.38187	0.38099	0.38011	0.37923
0.680	0.37834	0.37746	0.37657	0.37567	0.37478	0.37388	0.37299	0.37209	0.37118	0.37028
0.690	0.36937	0.36847	0.36755	0.36664	0.36573	0.36481	0.36389	0.36297	0.36205	0.36112

Continued overleaf

TABLE 12 (contd)

	0.000	0.001	0.002	0.003	0.004	0.005	0.006	0.007	0.008	0.009
0.700	0.36020	0.35927	0.35834	0.35740	0.35647	0.35553	0.35459	0.35365	0.35271	0.35176
0.710	0.35081	0.34986	0.34891	0.34796	0.34700	0.34604	0.34508	0.34412	0.34316	0.34219
0.720	0.34123	0.34026	0.33928	0.33831	0.33733	0.33636	0.33538	0.33440	0.33341	0.33243
0.730	0.33144	0.33045	0.32946	0.32846	0.32747	0.32647	0.32547	0.32447	0.32347	0.32246
0.740	0.32145	0.32044	0.31943	0.31842	0.31740	0.31639	0.31537	0.31435	0.31332	0.31230
0.750	0.31127	0.31024	0.30921	0.30818	0.30715	0.30611	0.30507	0.30403	0.30299	0.30195
0.760	0.30090	0.29985	0.29880	0.29775	0.29670	0.29564	0.29459	0.29353	0.29247	0.29140
0.770	0.29034	0.28927	0.28820	0.28713	0.28606	0.28499	0.28391	0.28283	0.28175	0.28067
0.780	0.27959	0.27850	0.27742	0.27633	0.27524	0.27414	0.27305	0.27195	0.27086	0.26976
0.790	0.26865	0.26755	0.26645	0.26534	0.26423	0.26312	0.26201	0.26089	0.25978	0.25866
0.800	0.25754	0.25642	0.25529	0.25417	0.25304	0.25191	0.25078	0.24965	0.24851	0.24738
0.810	0.24624	0.24510	0.24396	0.24282	0.24167	0.24052	0.23938	0.23823	0.23707	0.23592
0.820	0.23476	0.23361	0.23245	0.23129	0.23012	0.22896	0.22779	0.22663	0.22546	0.22429
0.830	0.22311	0.22194	0.22076	0.21958	0.21840	0.21722	0.21604	0.21485	0.21367	0.21248
0.840	0.21129	0.21010	0.20890	0.20771	0.20651	0.20531	0.20411	0.20291	0.20170	0.20050
0.850	0.19929	0.19808	0.19687	0.19566	0.19444	0.19323	0.19201	0.19079	0.18957	0.18835
0.860	0.18712	0.18590	0.18467	0.18344	0.18221	0.18098	0.17974	0.17851	0.17727	0.17603
0.870	0.17479	0.17355	0.17230	0.17106	0.16981	0.16856	0.16731	0.16606	0.16480	0.16355
0.880	0.16229	0.16103	0.15977	0.15851	0.15724	0.15598	0.15471	0.15344	0.15217	0.15090
0.890	0.14962	0.14835	0.14707	0.14579	0.14451	0.14323	0.14195	0.14066	0.13938	0.13809
0.900	0.13680	0.13551	0.13421	0.13292	0.13162	0.13032	0.12902	0.12772	0.12642	0.12512
0.910	0.12381	0.12250	0.12119	0.11988	0.11857	0.11726	0.11594	0.11463	0.11331	0.11199
0.920	0.11067	0.10934	0.10802	0.10669	0.10536	0.10403	0.10270	0.10137	0.10004	0.09870
0.930	0.09736	0.09602	0.09468	0.09334	0.09200	0.09065	0.08931	0.08796	0.08661	0.08526
0.940	0.08391	0.08255	0.08120	0.07984	0.07848	0.07712	0.07576	0.07439	0.07303	0.07166
0.950	0.07030	0.06893	0.06756	0.06618	0.06481	0.06343	0.06206	0.06068	0.05930	0.05792
0.960	0.05653	0.05515	0.05376	0.05237	0.05099	0.04960	0.04820	0.04681	0.04541	0.04402
0.970	0.04262	0.04122	0.03982	0.03842	0.03701	0.03561	0.03420	0.03279	0.03138	0.02997
0.980	0.02856	0.02714	0.02573	0.02431	0.02289	0.02147	0.02005	0.01863	0.01720	0.01578
0.990	0.01435	0.01292	0.01149	0.01006	0.00863	0.00719	0.00575	0.00432	0.00288	0.00144

APPLICATIONS AND RESULTS

Diversity is less interesting as a statistical concept than as an expression of a dynamic situation, resulting from processes such as colonization and extinction, and from the interplay of the different rates that characterize the demography of the coexisting populations. Diversity decreases when individuals of rare species are substituted by individuals of species that were already more common, or when one or a few species multiply with rapidity. Such events may happen when ecosystems are subjected to changing conditions or to stress, for instance after an episode of fertilization. Upwelling areas, as well as polluted areas, usually show a relatively low diversity, at least in their centres. This is explained only partly by the fact that a rather limited number of species can thrive in the conditions prevailing in such places. The rest of the explanation must take into account the intense flow of energy, expressed in high rates of multiplication, mortality and dispersal. As conditions remain constant for a while, multiplicity of niches and accumulation of species due to an evolutionary or successional past are reflected in a high diversity.

The distribution of the measures of diversity, using the proposed indices, is bell-shaped and has an upper limit, independently of how large the total number of species is. Interaction between species in a community tends to restrict the range of variation of diversity. Only in the showcases of a museum, or in slides with selected diatoms, could very high values of diversity be computed. In natural communities, numerical values of the Shannon–Weaver index of diversity rarely exceed 5 bits per individual.

In practice, diversity is computed from lists of the species abundances. The computation of the index after Shannon and Weaver is made easy by a table giving the function $-p_i \log_2 p_i$ for the values of p_i between 0 and 1 (Table 12). It suffices to add the values of the function for the relative abundances (N_i/N) of each one of the species. Usually, the diversity is computed over subsets of the whole ecosystem, more or less arbitrarily extracted. Such subsets are defined by taxonomic affinity (phytoplankton, diatoms, bacteria, copepods, etc.) and/or by methods of sampling (net plankton, Utermöhl plankton). If reference or extrapolation to the diversity of the whole community is made, it is based on the assumption, rarely proved, that diversity calculated for a sample or group of selected organisms reflects the total diversity. But in marine phytoplankton, to give an example of the risks of such assumptions, diversity of the populations of diatoms is poorly correlated with diversity of total phytoplankton. This is because the maximal development and diversification of diatoms depends on events (mixing or appropriate levels of turbulence) characteristic of periods when diversity of other groups is low (Margalef, 1961b).

Many measures of diversity in phytoplankton communities in different places and times have been published (Margalef, 1956a, 1958, 1966, 1969a, 1974; Patten, 1962; Hulburt, 1963, 1964; Patten et al., 1963; Zaika and Andryuschenko, 1969; Platt and Subba Rao, 1970; Travers, 1971; Fedorov,

K

1973; Legendre, 1973; Moss, 1973; Heip, 1974—a list by no means complete). Any attempt to extract some meaning from such data must be subjected to a critical appraisal based upon the considerations presented above. Most papers use the Shannon–Weaver index. This index will be used in the following comments. Diversity of phytoplankton, in bits per cell, is usually between 1 and 2·5 in coastal waters, being especially low in estuarine, polluted or upwelling areas. Values from 3·5 to 4·5 are most frequently measured in the oceanic plankton, but strong proliferations of coccolithophorids or other organisms may be the cause of locally low values of diversity. Although diversity is usually low in the centres of upwelling areas, horizontal mixing leads to the frequent observation of rather high diversities around the centres of strong vertical flow. In most of the oceanic areas of low productivity, a great number of species can be recognized, with rather low and more uniform population densities, resulting in diversities close to 5. But, even in such oligotrophic areas as the Sargasso Sea, temporal or local predominance of one or a few species results in frequent reports of lower diversities.

CONSIDERATION OF SPACE

Several difficulties associated with the uncritical use of measures of diversity become apparent when the occupation of space by the communities is considered. A 'point diversity' can be calculated on the composition of a 'small' sample, disregarding the spatial distribution of individuals in the sampled space. If the distribution of the probability of occurrence of the different species remained everywhere the same, diversity would be uniform, except for very small samples. In the experience of the author, identifying and counting at least 300 individuals in each one of the samples extracted from a well-mixed mass of water and plankton provide reliable values of diversity.

But in nature there are local differences in the probability of occurrence of the different species, as well as in the probabilities of their coexistence. Usually the numerical values of diversity increase when the sample is enlarged, and these values, as well as their increase with sample size, are different from place to place. A number of different values of diversity can be computed over enlarged samples, or from overlapping samples, in the same way as information in statistical terms can be measured over different segments of a written text, overlapping or not overlapping.

To stretch the concept of diversity to that of spectrum of diversity is unavoidable. This implies that the meaning of any measure of diversity should be considered in relation to the size of the sample, and that, at a further stage, change in the value of diversity should be related to the way and direction in which the sample of the community under study is being enlarged. It is a difficult task to compare diversities. Communities of very different organization, as will be commented later, can produce the same diversities in samples of the

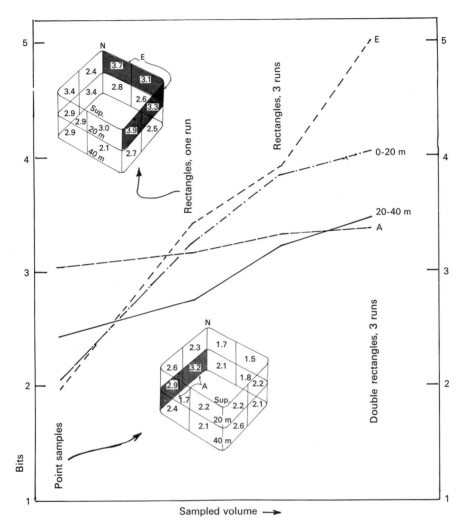

Figure 50
Distribution of the diversity of phytoplankton in the Mediterranean Sea, 39°49′ N, 0°8′E,
31 July 1968. The 139 samples of 100 ml each were taken on the sides of a square prism, 2 miles
wide and 40 m deep and analysed under the inverted microscope after sedimentation; the sampled
space has been divided into sixteen rectangles. Diversity spectra (bits per cell, Shannon–Weaver's
index) are given in the spaces A and E (see text) and for the layers 0 to 20 and 20 to 40 m depth.
The region of the spectra where diversity in turbulent water A is higher than diversity in
stratified water E, at left, is referred to as 'Hutchinson's paradox'. Lower inset: average diversity
per point sample. Upper inset: diversity computed for the whole of each of the sixteen rectangles,
assuming that each sample was representative of the space around it. (Redrawn and slightly
modified from Margalef, 1969a).

K*   259 same size. Any consideration of diversity in relation to properties of the organization of the ecosystem, such as the way stability is achieved, has to be based on the study of spectra of diversity and not only of point diversities. Although the way samples are collected and treated usually breaks the natural continuity in the spatial distribution of populations, there are seldom more substantial reasons for breaking down a continuous spectrum into arbitrary sections. However, this has been done often, and writers speak of point diversity or *alpha* diversity, referring to small samples, and *beta* diversity or pattern diversity, computed over collections of samples from a heterogeneous area (Pielou, 1969).

Figure 50 refers to a concrete distribution that may prove typical in marine plankton. A considerably heterogeneous distribution was observed over 2 miles. The water in region A was fertile, turbulent and rich in diatoms, as a result of internal waves breaking over the sloping bottom; region E was occupied by stratified water, populated by many dinoflagellates (Margalef, 1969a). The two diversity spectra, corresponding to the regions A and E, are different. Analogous differences, although less important, are observed between the diversity spectra obtained respectively in the top layer (0 to 20 m) and in the deeper layer (20 to 40 m). In turbulent water, populated by rapidly growing plankton, the average diversity of the studied samples (100 ml of water) is not much lower than the diversity computed for a much larger region. As all spectra must start at zero, diversity increases rapidly with volume at the lower end of the spectrum. In stratified water, swimming organisms with a lower rate of increase are common. Here, diversity computed for the whole region is much higher than the average diversity of isolated samples, and the spectrum of diversity is uniformly steep. Turbulence mixes developing populations over small scales of the spatial range, contributing to the 'paradox of plankton', as named and discussed by Hutchinson (1961), that is, to the apparently excessive diversity of planktonic populations. Notice that the lines of different spectra eventually cross, implying that samples of the same size can produce the same diversities in communities of completely different organization.

8.4 Associations

Louis Legendre
and Pierre Legendre

There are many questions that arise from the very concept of phytoplankton associations. Three of these will be discussed briefly as an introduction to the subject: (a) the definition of an association of species; (b) the identification of phytoplankton associations; and (c) the ecological significance of associations.

Definitions of ecological associations are not always operational. The concept of association generally applies to any group of species, occurring together in some significant way, not necessarily implying any definite positive interaction among constituent species in relation to their environment. In other words, an association is simply a group of species (or other taxa), recognized in accordance with a given set of rules.

Many procedures have been developed to identify associations of species. Strategies available are extremely varied and have evolved according to general trends in science. Initially dependent on empirical approaches, associative methodology is presently taking full advantage of computer facilities. Nevertheless, all procedures, whatever their crudeness or their sophistication, have two purposes: first, to identify those species which co-occur, and second, to reduce the chances that such co-occurrence is fortuitous. In order to do so, phytoplankton must be sampled in accordance with some experimental pattern—geographical, vertical or temporal—and groups of species which are 'stable' along the given axes are recognized as associations. The criterion of association stability is the recurrence of a group of species, following the sampled axes.

Ecological significance of associations is a subject still open. For Fager and McGowan (1963), associations 'are composed of species that have similar reactions to properties of the environment'. According to Legendre (1973), the characteristic of phytoplankton associations is their internal stability along the given axes, which means that species are responding in a related way to environmental changes. Such a co-ordinated response implies that variations of associated species are primarily dependent on the same environmental control. The concept of 'associations' is therefore a simplification of the responses of individual species to the environment. The recurrence of associations is an indication that such an abstraction corresponds to natural properties of the interaction among species and with their environment. Such interaction might possibly be of evolutionary significance.

According to Fager (1963) 'recurrent organized systems of organisms' are characterized by structural similarity in terms of species presence and abundances. Following that concept, a simple and operational definition is proposed:

a phytoplankton association is a recurrent group of co-occurrent species (or other taxonomic categories).

In the following pages, rules and procedures established to recognize co-occurrence of species and recurrence of groups will be summarily stated, referring to the literature for detailed descriptions. In conclusion, practical guidelines for the choice of a proper method will be set forth.

ASSOCIATIVE METHODOLOGY

The process of grouping usually involves two main steps, explicitly stated or not: first, one establishes the degree of similarity (or distance) between the objects to be grouped, through some coefficient or other evaluation method which seems appropriate to the data at hand. Then one proceeds to find the groups by applying a clustering rule on the association matrix, and by deciding whether or not a given pair of objects obeys the clustering rule at the stated clustering level. (For example: are the two objects similar enough to be included in the same cluster?) These steps will become obvious as the discussion proceeds.

Biogeographical clustering

If one represents the phytoplankton data collected at various sampling localities in the usual form of a rectangular array of numbers (often referred to as a matrix), the various species observed or counted may be made to correspond to the horizontal rows while the columns represent the samples (Fig. 51). Two analytical approaches are then available. One may be interested in the relationships between samples in terms of their species composition. This is referred to as a Q study and it usually involves the calculation of a coefficient of association (similarity or distance) between all pairs of samples. Alternatively, one may be interested in the interrelationships between species in terms of their co-occurrence or correlations of their abundance fluctuations. This type of analysis of the data matrix is termed R analysis and usually involves calculation of co-efficients of association between all pairs of species. Most of the clustering procedures described in the following sections are appropriate to the study of either Q or R association matrices.

The biogeographic approach to association identification such as described in Chapter **8.1.1** is clearly a Q study of the affinity between sampling localities, often carried out on a large scale. When groups of localities are established, the species inhabiting them may be described as biogeographic communities. This was the approach of earlier workers, such as Gran (1902) whose 'plankton elements' supplanted Cleve's (1897, 1903) 'plankton types'. Modern versions may be found in Smayda (1958), Wimpenny (1966b) and Hasle (1969). A similar approach on a small scale is presented by Grant and Kerr (1970), where the groups identified after two Q-mode partitions of samples were characterized by discriminating species.

Presence–absence data

Numerical phytoplankton data can be divided into (a) presence–absence or binary data, and (b) quantitative multistate data, often called quantitative data. This follows the use established by Sneath and Sokal (1973).

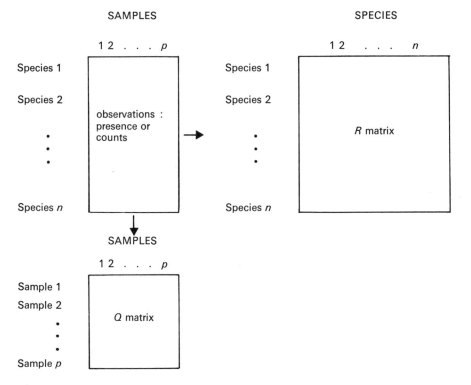

Figure 51
The basic data matrix gives rise to a *Q* matrix when similarities are calculated between samples, or to an *R* matrix when some measure of association is calculated between species.

Quantitative data (counts, transformed or not), when available, are not always suitable for association analysis. First, 'the phytoplankton does not afford us such evident and clear examples of associations as... in the zooplankton' (Wimpenny 1966*b*). Second, the species counts may not reflect the proportions present in the water, owing to poor sampling, preservation, identification or counting. Last, the behaviour of phytoplankton patches may overshadow the quantitative structure. For these reasons, most of the authors feel more at ease with presence–absence type of data, and the problem of modelling a basic definition of an association through the use of such binary data has been solved in various ways in the literature.

263

Binary similarity measures. Comparing species for their presence or absence in various samples first involves counting the number of joint occurrences, of double absences, and of singletons. The results are best summarized in a 2 × 2 frequency table such as:

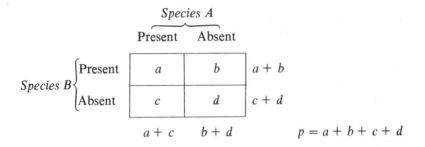

where a is the number of samples where both species occur, c and b are the number of samples where species A and B respectively occur without the other, d is the number of double absences, and p is the total number of samples. The coefficients used in association studies consistently exclude d, the term of negative matches, since the significance of an absence is uncertain: the environmental conditions may be adverse, or the species may not have been sampled as a consequence of patchiness, microdistribution or rarity. The next step is to measure, on this basis, the degree of association between species. Finding the groups of co-occurring species comes last, and this step will be discussed in the next section.

Measures of association stand for both similarity (S) and distance (D) coefficients, which are often the one-complement of each other $(D = 1 - S$ when $0 \leqslant S \leqslant 1)$. Association coefficients used for phytoplankton ecology may be simple or probabilistic. Probability of species matches is associated to probabilistic coefficients, while simple coefficients give a measure of association without any statement about its statistical significance. A probabilistic coefficient is preferred when an 'objective' limit of association extension is sought, or when a probabilistic clustering model is used. However, simple coefficients are more often used for their computational simplicity, and a lower-bound value is usually set for the delimitation of associations.

The best-known of the simple coefficients has been described by Jaccard in 1908. It compares the number of joint occurrences a to the total number of samples where either one of the species was found:

$$S = \frac{a}{a + b + c}.$$

This coefficient has been used in marine phytoplankton studies by Thorrington-Smith, 1971 (in the distance form $D = 1 - S$ proposed by Holloway and Jardine, 1968), and by Reyssac and Roux (1972).

Many other similarity coefficients are available in the biological literature, several of which have been reviewed by Sneath and Sokal (1973). However, most of these have not been used in marine phytoplankton association studies.

A more sophisticated coefficient has been proposed by Fager and McGowan (1963) in connection with zooplankton studies, and has been used by Venrick (1971) on marine phytoplankton data. It replaces the earlier probabilistic coefficient of species affinity proposed by Fager in 1957. It consists of the geometric mean of the proportion of joint occurrences, with a correction for sample size:

$$S = \frac{a}{\sqrt{(a + b)(a + c)}} - \frac{1}{2\sqrt{a + c}} \qquad (c \geqslant b).$$

This coefficient is preferable due to its mathematical completeness. A closely related form has been used by Margalef (1966).

A probabilistic criterion of plankton associations has been used with success by Krylov (1968), based on χ^2. As a criterion of similarity for two species, the probability is calculated that the number of joint occurrences would be as large as observed, under the null hypothesis of random and independent distribution of the two species over the samples. The similarity measure used is the probability associated with χ^2 with 1 degree of freedom, and χ^2 may be calculated as:

$$\chi^2 = \frac{p[|ad - bc| - (p/2)]^2}{(a + b)(c + d)(a + c)(b + d)}.$$

If p is smaller than 20 or if a, b, c or d is smaller than 5, then the similarity must be evaluated through Fisher's exact probability formula instead of χ^2. This formula is described in most textbooks of statistics, and also in Finney (1948). But in any case, the similarity is set at zero when the expected frequency of joint occurrences $(a + b)(a + c)/p$ is larger than or equal to the observed frequency a, since only positive associations are of interest. This measure of affinity of two species, although computationally more elaborate, should be used whenever an objective threshold value is sought, for instance $\mathbf{p} \leqslant 0.05$, as is the case with some of the association-forming clustering methods hereafter.

Clustering procedures. The method of grouping species most in use is linkage clustering. It is easy to understand, it can easily handle large numbers of species, and it is available in many computer packages. The basic idea in linkage clustering is as follows. On the basis of the matrix of similarities (or distances) between all pairs of species, a list of the pairs formed along a scale of decreasing similarity values can be written. The first pairs in this list are those which will cluster first, and following the list, as the similarity criterion is relaxed, more pairs are formed and the groups grow, up to the point where all species are included in the same group.

In this process of agglomeration, different strategies may be followed, i.e.

265

in single-linkage clustering, a new species becomes a member of a group at similarity level *C* if it has a similarity of at least *C* with at least one species which is already in the group. In complete linkage, the species would have to have a similarity of at least *C* with all the other species of the group in order to be included. Any amount of intermediate linkage can also be chosen as grouping strategy. A formal presentation of single-linkage clustering, based on graph theory, can be found in Legendre and Rogers (1972). The possible variations on the theme of clustering strategies have been well reviewed in Sneath and Sokal (1973).

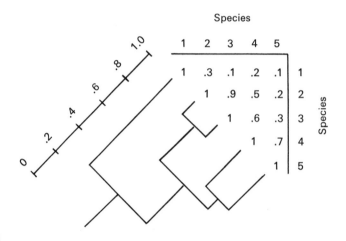

Figure 52
Fictitious example of single-linkage clustering on the similarity matrix to form a hierarchical grouping of the species. The highest values of similarity for each species, next to the 1-diagonal of the similarity matrix, form the decreasing similarity scale at the top of the dendrogram.

It may be noted that the philosophy behind single-linkage clustering is the same as that of dendrograms (Fig. 52), used on phytoplankton data, for instance, by Reyssac and Roux (1972), and the same also as in the classical phytosociological method of re-ordering the rows and columns of the similarity matrix to bring together species with close affinity. Examples of this approach are discussed later in this chapter. Thus single-linkage clustering is the preferable method, especially with large numbers of species, since it is the easiest one to translate in computer algorithms. Single-linkage clustering of a Jaccard similarity matrix has been used with success, for instance by Thorrington-Smith (1971) in the determination of phytoplankton associations in the West Indian Ocean. The result of the clustering activity was expressed in a dendrogram, in this case.

A method of agglomerating species around 'typical' species, designed in the first ages of numerical taxonomy by Rogers and Tanimoto (1960), has also been used with success on marine phytoplankton data by Margalef (1966), and

266

following him, by Reyssac and Roux (1972). But similar results could be obtained more easily by single-linkage clustering.

Another interesting approach is that of Fager (1957), which is actually a form of complete-linkage clustering associated with complementary ecological rules of thumb. This procedure has been made operational in a computer programme (Fager and McGowan, 1963), but any agglomerative clustering programme could be used instead, refining the groups of associated species by hand, following the four criteria which are presented in order of decreasing importance:

1. One is looking for completely linked clusters of species at a pre-stated similarity level. With a probabilistic index (for instance, Krylov, 1968), one can look for completely linked clusters at $S = 1 - p \geqslant 0.95$, for example. With a non-probabilistic index, a preset value may be used: Fager and McGowan used $S \geqslant 0.5$ as a clustering bound.

2. In this algorithm, one does not want the final clusters to share species. So between two alternative partitions of the species, one should choose to form groups containing the largest number of species. Krylov has added an extra limitation: no group should bear less than three species.

3. Between alternate partitions into groups with the same number of species, one should maximize the number of independent groups formed.

4(a). Between two or more groups with about the same number of species and with species in common, the group found in the greatest number of samples is selected.

4(b). Krylov (1968) prefers to replace requirement 4(a) with the following: among alternative species, the one with the lowest total of positive associations with all the others, is to be included in the group.

On this basis, Fager (1957), Fager and McGowan (1963) and Krylov (1968) have found useful and repeatable groupings of plankton data. Venrick (1971) explained an interesting step, carried out by Fager's computer programme, a step which answers an important problem of biological associations. After recognizing independent but completely linked groups of species as associations by complete linkage, he related associate species to one or several groups by single-linkage analysis. These species do not have to be associated to all the members of an association; and they may be associates to several groups, which reflects the organization of natural communities well.

Quantitative methods

Most often, production of quantitative data is not directly oriented towards association studies, but is related to other fields of ecological research, such as diversity or population studies. Nevertheless, it is often felt that such valuable material as quantitative phytoplankton data might be subjected with profit to associative procedures.

Quantitative similarity measures. Many researchers have applied parametric procedures to the subject of associations, since those statistics are primarily designed to handle quantitative data (cell counts). The most natural way is to start with the correlation matrix, where the similarity measure is the parametric correlation coefficient (Pearson's *r*). According to Fager (1957), associations must group those species which are 'a nearly constant part of each other's biological environment' and yet, even if two species follow the same occurrence pattern but do not fluctuate in a similar fashion, 'a correlation coefficient will indicate no relationship even though they are a constant part of each other's biological environment'. This drawback of correlation coefficients stresses the fact that low correlation values do not have any meaning whatsoever. Indeed, low correlation does not mean that two species are either poorly or strongly related: it means absolutely nothing and that is a serious inconvenience for further clustering procedures. As phytoplankton data are highly heterogeneous (large number of noughts), many pairs of species appear strongly correlated only because they are frequently absent together and, according to Margalef and González Bernáldez (1969), 'one of the major criticisms... is that the value of the correlation coefficients is excessively increased among the less frequent species'. There are two procedures to temper the impact of heterogeneity on the correlation matrix: (a) elimination of highly heterogeneous species, which results in associating only the most frequent species; and (b) calculation of correlation coefficients between pairs of species, using only those samples where both species occur together (some computer packages are generating similar results through replacement of noughts by the average abundance of the species). Non-parametric correlation coefficients (Kendall's τ) may be calculated for ranked data or data grouped in classes of abundance.

Another approach proceeds from a different viewpoint. Legendre (1973) investigated phytoplankton associations by means of a probabilistic similarity index, developed by Goodall (1966); given a set of *n* species, observed at *t* stations, the similarity index for a pair of species is defined as the complement $(1 - \mathbf{p})$ of the probability $\mathbf{p}$ that two randomly choosen species will be at least as alike as the two species under consideration. An operational description of a version of Goodall's probabilistic index is given by Orlóci (1975), together with a BASIC computer programme, and the proper clustering procedure is described below.

Clustering procedures. Clustering procedures, already described for presence–absence data, are also appropriate for quantitative phytoplankton data. Nevertheless, some of these procedures, as well as a special probabilistic agglomeration, are discussed hereafter in the quantitative context.

Associations may be extracted from a correlation matrix through various procedures. The simplest one is to select the order in which the species appear in the matrix, by trial and error, in such a way that as many as possible of the high positive correlations appear as close as possible to the principal diagonal of the matrix. Such a procedure was followed by Colebrook and Robinson (1964) and by Ibanez (1972), in both cases to study plankton associations from the North Sea. An example is given in Table 13 where two biologically meaningful associations are produced, since *Hyalochaete* and *Phaeoceros* are two subgenera of *Chaetoceros*, and all species of *Ceratium* are grouped together.

TABLE 13. Matrix of the correlation coefficients between the annual fluctuations of abundances of phytoplankton species in the North Sea (area D1). Correlations are given to one place of decimals and multiplied by 10; negative correlations are indicated by a bar. (Adapted from Colebrook and Robinson, 1964, table I and pl. XXV.)

2	3	4	5	6	7	8	9	10	11	12	13	14	
0	5	3	2̄	1̄	4̄	2	1	2̄	3̄	5̄	1̄	2	1 *Thalassiosira* spp.
	3	5	4	1̄	5	2	2	0	5	4	6	3	2 *Chaetoceros* sg. *Hyalochaete*
		7	4	1	2̄	1̄	1̄	2̄	2̄	2̄	1	1	3 *Chaetoceros* sg. *Phaeoceros*
			4	4̄	3	1	1̄	3	1̄	2̄	2	1	4 *Rhizosolenia imbricata* var. *shrubsolei*
				4	0	3	0	2̄	0	1	0	1	5 *R. alata* var. *alata*
					1	1̄	3	1	0	1	1	2	6 *Thalassionema nitzschioides*
						7	7	4	5	3	6	2	7 *Rhizosolenia styliformis*
							9	8	7	6	7	4	8 *Ceratium furca*
								8	7	6	7	5	9 *C. macroceros*
									8	8	6	6	10 *C. tripos*
										9	8	8	11 *C. horridum*
											8	9	12 *C. longipes*
												7	13 *C. fusus*
													14 *Biddulphia sinensis*

Ordering species in the matrix may be done by hand, but an iterative procedure has been known since Beum and Brundage (1950). Other approaches are producing results equivalent to maximizing the diagonal of the correlation matrix: Margalef and González Bernáldez (1969) for Caribbean waters, as well as Reyssac and Roux (1972) for the Ivory Coast, produced phytoplankton associations through centroid agglomeration with reasonable success. There are modern techniques producing equivalent results, capable of handling large correlation matrices on computers: those are the linkage clustering methods, described above.

Clustering the correlation matrix is a simple and attractive approach to the problem of phytoplankton associations and this procedure has been demonstrated to be useful in marine ecological problems. Obviously, the fact that the less frequent species, as a consequence of their heterogeneity, might exhibit high correlations is to be kept in mind and carefully checked. However, similarly occurring species in non-linear relation may show weak correlation,

269

the associations thus produced being minimal ones. Clustering the correlation matrix is therefore recommended as a practical procedure.

Factorial methods were successfully used in marine phytoplankton ecology, as demonstrated for instance by Ibanez (1972), and a procedure to calculate the significance of factorial axes was described by Frontier (1976).

Figure 53
Position of phytoplankton species from the Caribbean on 3rd and 4th principal components, showing hyperspheroidicity and lack of preferential directions. (Adapted from Margalef and González Bernáldez, 1969.)

However, this chapter is devoted exclusively to those methods which efficiently identify marine phytoplankton associations. Principal component analysis may appear to be a sophisticated method of extracting associations from a correlation matrix, but Margalef and González Bernáldez (1969), Ibanez (1972) and Reyssac and Roux (1972) have attempted to produce phytoplankton associations through principal components, without any real efficiency. In most cases, the hyperspheroidicity (tendency for all species to remain grouped together in all dimensions) and the lack of preferential directions prevented a clear-cut definition of groups of associated species (Fig. 53). Correspondence analysis (Reyssac and Roux, 1972) suffered from similar weaknesses with regard to phytoplankton associations. Factorial methods, therefore, do not seem to be an efficient association-forming procedure with marine phyto-plankton. A similar conclusion was reached by Whittaker and Gauch (1973) for terrestrial plants.

The last procedure takes advantage of Goodall's probabilistic similarity index. As the probabilistic index is the complement of a probability, it is possible to define a statistical approach to the formation of clusters of species. Clusters are formed by probabilistic agglomeration around statistically significant pairs of species. The largest cluster recognized is removed from the original set of species and residual species are then clustered with recomputed probabilities. The final results are associations of species significantly similar among themselves, and species not significantly associated to others are not included in any cluster. The whole procedure is described in Legendre (1973), where a flow chart is given.

CONCLUSION

In concluding this exposé on associative procedures, a few points should be stressed, these remarks being a warning against excessive enthusiasm about numerical methods.

It is sometimes felt that valuable information might be extracted from poor data, through sophisticated mathematical procedures. Such an approach is obviously erroneous, since the information content of data cannot be increased above its original level by any means. Biased data should be treated, when required, only with the crudest available methods, and resulting associations should be regarded with suspicion, remembering the wise saying 'garbage in, garbage out!'

At the same time, unbiased data may well be of different precision levels: various procedures have been designed to handle presence–absence data, numerical data, etc. Fairly often, plankton data are too precise, considering the information extracted, and a great deal of energy is consumed to achieve useless precision. For instance, Ibanez (1974) demonstrated that 3-state zooplankton data (absent or rare, present, abundant) are as informative as numerical data in principal component analysis. More information about natural populations might therefore be obtained, in some instances, by estimating more samples in a cruder fashion.

The ecological situation under investigation is of importance in choosing a procedure to identify associations. Phytoplankton populations, in areas of the oceans dominated by strong environmental gradients, should not be approached in the same way as populations from more complex environments. Strong environmental gradients generally lead to better-defined population patterns, which do not require highly sophisticated mathematical analyses. In other words, there is no point in 'over-identifying' associations: a straightforward procedure, when applicable, is better than a very involved one. There are even

271

dangers in the indiscriminate use of complex methods to investigate simple problems. Frontier (1974) discusses the use of principal component analysis in plankton ecology: when unjustified, such a procedure may lead to seemingly impressive results which only reflect, in actual cases, preconceptions of the investigator transposed in the sampling design. However, sophisticated methods are extremely useful when dealing with ecological situations where no population pattern clearly emerges, and exciting achievements may be attained when they are used with care and mastery.

Finally, using methods of a level higher than necessary is a gross loss of energy ... and often of money. Table 14 is designed to provide a practical guide to the choice of a proper association procedure.

TABLE 14. Characterization of associative methods

Method		Discriminative characteristics
BINARY DATA (presence–absence)	*1. Similarity measures*	
	Jaccard	For single-linkage clustering
	Fager and McGowan	For complete-linkage clustering
	Krylov (χ^2)	For complete-linkage clustering with probabilistic threshold
	2. Clustering procedures	
	Single linkage	Associations at unspecified level
	Complete linkage (Fager)	Clusters of highly associated species with clouds of related species
QUANTITATIVE DATA (cell counts)	*1. Similarity measures*	
	Correlation coefficients:	
	Kendall (τ)	Semi-quantitative data
	Pearson (r)	Quantitative data; well-structured environment
	Probabilistic index (Goodall, Orlóci)	Any type of data; for probabilistic agglomerative clustering
	2. Clustering procedures	
	Clustering the correlation matrix	Associations at unspecified level (single-linkage clustering)
	Probabilistic agglomeration	Densely associated species, clustered according to a probability criterion
	Factorial methods	Of questionable value in identifying phytoplankton associations

8.5   From phytoplankters to biomass

Theodore J. Smayda

Cell size differs greatly between species of phytoplankton; cell diameter varies 1,000fold from about 1 µm (*Micromonas pusilla*) to >1,000 µm (*Ethmodiscus rex*), and cell volume a millionfold. The mean cell volume of diatom species collected from the Gulf of Panama (Smayda, 1965) ranged from about 50 µm³ (*Nitzschia delicatissima*) to 12 × 10⁶ µm³ (*Rhizosolenia acuminata*). The routine estimation of standing stock as cell numbers is therefore an imprecise and sometimes even inadequate measure of phytoplankton abundance. Determinations of proximate constituents such as levels of chlorophyll or detrital plus viable carbon are likewise inadequate measurements of total phytoplankton abundance for use in studies on certain aspects of phytoplankton-grazing relationships, on nutrient dynamics, including turnover, on species or size-class contributions to primary production (Paasche, 1960; Smayda, 1965, 1966), etc. Some representative publications demonstrating the coupled effect of cell abundance and cell size on grazing (Frost, 1973) and the effect of cell size on various metabolic and cellular processes (Paasche, 1973*b*; Banse, 1976; Taguchi, 1976; Durbin, 1977) also illustrate the inadequacy of cell numbers alone as a measure of phytoplankton standing stock, and the need to consider cell size and biomass.

CELL VOLUME

Cell abundance data can be transformed into more useful and appropriate form by taking into account differences in cell size and estimating phytoplankton biomass indirectly from phytoplankton counts after determination of cell volume (see Paasche, 1960; Smayda, 1965). The calculation of species cell volumes to estimate phytoplankton community biomass is more complicated than commonly believed. The use of cell-volume data calculated for the same species from other localities, or from different seasons, or from one sample for use over an extended time series is strongly discouraged. Considerable intra-specific variability in cell size (particularly for diatoms) and hence, cell volume, characterizes the phytoplankton. Potentially significant errors may accompany use of such extrapolated cell-volume data. Therefore, estimations of species cell volume should be based on the investigator's own material.

The cell shape of the various phytoplankton species also varies greatly. Selection of the equivalent geometric shapes for the species in question requires careful attention. It determines the formulae used to calculate cell volume and, therefore, the linear measurements needed per cell. There is a great temptation

273

to assign simple geometric shapes, such as the sphere, cylinder, elliptical cylinder, cone, etc., to ease the tedious and laborious process of calculating cell volume. However, many species of phytoplankton are characterized by complicated cell shapes, particularly dinoflagellates and certain large, asymmetrical diatoms. Kovala and Larrance (1966), for example, used 18 different shapes. Eleven linear measurements per cell were required for certain species of *Peridinium*! It is sometimes necessary to assign two or more different geometric shapes for different portions of a cell to calculate its cell volume more accurately. It is recommended that investigators first make representative sketches of the species to be measured before assignment of the equivalent geometric form. This will help to assess the appropriate shape and be useful subsequently as a guide to ensure that the appropriate linear measurements are being made.

The appropriate dimensions of at least 25 randomly selected cells are then measured, and the volume of *each* of the measured cells calculated, from which the mean cell volume is derived. The mean cell volume should *not* be calculated from the average linear dimensions of the individual cells.

From the cell count data for each species transformed in this way, the cell volume (biomass) for the total population is calculated:

$$\sum_{i=1}^{m} V = (V_1)(N_1) + (V_2)(N_2) + \ldots (V_m)(N_m) \tag{1}$$

where V is the total cell volume, m equals the number of species found, i represents species i ($i = 1, 2, \ldots, m$) in terms of its mean cell volume (V_i), and N_i represents the number of individuals of species i. The mean volume estimate is then transformed into a 'live weight' estimate after assuming a density of 1·1 (Holmes *et al.*, 1969). Examples of such calculations and related procedures will be found in Smayda (1965).

A major shortcoming of this indirect estimation of biomass, aside from being labour-intensive and tedious, is that transformation of the numerical abundance of each species into biomass units may exaggerate the overall importance of the larger-celled species to community dynamics relative to the smaller species, the converse (see Paasche, 1960) of that characterizing use of cell-count data. For example, the large diatom *Rhizosolenia acuminata* accounted for only 80 cells per litre of the total diatom population of 145,280 cells per litre in a sample collected in the Gulf of Panama (Smayda, 1965), but comprised approximately 60 per cent of the total diatom biomass of 1·6 mg per litre. Clearly, neither estimate of standing stock adequately represented the importance of this taxon in that community.

CELL AREA AND PLASMA VOLUME

There is an additional complication. Lohmann (1908) long ago communicated that total cell volume is frequently an inadequate estimate of biomass because

274

it includes the cell vacuole, which contains relatively non-nutritious cell sap. This would be especially characteristic of larger species whose larger vacuoles presumably contain a greater volume of cell sap. Large cells or species, accordingly, might be termed 'watery' when compared to smaller cells or species. Lohmann suggested that the cell *plasma volume* is a more correct estimate of biomass. Plasma volume, essentially a measure of total cell volume minus vacuole volume, has been used subsequently to Lohmann by Hickel (1967) and Zeitzschel (1970) to estimate phytoplankton abundance.

Smayda (1965) and Strathmann (1967) discuss the complicated calculations and assumptions associated with plasma-volume estimates. The most subjective aspect is to establish the ratio of vacuole volume to total cell volume in species of differing cell size. The percentage of total cell volume represented by the vacuole probably increases with cell size, but a relationship between this ratio and total cell volume (or cell diameter) has not been established, should a positive relationship exist. The plasma volume (PV) primarily represents the sum of (a) the thin cytoplasmic layer (cyt. lay.) adhering to the inner cell wall (the chloroplasts are embedded here), (b) the thin protoplasmic bridge (prot.br.) strands traversing the vacuole and supporting the nucleus, and (c) the volume of the nutritious component of the vacuolar sap which Lohmann arbitrarily assumed equalled 10 per cent of the cytoplasmic layer volume. Plasma volume is calculated following the steps outlined for calculation of total cell volume; in fact, it does not entail much additional work.

Lohmann calculated plasma volume as the product of the cell surface area and parietal cytoplasmic-layer thickness (1–$2\,\mu m$), to which he added the volume of the protoplasmic bridge (see Smayda, 1965, fig. 2 or Smayda, 1970, fig. 5) assuming it to be a sphere, and then arbitrarily added 10 per cent of their sum to represent the nutritious portion of the vacuolar cell sap:

$$PV = 1 \cdot 10\,[(\text{surface area, } \mu m^2)(1\text{–}2\,\mu m) + (\text{prot.br., } \mu m^3)] \qquad (2)$$

Smayda (1965) used a simplified procedure to calculate plasma volume:

$$PV = (\text{surface area, } \mu m^2)(\text{cyt.lay., } 1\text{–}2\,\mu m) + (0 \cdot 10)(\text{total cell volume, } \mu m^3)$$
$$(3)$$

A key measurement in such calculations is the thickness of the parietal cytoplasmic layer. It is neither possible nor practical to detect and measure this layer during the routine processing of natural populations. Shrinkage of this layer also complicates its measurement in preserved cells. Lohmann determined that the thickness of this layer in diatoms (using *Coscinodiscus* and *Rhizosolenia*) was 1 to $2\,\mu m$ depending on cell size, the range used by subsequent workers. Selection of the cytoplasmic layer thickness is very subjective (Smayda, 1965; Strathmann, 1967), even in applying Lohmann's estimates. Clearly, plasma volume cannot exceed total cell volume, and both volumes are equal in cells without vacuoles. If plasma volume is to be estimated for a cell whose surface-to-volume ratio ($\mu m^2 : \mu m^3$) is near unity, then a cytoplasmic layer thickness not

275

exceeding 1 µm must be assumed. Otherwise, the calculated plasma volume would exceed the total cell volume. Smayda (1965) has recommended that the cytoplasmic layer thickness used in such calculations be based on the cell-area to cell-volume ratio (Table 15). It must be stressed that these proposed guidelines are not based on direct observations and measurements.

TABLE 15. Suggested cytoplasmic-layer thickness, in µm, to be used in the calculation of plasma volume (PV) of diatoms of different cell-surface-area to cell-volume ratios ($\mu m^2 : \mu m^3$) (from Smayda, 1965)

Surface: Volume ($\mu m^2 : \mu m^3$)	>0·90	0·51 to 0·89	0·35 to 0·50	<0·35
Cytoplasmic-layer thickness (µm)	PV = total volume	1·0	1·5	2·0

A conspicuous feature of Equations 2 and 3 is that the vacuole volume is *not* estimated directly in the calculation of plasma volume. A more appropriate estimate of plasma volume, theoretically, would be to determine the total cell volume (CV) and cell-vacuole volume (VV), subtract the latter from the total cell volume, and include 10 per cent of the vacuole volume as an approximation of the nutritious portion of the cell sap:

$$PV = [(CV - VV) + (0·10)(VV)] \tag{4}$$

The thickness of the cytoplasmic layer would have to be estimated in order to establish the linear dimensions of the vacuolar space. For example, for a cylindrical cell of width a and length l, and with an assumed cytoplasmic layer thickness of 2 µm, the estimated vacuolar width (a') and length (l') is ($a - 2$ µm) and ($l - 2$ µm), respectively.

A comparison of estimated plasma volumes of some arbitrarily chosen cell sizes and shapes when calculated by Equations 3 and 4 is presented in Table 16. In every instance, volumes calculated by Equation 3 exceed by about 30 per cent those calculated by Equation 4 for all cells with a $\mu m^2 : \mu m^3$ ratio of ⩾0·30 when the cytoplasmic-layer thickness suggested (Table 15) as appropriate to the cell-area to cell-volume ratio is used. For larger cells (0·20 to 0·30), volumes derived from Equation 3 are only 10 to 20 per cent greater, and the difference is less than 10 per cent for even more voluminous cells. Calculation by Equation 10 (see below) of the plasma-volume carbon content for cells having a $\mu m^2 : \mu m^3$ ratio of ⩾0·30 (Table 16) reveals that the carbon expected for the Equation 3 plasma-volume estimates exceeds by about 20 to 25 per cent those based on Equation 4. Interestingly, the differences between plasma volumes estimated using Equations 3 and 4 are usually only ±15 per cent when the cytoplasmic-layer thickness used in Equation 3 is selected from Table 15, and a constant thickness of 2 µm is used in Equation 4.

There are no quantitative grounds on which to justify preferential use of either Equation 3 or 4. Both are easier to execute than Lohmann's procedure

276

TABLE 16. Comparison of plasma volume (PV) estimations for different-sized phytoplankton cells and variable cytoplasmic-layer (CL) thickness when calculated by Equations 3 and 4

Shape	$\mu m^2 : \mu m^3$	CL (μm)	Plasma volume (μm^3)		$\dfrac{\text{Equation 3}}{\text{Equation 4}} \times 100$ (%)
			Equation 3	Equation 4	
Cube	0·60	*1·0	700	539	130
10 μm		1·5	1,000	679	147
(1,000 μm^3)		2·0	PV > CV	806	—
Sphere	0·61	*1·0	366	282	130
10 μm		1·5	523	361	145
(523 μm^3)		·2·0	680	421	162
Sphere	0·06	1·0	$84·0 \times 10^3$	$80·0 \times 10^3$	105
100 μm		1·5	$99·5 \times 10^3$	$93·5 \times 10^3$	106
(524 × 10^3 μm^3)		*2·0	$115·0 \times 10^3$	$106·7 \times 10^3$	108
Sphere	0·006	1·0	$55·5 \times 10^6$	$55·2 \times 10^6$	101
1,000 μm		1·5	$57·0 \times 10^6$	$56·6 \times 10^6$	101
(524 × 10^6 μm^3)		*2·0	$58·6 \times 10^6$	$58·0 \times 10^6$	101
Cylinder	1·11	1·0	PV > CV	200	—
4 × 20 μm		1·5	PV > CV	239	—
(251 μm^3)		2·0	PV > CV	PV > CV	—
Cylinder	0·77	*1·0	491	362	136
6 × 20 μm		1·5	PV > CV	457	—
(565 μm^3)		2·0	PV > CV	510	—
Cylinder	0·50	1·0	942	757	124
10 × 20 μm.		*1·5	1,335	982	136
(1,571 μm^3)		2·0	PV > CV	1,164	—
Cylinder	0·30	1·0	2,512	2,161	116
20 × 20 μm		1·5	3,454	2,810	123
(6,283 μm^3)		*2·0	4,396	3,328	132
Cylinder	0·22	*2·0	16,965	14,044	121
20 × 100 μm					
(31·4 × 10^3 μm^3)					
Cylinder	0·20	1·0	12,566	10,463	120
40 × 20 μm		1·5	10,053	8,669	116
(25·1 × 10^3 μm^3)		*2·0	7,539	6,747	112

* = Value of cytoplasmic layer based on Table 15.
CV = Total cell volume.

(Equation 2). In any case, it must be emphasized that the investigator will probably not establish the true values for either the thickness of the cytoplasmic layer or the proportion of organic carbon (nutritious material) in the vacuole in the material being processed. Moreover, subsequent calculations of carbon, such as by Equation 10, are also based strictly on empirical relationships, rather than representing the true relationship between cellular carbon and cell-plasma volume. Given these uncertainties and the general problem of carbon content, including its cycling and measurement in natural phytoplankton populations, the 20 to 25 per cent difference in carbon levels estimated from

277

L

plasma volumes calculated by Equations 3 and 4 for most phytoplankton cell sizes does not appear to be a serious problem. The investigator may, therefore, adopt whichever procedure (Equation 3 or 4) is more suitable to his material. When working with monospecific algal cultures, direct determination of carbon content is recommended. If impracticable, then a serious effort to determine the mean thickness of the cytoplasmic layer by appropriate microscopy is recommended before proceeding to plasma volume and derived carbon estimations.

While the estimation of plasma volume is based on somewhat arbitrary assumptions, it would appear to provide a more accurate estimate of 'metabolically active' tissue than does total cell volume, at least for the larger species (Smayda, 1965). This conclusion is strengthened by observations (Paasche, 1960; Smayda, 1965) which suggest that the cell surface area is a more adequate measure of phytoplankton standing stock than cell volume when relating rates of primary production to standing stock abundance. Cell surface area provides a fairly good estimate of the plasma volume, inasmuch as the latter is primarily a function of cell surface area. In fact, these two estimates of standing stock size would appear to be interchangeable, either estimate providing a measure of the assimilative and photosynthetic surfaces as well as metabolically active tissue.

CELL CARBON AND NITROGEN

While the above transformations provide approximations of biomass, the more desirable and useful estimate is often the carbon or nitrogen content of this biomass. Because it is presently not possible to measure *in situ* levels of living algal carbon and nitrogen, nor to fractionate natural communities into constituent species for such analyses, the relationship between cell carbon and cell volume has been sought (Mullin *et al.*, 1966; Strathmann, 1967) in order to estimate phytoplankton carbon in seawater from preserved phytoplankton samples. Mullin and co-workers established that the carbon content per μm^3 of phytoplankton cell volume varies inversely with cell volume, whereas no dependence of the ratio of cell carbon to cell surface area (μm^2) on cell size could be demonstrated. From these relationships, they developed an equation to estimate organic carbon in phytoplankton from cell volume. Strathmann has extended these observations and provided the following equations allowing the estimation of cell carbon from cell volume:

$$\log_{10} C = 0.758 \,(\log_{10} V) - 0.422 \quad diatoms \tag{5}$$

$$\log_{10} C = 0.866 \,(\log_{10} V) - 0.460 \quad other\ phytoplankton \tag{6}$$

with V representing total cell volume (μm^3) and C the amount of carbon as picograms per cell. Since diatoms have a lower carbon content per unit cell volume, they are usually treated separately from other groups of phytoplankton.

278

Some minor modifications in constants have been reported since Strathmann's formulation. Eppley *et al.* (1970) and Eppley (in Anon., 1974) give:

$$\log_{10}C = 0.76\,(\log_{10}V) - 0.352 \quad diatoms \tag{7}$$

$$\log_{10}C = 0.94\,(\log_{10}V) - 0.60 \quad other\ phytoplankton \tag{8}$$

and Taguchi (1976):

$$\log_{10}C = 0.74\,(\log_{10}V) - 0.58 \quad diatoms. \tag{9}$$

Strathmann found that plasma volume provides a more precise estimate of cell carbon in diatoms than does cell volume. Based on calculations using 1 μm as the cytoplasmic-layer thickness, he obtained:

$$\log_{10}C = 0.892\,(\log_{10}PV) - 0.61 \tag{10}$$

This additional information suggests that calculation of plasma volume, with subsequent estimation of its carbon content, represents the best of the available methods for indirect estimations of phytoplankton biomass.

Suitable equations allowing for indirect estimations of other proximate constituents are presently lacking. Taguchi (1976) reported the ratio of carbon to nitrogen to be significantly dependent on cell size, with smaller cells being relatively nitrogen-rich, and related the ratio of cell C to cell N ($C:N$), by weight, to cell volume (V), μm³:

$$\log_{10}(C:N) = 0.72\log_{10}V + 1.31 \tag{11}$$

However, this equation yields values for the $C:N$ ratio considerably in excess of the approximately 6:1 ratio normally found for actively growing populations.

The foregoing equations should be accepted only as approximations, and are probably most applicable to cells in exponential growth phase. There is evidence that nutrient-limitation (Strathmann, 1967) and possibly temperature (see Eppley, 1972) influence cell carbon and nitrogen content. Banse (1977) presents a good appraisal of the general problems associated with measuring biomass of natural phytoplankton populations, and of the use of conversion factors to estimate algal carbon and nitrogen levels. Despite such shortcomings and the tremendous investment of time required to transform numerical data into biomass estimates of phytoplankton standing stock, judicious use of particle size counters, computer programmes, and well-conceived investigations partially facilitate such efforts where appropriate.

8.6   Data storage and retrieval

J. M. Colebrook

The extent of the facilities available for the preparation, storage, manipulation and processing of data depends on the sophistication and size of the computer system used. In addition, the nature and structure of phytoplankton data varies considerably depending on the objectives and methods used for surveys.

This diversity of resources and requirements makes it very difficult to provide a comprehensive presentation of the processes involved in the storage and retrieval of phytoplankton data. What follows, therefore, should be regarded as a rough guide to desirable practices rather than as recommendations. An example of practical application may be found in Colebrook (1975a).

There is no internationally agreed format for the exchange of phytoplankton data nor is there any agreed system of numerical labels for species or other taxonomic categories. Exchange procedures are, however, being considered for some of the more reproducible observations such as chlorophyll a concentrations and productivity by ^{14}C. Also, the recommended inventory systems for reporting data, ROSCOP II and ROMBI are relevant to phytoplankton data.

Before embarking on the development of an automatic data processing system it is important to consider whether this is the most efficient and economical method of getting the required results from a data set. A good criterion to use is the extent to which operations will be repeated. It is probably wise to use automatic data processing for a survey involving a series of cruises producing comparable data or in situations where a large data set will be searched repeatedly.

In this account it will be assumed that: (a) observations are derived from a survey producing data in a multivariate structure involving appreciable numbers of species or other taxonomic categories, and/or stations, and/or depths, and/or time intervals; (b) the computer system to be used is file-oriented or has, at least, facilities for reading, printing, storing and editing files of data and information; (c) a file is a named and defined set of data consisting of a set of records and each record is the image of an 80-column punched card; and (d) the data will be stored in an archive consisting of a set of files.

DATA RETRIEVAL

The objective of any system for storing phytoplankton data is to facilitate its retrieval and subsequent processing. The retrieval process normally involves the establishment of one or more arrays of data representing a specified subset

280

of the data in the archive. The format and structure of the stored data should, therefore, be determined by the requirements of the retrieval procedures rather than by the easiest route through data preparation and the establishment of an archive. A data file is, or preferably should be, established once only, while data will be extracted from it on numerous occasions.

For normal currently available computer systems, the basic retrieval processes, listed in order of ease and efficiency are:
1. Selection of one or more files.
2. Sequential selection of one or more blocks of records from a file.
3. Sequential selection of a set of non-consecutive records from a file.
4. Sequential selection of one or more elements from a record.
5. Non-sequential selection of blocks of records.
6. Non-sequential selection of a set of records.
7. Non-sequential selection of a set of elements.

Archives should, therefore, be designed so that the more commonly used retrievals involve the processes near the top of this list and processes 5, 6 and 7 should be avoided as far as possible. At the same time, it should be recognized that an archive may well be required to serve a retrieval process that was not visualized in the original design of the system.

At one extreme, identification parameters can be associated with each observation in a single record. For example, date, time, position, depth, species and count would be successive elements in a single record. Such a structure provides fully flexible, but not very efficient, retrieval (involving processes 1 and 4). At the other extreme, all identification parameters can be grouped at the head of the file, providing an inventory, and the observations grouped in the form of one or more arrays. In retrieval, the identification parameters are searched and a set of row and column numbers are produced corresponding to the required observations. No flexibility is lost using this scheme and, although some retrieval patterns may be less efficient than with fully labelled observations, suitable arrangement of the arrays to fit the more commonly used forms of retrieval offers considerable gains in efficiency by the use of processes 2 and 3.

In practice it may be convenient to adopt a compromise structure with, for example, a file containing all the data for a set of cruises, each being organized into inventory and observation sections.

The inventoried-file system is not necessarily the correct solution for every data-storage problem. Most phytoplankton data sets that are large enough to warrant computer storage and processing are usually still relatively small compared with many data-processing tasks. However, they tend to be more complex in structure, frequently involving data for taxonomic categories over and above the date, time, position and depth coordinates common to most field-survey data. Such data are ideally suited for storage in inventoried files. It is not unusual for a survey to contain data for a large number of species while individual cruises or samples may contain data for a relatively small subset. Much time can be saved in searching an inventory, as opposed to the whole data set, for the presence or absence of a particular species.

DATA FORMATS

This refers to the arrangement and number of digits required to represent the observations. Relevant computer manuals should be consulted as to the options available but, in general, numbers can be stored as either integers or real numbers (i.e. with a decimal point) and these can be either fixed-point or floating-point numbers. Fixed or free formats can be used. Considerable economy of storage space, and hence retrieval times, can be achieved by the selection of the most compact format compatible with the accuracy of the observations. Consider, for example, the particular case of counts of species. If subsampling is involved in the counting process, then the counts may be expressed in relation to the smallest fraction regularly used in the subsampling. This may avoid trailing zeros or probably insignificant last digits. The use of a coding system might be considered, for example, $\log(\text{count} + 1)$ expressed to two decimal places and multiplied by 100 provides a three-digit integer representation of counts with a range sufficient to cover most situations and with errors which are probably small compared with sampling and subsampling errors.

DATA PREPARATION

This normally involves the preparation of manuscripts of the data and manual key-punching of these into a machine-readable form (punched cards, paper tape or magnetic tape). The preparation of a single manuscript format for data that will satisfy the needs of both sample analysts and key-punch operators does present problems but is desirable, to avoid copying errors. At least, the first product of processing should be a listing of the data in a form which looks like the original manuscript in order to simplify checking for errors. The programme which produces this listing can also incorporate some editing procedures; for example, species labels can be checked against a catalogue and can also be checked for duplication.

There is no reason why a set of data involved in a single data-preparation task should represent the entire contents of an archive file nor need it be in the required structure or format. At this stage, convenience in data preparation, including the preparation of manuscripts, key-punching, editing and checking, should decide the extent and structure of the data.

Establishing an archive file can be a separate process requiring its own programme.

9

Comments on related fields

9.1 About bacterioplankton

John McNeill Sieburth

The procaryotic biomass of the plankton known as bacterioplankton is a large and active component of the plankton. The fact that this population of free cells is in the ultrananoplankton size range ($< 2\ \mu$m) means that for all practical purposes they do not present themselves to the phytoplanktologists and are usually ignored by them. It may therefore come as a surprise to non-bacteriologists who have not taken the data of Sorokin (1971*a*, 1971*b*) seriously, that this procaryotic biomass can equal some 30 per cent of the protist biomass in the photic zone and some 40 per cent of the total microbial biomass in the aphotic zone (Sieburth *et al.*, 1977; Lavoie and Sieburth, submitted for publication).

Thanks to Sheldon (1972) who demonstrated that the perforated poly-carbonate filters (Nuclepore) act as screens or sieves, and to Salonen (1974) who showed that bacteria can be efficiently separated from the phytoplankton by 2-μm Nuclepore filters, we can now obtain bacterioplankton in the ultra-nanoplankton size fraction just as nanoplankton and net plankton are obtained by appropriate techniques. Bacteriologists have been slow to realize that most of the bacteria suspended in the water exist as free cells (Wiebe and Pomeroy, 1972) and that they can be selectively obtained by Nuclepore filtration (Sieburth *et al.*, 1977). A preparation of selectively filtered bacterio-plankton is shown in Figure 54.

It may also come as a surprise to some phytoplanktologists and even some bacteriologists that the bacterioplankton are more than just little nondescript bags of enzymes. They have a variety of distinctive morphologies and life cycles and to some degree can be studied like the phytoplankton. The smallest forms are at the limit of resolution of the light microscope, having a width approaching that of the head of a bacteriophage particle. Bacteriophage particles are also relatively abundant and widespread in near-shore waters (Ahrens, 1971; Zachary, 1974) and even in natural populations of bacterioplankton, as shown in Figure 54, can be clearly and frequently seen attached to the outside of bacteria and inside the ghosts of host cells. Phage may not be the only factor, besides protozoan ingestion, controlling bacterioplankton populations. Comma-shaped bacteria of the proper size and with a single-sheathed polar flagellum are indicative of the parasitic and lytic genus *Bdellovibrio* which are also quite host-specific for other bacteria and are present in significant numbers (Taylor *et al.*, 1974; Marbach *et al.*, 1976). The presence of bacterial 'horseshoe' and 'ring' forms indicates that members of the genus *Microcyclus* (Raj, 1977) may be quite prevalent in the marine environment.

As water passes through the pores of a filter, streamlines are formed and the bacterial cells apparently line up longitudinally along these streamlines. Therefore the width of the cells and the diameter of the pores determine which cells will pass. Even in preparations passing through 2-μm porosity Nuclepores, the most striking feature is that cells 6 μm in length or longer are quite common. Hoppe (1976) showed that 80 per cent or more of the cells in the bacterioplankton were active, passed a 0·6 μm filter, and could not be cultured on an agar medium, while only a small fraction of the population of wider cells retained by a 0·6-μm filter yielded colony-forming units (CFUs) on an agar medium. Hoppe also found that CFUs rapidly decrease in population with distance from shore while the population of the active small cells remains relatively high. These larger cells in the bacterioplankton which give CFUs on solid media appear to be mainly epibacteria associated with all surfaces including suspended organic debris such as the dominant remnants of faecal pellets. The epibacteria are apparently present in the plankton in small but ever-present numbers as strays, transients and as a threshold inoculum.

These epibacteria appear to be adapted for motility, attachment to surfaces by fibrils (Marshall, 1976) and the utilization of solid substrates by the liberation of exoenzymes. However, the smaller cells free in the water appear to be the autochthonous planktobacteria, refractory to growth on agar media, which utilize the small, transient and regularly leaked dissolved organic carbon (DOC) in the photic zone which they rapidly reduce to threshold levels (Sieburth *et al.*, 1977). These cells may be thought of as the centric diatoms and phytoflagellates living free in the euphotic zone, while the larger epibacteria may be thought of as the pennate diatoms sliding over and attached to solid surfaces. The planktobacteria are apparently adapted for an existence on dissolved organic matter, while the epibacteria are apparently adapted for an existence on particulate organic matter.

Selectively filtered bacterioplankton dominated by indigenous planktobacteria, free of eucaryotic micro-organisms and with a very small load of organic debris, can now be used to supply the information sought by the big-picture planktologists, such as Strickland (1971). What is the biomass and what are the growth rates of oceanic bacterioplankton? The total bacterial biomass of both the planktobacteria free in the water and the epibacteria on organic debris can be estimated with the limulus amoebocyte lysate (LAL) test for the quantification of the lipopolysaccharides occurring in the cell walls of gram-negative bacteria (Watson *et al.*, 1977). However, the biomass of the bacteria free in the

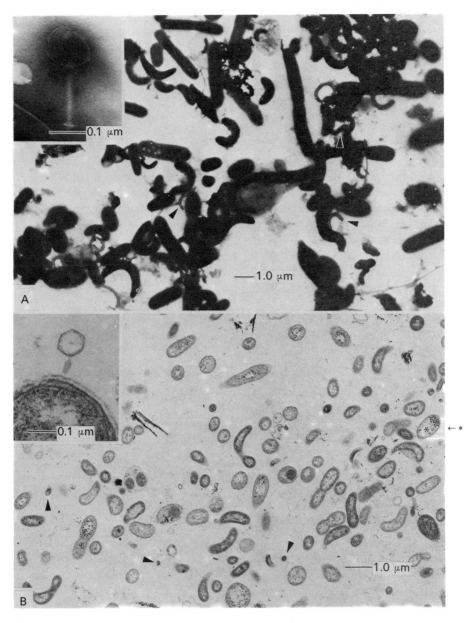

Figure 54
Electron micrographs of natural populations of bacterioplankton (from Narragansett Bay, Rhode Island) that pass a 2-μm porosity Nuclepore membrane (arrows denote minibacteria <0·2 μm in width): (A) negatively stained preparation; (B) thin section of same sample; note phage-infected cell (*). Inserts in A and B show frequently seen bacteriophage in same samples. (Electron micrographs by Paul W. Johnson.)

L*

water can also be determined by assaying these selectively filtered populations for adenosine triphosphate (ATP) (Sieburth et al., 1977; Lavoie and Sieburth, submitted). We can also assay these preparations for adenylate energy charge (Wiebe and Bancroft, 1975) as well as organic carbon and get ATP: carbon ratios directly. Both values serve as indices of the physiological state of the natural populations of bacterioplankton. Dialysis or diffusion culture (Schultz and Gerhardt, 1969), which has been used to study pure cultures of phytoplankton (Jensen et al., 1972; Prakash et al., 1973), is also suitable for caging natural populations of bacterioplankton (Lavoie, 1975). Growth patterns which are controlled by diel changes in dissolved organic matter can now be observed with natural populations in situ (Sieburth et al., 1977). We also have sensitive and reliable procedures for analysing monomeric carbohydrate (Johnson and Sieburth, 1977), total carbohydrates (Burney and Sieburth, 1977) and estimating polysaccharides by difference. These determinations coupled with DOC values have shown that some 50 per cent of the labile DOC is due to carbohydrate (Johnson et al., submitted; Burney et al., submitted). By following the diel changes in these materials we can now study the role of the indigenous planktobacteria in the in situ utilization of DOC and carbohydrates which transiently accumulate during phytoplankton production and protozooplankton grazing on this biomass.

Bacterioplankton and protozooplankton are partners with the phytoplankton as all three trophic modes of the microbial plankton are interdependent. Despite the small physical dimensions of the planktobacteria or bacterioplankton, in oceanic waters they are numerous (10^5 to 10^6 cells ml^{-1}), have a biomass equal to 30 to 40 per cent of the protist biomass, and have marked and rapid activities (periods of growth with four-hour doubling times) (Sieburth et al., 1977 and submitted). Bacterioplankton cannot be ignored in either environmental or pure-culture studies. The smaller cells are not visible with light microscopy and they will not grow on artificial media, yet their presence will have a marked effect on the responses of phytoplankton in multicultures.

The microbiological community now has an arsenal of useful tools and procedures that provide alternatives to cultural procedures, including ^{14}C bottle experiments. These include selective filtration with perforated polycarbonate filters, reverse-flow filtration, epifluorescent microscopy, autoradiography, electron microscopy, ATP, LAL, carbon, plant pigment, and carbohydrate analyses, as well as diffusion culture, with which to study the microbial plankton. These techniques can be used in concert to obtain, characterize, quantify and study the influence and dependence of bacterioplankton, phytoplankton and protozooplankton upon one another. There is much too much for one man, one team or even one shipload of investigators to handle competently. Meaningful hour-by-hour studies on the microbiology of plankton through diel periodicity in different water masses will require interdisciplinary and even multi-ship studies that will push our technologies and enterprise to the limit. The time is now. Only interdisciplinary teams working together can solve the apparent discrepancy between the high values of bacterial biomass and productivity and the low values of offshore primary productivity (Sieburth, 1977).

ADDENDUM

During the continuing study of the bacterioplankton in the picoplankton size range (0·2 to 2·0 μm; Sieburth *et al.*, in press) by Paul W. Johnson and myself, we have observed occasional chroococcoid cyanobacteria in oceanic plankton as well as in nearshore waters since 1975. These forms comprise some 0·3 per cent of the bacteria of Narragansett Bay and at 100 metres in the Sargasso Sea (November 1977) some 20 per cent of the cells had the ultrastructure of chroococcoid cyanobacteria $\geqslant 0·4$ μm in diameter. Samples of water from 50 and 100 metres in the Caribbean Sea (April 1978) contained $0·5 \times 1$ μm orange fluorescing cells, probably cyanobacteria as numerous as 13,500 ml^{-1} and yielded enrichment culture of similar cells. Stanley Watson and colleagues at Woods Hole who have also been using epifluorescence to enumerate oceanic bacterioplankton have noted the occurrence of small orange autofluorescing cells at concentrations to 10^4 ml^{-1} in upwelling waters off the coast of Africa and Peru. During May 1978 the connection between these counts and the isolation of phycoerythrin containing chroococcoid cyanobacteria isolated by Robert Guillard on other cruises was also made by these investigators (Stanley Watson, pers. comm.).

Bacterioplankton therefore contains in addition to the heterotrophic planktobacteria and epibacteria a less numerous but sometimes larger biomass of 'oxyphotobacteria' (Gibbons and Murray, 1978: *Int. J. Syst. Bacteriol.*, vol. 28, no. 1, p. 1–6) whose primary productivity may explain the difference in order of magnitude when compared with our studies on the release and uptake of dissolved organic matter.

9.2　About microzooplankton

John R. Beers[1]

Pelagic ecosystems in the marine environment contain a broad spectrum of living organisms. For practical purposes the organisms comprising the pelagic biosphere are often separated into somewhat artificial groups for study based on characteristics such as size differences, taxonomic position and nutritional type. The major division of organisms in the planktonic biosphere into phytoplankton and zooplankton recognizes differences in nutrition. The phytoplankton ideally are the photosynthetic (i.e. autotrophic) planktonic forms while the zooplankton embrace the plankton having particle ingestion (i.e. phagotrophy, a form of heterotrophy) as their principal mode of feeding.

The microzooplankton is a size-class designation and is that component of the zooplankton encompassing the smaller forms of phagotrophs. Most commonly, in recent years, microplankton, of which the microzooplankton would be the animal component, has been defined as forms in the size range of 20 to 200 µm (Dussart, 1965, 1966; see discussion in Parsons and Takahashi, 1973b). As a working definition, the microzooplankters have been considered as the animal forms which are of the size and configuration that allows them to *pass through* approximately 200 µm mesh netting (e.g. Anon., 1968b; Beers and Stewart, 1969, 1971). Hence, this use of the term microzooplankton includes, in addition to animal plankters in the approximate 20 to 200 µm range, those in the range 2 to 20 µm which are often categorized as 'nano' forms in schemes of size classification.

Taxonomically, the microzooplankton embodies most of the pelagic protozoans and a variety of metazoans such as naupliar stages of many copepods. Some meroplanktonic animals (e.g. larval stages of some molluscs) are of the size to occur in microzooplankton samples. Also, benthic taxa of the proper size and forms which inhabit surfaces of various suspended materials may be found at times in microzooplankton samples, especially from coastal neritic waters.

Protozoan microzooplankton may include ciliates, sarcodinans and flagellates. Planktonic ciliate populations are frequently dominated by species of the suborders Oligotrichina and Tintinnina of the Oligotrichida (see proposed classification scheme of Corliss, 1974), although free-living members of other groups may be important at times, especially in coastal waters (e.g. *Tiarina*, prorodontine gymnostome; *Didinium*, haptorid gymnostome). Sarcodinans include naked forms (e.g. amoebae), often most abundant in association with regenerating organic materials, as well as forms with tests (e.g. Foraminifera)

---

1. This work was supported by the Biological Oceanography Program of the Oceanography Section, National Science Foundation (United States), Grant OCE 76-21655.

and those with skeletons (i.e. Actinopods: Radiolaria and Acantharia). The hard parts of various sarcodinans may include long spines and other radiating structures which effectively increase the organism's dimensions and, hence, may affect its classification as a microzooplankter. The sediment populations of hard parts of calcareous foraminifera and siliceous radiolaria have been extensively studied by palaeontologists because of their use as palaeoecological indicators. At the same time, juvenile stages with only rudimentary development of the hard parts that may be found abundantly in the populations of living microzooplankton have been largely ignored because of their generally poor representation in the sediment samples studied. The most prominent of the phagotrophic planktonic flagellates are dinoflagellates, but abundant populations of small sessile forms (e.g. choanoflagellates) can be found in association with other plankton organisms. A similar association is seen in ciliates such as *Vorticella* (sessiline peritrich) whose mature forms may live attached to the bodies of various plankters but which have motile free-swimming stages (i.e. the telotroch) by which the species can be dispersed. In addition to the recognized zooflagellates, the importance of small non-pigmented flagellates that may be specifically phagotrophic in nutrition and not simply utilizing dissolved organic materials, although a subject of some speculation, has not been rigorously examined.

In plankton samples, whether collected by water-bottle, pump or net, there is a mixture of both phytoplankton and zooplankton, and there is no practical method of physically separating all the components of the two groups. There is a distinct overlap in the size spectra of the microzooplankton and phytoplankton with protozoans extending down to 10 μm 'equivalent sphere diameter' (e.g. *Uronema* sp.; Hamilton and Preslan, 1969) and lower. Furthermore, separation of phytoplankton and microzooplankton on the basis of major taxonomic group or nutritional type is also not feasible since in some groups (e.g. the Dinoflagellata) there are autotrophs and phagotrophs plus, in addition, other heterotrophic forms which derive nutrition from dissolved organic materials. Further, organisms may have different means of meeting nutritional requirements at various times in their life cycle. Also, forms belonging to principally animal-like major taxonomic groups may be mainly, or wholly, plant-like in nutrition, e.g. functional chloroplasts of the gymnostome ciliate *Mesodinium rubrum* (Taylor *et al.*, 1971; Hibberd, 1977) and other ciliates (Blackbourn *et al.*, 1973).

The study of microzooplankton in marine food-web dynamics has been receiving increasing attention as an understanding of their role is developed. Microzooplankters may serve as links between the populations of small phytoplankton (e.g. the so-called 'monads and flagellates') and bacteria, on the one hand, and the larger zooplankton and nekton on the other. As primary consumers of autotrophic and heterotrophic production, the microzooplankton occupy an important position with respect to the elevation and distribution of materials, including various types of pollutants, which may enter the base of the food chain or be synthesized through the activities of the producers.

FIELD STUDY OF MICROZOOPLANKTON

Considering the total size spectrum of the microplankton, the abundance (numbers and/or biomass) of microzooplankton in all marine environments is almost always less than that of phytoplankton, although animals may predominate in the larger size classes of the microplankton. While microzooplankters, especially sarcodinan protozoans, can be found at most depths within the water column, their abundance is greatest in the euphotic zone, with a generally rapid decrease below the compensation depth (e.g. Beers and Stewart, 1969; Beers et al., 1975). Within the euphotic zone, the abundance of microzooplankton tends to be positively correlated with that of the phytoplankton, although the ratio of the two may not be constant. While the general principles described for phytoplankton sampling (Chapter 2.1) apply to microzooplankton, a larger sample is usually required for study of the latter to achieve an equivalent level of confidence in counting because of their generally lower abundance relative to the phytoplankton.

Microzooplankton can be sampled using the same type of gear as for phytoplankton. Any sampling by water-bottle, however, would generally be restricted to the larger-size bottles in order to obtain sufficient numbers of the bigger microzooplankters for enumeration. An alternative is to take several bottle samples and pool the materials, but this is time-consuming and the samples would not come from the same precise depth. Large-volume water-bottles (i.e. volumes > 5 to 10 litres) are cumbersome to work with on board ship.

Sampling the microzooplankton with fine-mesh nets presents the same problems as for phytoplankton and, in general, should only be considered for qualitative purposes. Many microzooplankters have better-developed swimming capabilities than phytoplankton and, hence, their ability to avoid a slowly towed net is probably greater. In any case, the total population of microzooplankton cannot be sampled with nets since the smallest animals will pass through even the finest mesh netting.

The use of pumps for sampling the microzooplankton provides the potential for obtaining quantitative samples of the entire size spectrum of organisms. Furthermore, pump systems can allow for sampling any volume of material desired either at discrete depths or integrating over various horizontal, vertical or oblique intervals. The pumped water can be retained unconcentrated for study of the smaller, generally more abundant microzooplankters, or can be passed through various mesh-size cloths in order to concentrate the larger, relatively less abundant organisms in the bigger size classes of the microzooplankton. Pumping has been criticized as potentially damaging to organisms, especially large watery forms. However, with the appropriate choice of gear (see Chapter 3.2) the possibility of an objectionable level of damage to microzooplankton in the pumped water is negligible.

PRESERVATION AND STORAGE OF MICROZOOPLANKTON

As with the phytoplankton, the microzooplankton includes a wide spectrum of types with regard to their hardiness and keeping qualities. The process of fixation may be responsible for much of the distortion seen in preserved samples, but undoubtedly some damage is caused in the actual collection of samples, especially when the materials are concentrated on filter cloths. Many of the ciliate species which are without hard parts are also delicate and they become markedly distorted. Included in this category is the ubiquitous and often important group of oligotrichs. Problems with their fixation and preservation are similar to those met with in the study of naked flagellates.

A comprehensive consideration of zooplankton fixation and preservation, including specific recommendations for several microzooplankton groups, has recently been published (Steedman, 1976) as No. 4 (*Zooplankton Fixation and Preservation*) in this series of Unesco Monographs on Oceanographic Methodology. In general, the principles of fixation, etc., discussed are also pertinent to microzooplankton, no single fixative yielding the best results for all taxonomic groups. Special procedures have to be used for the preservation of some taxa (e.g. Foraminifera, Acantharia, see below). Formaldehyde is recommended as the best general reagent for fixation and preservation of mixed marine zooplankton samples for study of taxonomy and morphological characters. A 2 per cent formaldehyde concentration in the fixing/preserving fluid media is suitable if a 1:9 ratio of volume of plankton to fluid is not exceeded; when it is, a 4 per cent formaldehyde concentration is recommended (Steedman, 1976).

A pH of approximately that of natural seawater (i.e. pH 8·2) is required for long-term preservation of some mineralized structures (e.g. calcareous foraminifera tests). To maintain this, materials such as borax (sodium borate) must be added to the preserving solution. Materials such as hexamethylaminetetramine which can give pH levels in excess of 8·4 to 8·6 should be avoided because of their detrimental effect on the preservation of other microzooplankters. Their use, for example, can result in the solubilizing of protein, which tends to swell and disrupt forms such as naupliar copepods (Steedman, 1976), and the potential dissolution of siliceous skeletons of radiolarians (Beers, 1976) and even calcareous structures when samples are stored in warm places (Steedman, 1976). Since borax is not a true buffer, it must be replenished if the pH drops; pH checks should be made at frequent intervals immediately following fixation, decreasing with time as the pH of the samples stabilizes.

Other 'additives' to the samples may be needed if all groups of microzooplankton are to be maintained. For example, the addition of strontium chloride to augment the naturally occurring strontium level in seawater is generally necessary if the skeletons of acantharians are to be preserved (Beers and Stewart, 1970). With the addition of such materials there is an increasing risk of objectionable precipitates developing in the preserved samples. These can diminish, and even destroy, the value of the samples for the study of the smaller microplankton.

291

Specific methods for fixing particular protozoan groups, including some procedures for cytochemical study, are considered in Steedman (1976). Cytological fixatives are, however, not usually satisfactory for general preservation.

LABORATORY STUDY OF MICROZOOPLANKTON

Depending upon which part of the microzooplankton size range is under analysis, samples are studied using compound microscopes or lower-magnification dissecting microscopes. In the size range that has to be examined at relatively high magnification, the abundance of phytoplankton and detrital material is generally larger than that of the animals, making microzooplankton enumeration and identification a tedious and time-consuming process. When the size of a taxon allows for a choice between compound and dissecting microscopy, the latter may be preferable since with this method it is possible to manipulate specimens into the most desirable position for identification.

The Utermöhl inverted-microscope procedure (see Chapter **5.2.1**) is a particularly useful form of compound microscopy for studying microzooplankton. Materials for study can be settled from up to as much as 100 ml of sample (either unconcentrated seawater or pre-concentrated material). For quantitative study the amount of material settled should not exceed that which can be examined on a single plane. The specific density of naked microzooplankters is low and, hence, settling times needed for inverted-microscope study are similar to those proposed for formaldehyde-preserved phytoplankton (see Chapter **5.2.1**). In those infrequent cases when it is not necessary to examine the entire chamber bottom, a standard sequence of alternating rows (i.e. every 2nd, 3rd, 4th, etc.) should be used to average over any non-random distribution. Some microzooplankters, such as the copepods, are particularly prone to trapping gas bubbles under their carapace when the sample is randomized by agitation and, instead of sedimenting, rise to the surface of the settling chamber.

In addition to study using inverted or dissecting microscopy, the smaller microzooplankton such as the non-loricate ciliates can be enumerated with the compound microscope using small-volume counting chambers such as Sedgwick–Rafter, Palmer–Maloney, etc. An advantage with the standard compound microscope is that phase-contrast illumination at higher magnifications is often easier to achieve than on the inverted microscope where the condenser working distance may limit such observation.

The use of stains such as rose Bengal, which for inverted microscope study can be added during the settling, to separate organic from inorganic materials generally speeds up the counting process. 'Vital' stains such as neutral red (Dressel *et al.*, 1972; Crippen and Perrier, 1974) to which the material is subjected prior to fixation, can be helpful in recognizing those individuals that were alive at the time of sampling. However, their use with field samples of plankton is complicated by the fact that the organisms may have been killed or be in a physiologically declining state as the result of the collecting procedure or the density of material in the sample and, hence, may not accept the stain as when in good condition. Also, the incubation time needed for stains to be taken up would allow for possible deterioration of the sample materials, especially where there has been damage during the sampling.

For some considerations, such as with planktonic food-web studies, it may be desirable to make a separate record of organisms observed as hard parts only without the protoplasmic structures (e.g. empty tintinnid loricae, radiolarian skeletons, etc.), although it is difficult to determine whether they were intact when collected since with many such taxa the attachment of the soft body elements to the hard parts is tenuous and subject to possible mechanical damage during sampling and any subsequent concentrating procedure.

Microzooplankton samples that are to be studied with a dissecting microscope can be treated as other zooplankton materials. Appropriate aliquots for examination can be provided by splitting, using, for instance, the 'Folsom' splitter (McEwen et al., 1954) or 'Motoda' box (Motoda, 1959). Materials may also be concentrated by such procedures as the reverse-filtration technique of Dodson and Thomas (see Chapter **5.4.1**).

Special care in transferring microzooplankton material to the examining vessel is necessary if it is only a fraction of the total sample. Because of the relatively large size and density of some microzooplankton compared to phytoplankton, their random distribution following agitation (i.e. shaking) of a sample is, at best, only of short duration. Subsamples can be removed by pipetting, but the *mouth* and not just the bore of the pipette should be large relative to the size of organisms being studied. The length of the pipette should be kept fairly short. Commercially available instruments such as Stempel pipettes can be used. The materials should be randomized just prior to removing a subsample; agitation with the pipette itself in a random motion so as not to set up any particular current pattern may be helpful.

Because the abundance of microzooplankton is often low relative to the phytoplankton, it may be considerably more time-consuming to enumerate them and, hence, a desired level of precision sometimes cannot be achieved realistically. A decision on the number of organisms to count has a direct effect on the precision of results. As discussed in Chapter **7.1.2**, subsample counting is only the last step of several from the point of collection, all of which have their own associated levels of variability which can affect the accuracy of the results. Unless information is desired regarding specific taxa, counts can be made at more inclusive levels of identification for purposes such as food-web studies where, for example, all species of like size in a given major group may reasonably be considered to be similar with regard to many of their activities and roles.

Special techniques have been developed for the study of some microzooplankton groups. For example, in the study of siliceous radiolarians where identification is on the basis of the morphology of hard parts, all organic matter in the samples, including that in other organisms and detritus, can be removed

293

by ignition (Sachs *et al.*, 1964; Sachs, 1965; Smith, 1967) or ultraviolet irradiation (e.g. Holmes, 1967; Swift, 1967). The remaining material can be mounted on 'strewn' glass slides. The study of such strewn slides is facilitated by using in conjunction with its examination a reference slide such as the 'England Finder' (described in Riedel and Foreman, 1961) with which the location of organisms can be determined and recorded for easy future relocating. Another method of sorting sarcodinans, such as foraminifera and radiolarians, from other plankters takes advantage of their high specific gravity relative to non-'shelled' plankton. Samples are introduced into saturated salt solutions (NaCl or $NaNO_3$) and while the 'shelled' sarcodinans sink the other forms remain on the surface (Bé, 1959; McGowan and Fraundorf, 1964). Juvenile sarcodinans do not have fully developed skeletons and, therefore, would not be expected to sink as rapidly as mature forms.

TAXONOMIC IDENTIFICATION OF MICROZOOPLANKTON

While some of the studies of Lohmann (e.g. 1908) in the early days of quantitative plankton work considered all elements of the microzooplankton, most considerations in the interim have been of isolated groups and only recently has there been a rebirth of interest in the total population as a component of pelagic food webs. Hence, as with the phytoplankton, literature references to microzooplankton taxa are scattered, and much of the material on the groups is relatively old. Considerable study has been made of some taxonomic groups (e.g. the tintinnid ciliates) whereas others suffer from a distinct paucity of information (e.g. the developmental stages of Radiolaria).

Taxonomic identification within many of the protozoan groups suffers from the fact that observation and study of the forms have generally been conducted in such a manner that natural relationships are not seen. This can lead to the grouping of species into highly spurious phylogenetic associations. Furthermore, the determination of species may also be complicated by such occurrences as the false identification of environmentally induced variants as distinct species and the naming of diverse stages in the life cycle of an organism as different species. While some progress has been made to discern true relationships, through more in-depth study including the use of sophisticated technical approaches on the cytological level (e.g. Hollande and Enjumet, 1960, on the cytology, evolution and systematics of the sphaerellid Radiolaria), it is probable that fully accurate definitions of species and the devising of satisfactory systems of taxonomic classification based on natural relationships will not be forthcoming until there is much more extensive laboratory culture and study of the living organisms.

Microzooplankton taxonomic groups are considered in varying degrees of detail in several general references on plankton taxa (e.g. Davis, 1955; Newell and Newell, 1963), including some which are most pertinent to specific geographical areas (e.g. Trégouboff and Rose, 1957, on the Mediterranean plankton; Wailes, 1937, 1939, 1943 in the series on Canadian Pacific fauna).

There are no comprehensive separate literature sources for phagotrophic flagellates, partly because of the difficulties in distinguishing these distinctly from other flagellates except in a few instances (e.g. *Noctiluca*, *Polykrikos*). Some references which contain considerations of phagotrophic flagellates include Biecheler (1952), Grassé (1952), and Pringsheim (1959).

The two major ciliate orders represented in pelagic protozoan populations are the Oligotrichina and Tintinnina. The most detailed account of oligotrich species is found in Kahl (1932) and includes descriptions based on live-animal examination and preserved specimens (e.g. Leegaard, 1915). The tintinnids stand out as one of the most intensively studied groups of marine ciliates. This undoubtedly results largely from the species of this group possessing distinctive vase-shaped houses, i.e. the loricae, in which the living organism resides and which preserve well in fixed samples and are used for identification purposes. Kofoid and Campbell (1929, augmented by Kofoid and Campbell, 1939, and Campbell, 1942) reviewed the already extensive literature on tintinnids to that time and described many new species. While their work still is generally used, it was criticized soon after publication for having given species status to forms which simply may represent ecotypic variation (e.g. Hofker, 1930). Numerous other investigators (e.g. Balech, 1959b, 1962; Halme and Lukkarinen, 1960; Margalef and Durán, 1953) have also noted the wide degree of variability in lorica morphology with transitional forms bridging the gap between two or more of the species set up by Kofoid and Campbell. Marshall (1969) provides a handy reference to the tintinnid species found in the Arctic Ocean and at latitudes above the tropics in the North Atlantic.

Amongst the pelagic sarcodinans the foraminifera are perhaps the best defined taxonomically, although in this group as well as the other sarcodinans, early developmental stages often cannot be identified to species because of their lack of distinctive features. Only about 30 species of planktonic foraminifera are recognized. Their test morphology and geographical distribution are well summarized in Bé (1967). Postuma (1971) considers the living planktonic foraminifera in his coverage of Mesozoic and Cenozoic planktonic representatives of this group. The Acantharia are considered in the monograph by Schewiakoff (1926), which includes some discussion of developmental stages. References to facilitate the routine identification of radiolarian specimens in microplankton samples are wanting largely because of a lack of knowledge of their developmental stages which often apparently predominate in samples. Trégouboff (1953) provides an overview of the group, but does not provide the detailed descriptions necessary for identifications at the species level. The study of Haeckel (1887), in which the extensive radiolarian material collected during the *Challenger* expedition is figured, is still perhaps the most comprehensive treatment of the group. However, the system of classification he developed for placing species into higher groups, and which has been used extensively since (e.g. Trégouboff, 1953), has been criticized for its artificiality by numerous investigators and attempts have been made to provide a classification based more on supposed natural relationships (see Riedel, 1971). For the identification of Radiolaria in plankton samples, Renz (1976), for example, provides figures of numerous forms which, although studied from the North Pacific, have considerably broader distribution.

9.3   About freshwater phytoplankton

Eva Willén and Torbjörn Willén

A lake with a small area and a moderate depth offers good conditions for most aspects of phytoplankton research. Usually, consideration of large-scale hydrological patterns is necessary only in the largest bodies of freshwater where the conditions resemble those in the sea, e.g. geostrophical currents and thermal bar effects. Although there are many similarities between marine and freshwater biotopes, the structure and dynamics of phytoplankton may be very different. Study of phytoplankton in lakes has become increasingly focused on practical problems in connection with eutrophication processes.

Phytoplankton species occurring in freshwaters are not the same as those in marine waters. Several genera and most groups, however, are represented in both biotopes. Blue–green and green algae are more abundant in freshwater as are taxa belonging to Euglenophyceae and Xanthophyceae. Within some genera a great number of taxa can be distinguished, e.g. the chlorophyte genus *Chlamydomonas* with about 450 species recorded mainly from freshwater lakes and ponds, while only a few taxa occur in marine waters. On the other hand, some genera such as the dinoflagellate *Ceratium* and the diatoms *Chaetoceros* and *Rhizosolenia* include numerous marine species but only a few living in freshwater. Within the group Coccolithineae marine members predominate. The group Silicoflagellineae only occurs in marine environments. The desmids (Conjugatae) are strictly confined to freshwater; this group is very rich in species —within the genus *Cosmarium* alone about 1,000 are described—and is used in different quotient systems to indicate oligotrophy (Hutchinson, 1967).

In contrast to oceanic conditions with a true plankton ('euplankton'), freshwater phytoplankton often contains benthic and littoral forms ('tychoplankton').

In both marine and freshwater biotopes there is a pronounced seasonal succession of phytoplankton. In temperate freshwater lakes a spring maximum of diatoms (which sometimes have already developed under the ice) is often followed by a summer minimum of chrysophytes and green algae. In late summer the blue–green algae appear. A minor peak of diatoms is common in the autumn. In polluted lakes the development of blue–green algae often begins after the spring circulation period and continues for several months. This is, however, only one example of deviations from a 'normal' succession, since many disparities exist in lakes due to trophic level, pollution, etc. In tropical lakes the phytoplankton maximum is often observed in winter while in polar regions and at high latitudes lakes have a single summer maximum. The seasonal periodicity, however, seems to be constant from year to year but the amplitude of changes in plankton numbers and biomass is usually considerable (Wetzel, 1975).

297

SAMPLING AND ANALYSIS

The uneven distribution of plankton often causes considerable problems in obtaining representative samples. This is especially evident during periods of mass development of blue–green algae. Aerial photography using infra-red sensitive film can be used to illustrate the momentary pattern of these surface-living algae (Reeves, 1975). Integration of samples over a limited area and/or depth is one way of removing the variability due to patchiness. In routine investigations of large lakes it is often necessary to choose a systematic method of sampling, bearing in mind that the largest error in any phytoplankton study is connected with sampling.

A representative water sample is essential for quantitative analysis. For counting purposes 100 to 300 ml will suffice, depending on plankton density. Water-bottles that open at a given depth, e.g. the Ruttner sampler (Ruttner, 1963), or reversing bottles are useful sampling equipment, as well as pumps and tubes. Nets of 10 to 25-µm mesh size may be used to provide material for the taxonomical checking. A survey of samplers, nets, centrifuges, etc., is given by Welch (1952) and Schwoerbel (1966) among others, and a valuable discussion on this subject may be found in Lund and Talling (1957).

The frequency of sampling depends on the scope of the investigation. Monthly sampling may be adequate in long-term routine programmes where gaps in the annual temporal variation are compensated by long-term series (Willén, 1975). In investigations over a shorter period of time the sampling should take place at least every 14 days. This interval may of course be too long if the study focuses on special problems such as migration or the detailed succession of species. When a general description of plankton composition and abundance is needed for some practical purpose, either in a single body of water or over a wide area, then a sampling frequency of four to five times a year may give sufficient information.

The preservative should be a liquid that does not destroy the organisms and at the same time facilitates sedimentation. The naked flagellates, which often are quantitatively important or dominant, are damaged by solutions containing formaldehyde. A formaldehyde solution does not allow blue–green algae to settle properly and large errors can be introduced into the evaluation of the total volume. The Lugol (I_2IK) preservative solution fulfils the above requirements. This solution, supplemented with acetic acid, has proved to be most durable in freshwater and lenient on the organisms (Hobro and Willén, 1977). However, because of its low pH (about 2·5) the storage time of material preserved from humic and acid lakes is short.

Taxonomical identification of phytoplankton should be made on living material whenever possible. This, however, does not apply to skeleton-bearing organisms, for which special preparations may be needed. An unpreserved sample decomposes within a short time, which makes preservation with a formaldehyde solution necessary even though nannoplanktic forms are destroyed.

Monographs useful in species determination are published in the well-known series by West and West (*A Monograph of the British Desmidiaceae*, 1904–23), Pascher (*Die Süsswasser-Flora Mitteleuropas*, 1913–32), Huber-Pestalozzi (*Das Phytoplankton des Süsswassers*, 1938–72) and Starmach (*Flora Slodkowodna Polski*, 1963–74); another classic piece of work is Rabenhorst's *Kryptogamen-Flora* to which several authors have contributed, namely Hustedt (1927–66) for diatoms, Schiller (1931–37) for dinoflagellates, Geitler (1932) for blue–greens, Krieger (1933–39) for desmids, Pascher (1937–39) for heterokonts and Kolwitz and Krieger (1941–44) for Zygnemales. A large number of valuable taxonomical revisions are listed in Bourrelly (1966–70) and Fott (1971).

In a quantitative evaluation of phytoplankton the counting procedure has many advantages, such as direct observations of the organisms and an estimation of the importance of the different species in a population. There are many counting methods (Lund and Talling, 1957; Vollenweider *et al.*, 1969; Schwoerbel, 1970). Preference should be given to a technique with as few subsamplings as possible and one should also avoid causing mechanical damage from centrifuging, filtration, etc. The Utermöhl methods fulfils these requirements (Utermöhl 1931, 1958) and is now widely adopted in freshwater plankton studies. It requires the use of an inverted microscope which enables counts to be made from a wide range of water volumes (sedimentation chambers) where observation of small delicate forms is possible.

The counting procedure is very time-consuming and simplifications must be made without reducing the accuracy. An acceptable solution, at least in routine work, is to count the most important species (Willén, 1976), the number of which varies in different bodies of water from 1–2 during periods of water-bloom to 10–15 in lakes with plankton of larger diversity. When selecting the species to be counted it is necessary to examine sedimentation chambers of different volumes. Counting a limited number of species may underestimate the total volume by only 3 to 5 per cent as calculated on annual mean values (Hobro and Willén, 1977). A fairly large number of individuals (50 to 100) ought to be counted, to get acceptable reliability (Lund *et al.*, 1958). Valuable comments on this problem are given in Chapter **7.1.2**.

When calculating volumes, the morphology of the algae must be thoroughly studied in order to enable choice of a suitable geometric formula. Large colonial forms, like *Microcystis*, present numerous difficulties in both counting and calculations of volume. One way of tackling the problem is to treat the sample with ultra-sounds, which split up the large colonies, thus making the counting of separate cells easier (Cronberg *et al.*, 1975).

PRACTICAL AND ECOLOGICAL ASPECTS

Lakes represent a very small fraction (0·009 per cent) of the total water content of the biosphere as compared to oceans (98 per cent). However, freshwaters, from many aspects, are of vital importance to man, a fact that has led to surveying programmes for their use and protection. In many cases the analysis of freshwater phytoplankton is connected with practical problems that involve, for example, the bloom of blue–green algae with accompanying problems of toxicity, etc., the development of certain diatoms that clog fishing nets and intakes of waterworks, or the presence of algae causing unpleasant odour and taste to drinking water. Most of these problems are discussed by, for example, Gorham (1964) and Palmer (1964).

Results of the qualitative analysis can be used either to calculate one or another quotient in order to range lakes in different trophic classes, or to disclose indicators of different pollution conditions. The Kolkwitz and Marsson (1908) system for classifying saprobic and katharobic zones according to the presence or absence of characteristic species has been modified or improved by several workers (Liebmann, 1962; Fjerdingstad, 1965). Values of diversity of freshwater phytoplankton populations are also used; the general tendency for species diversity is to decrease with increasing fertility of the lake (diversity methods and results are common to both marine and freshwaters, cf. Chapter **8**.3).

Studies of phytoplankton species and biomass are valuable when observing changes due to an increasing or decreasing nutrient supply and disturbances of aquatic ecosystems such as acidifications. Changes may also be studied using a single species, for example in algal-growth-potential tests where *Chlorella*, *Selenastrum*, *Chaetoceros* and *Phaeodactylum* are cultivated and counted by sensing-zone counters (Skulberg, 1975).

Owing to the time-consuming nature of quantitative plankton analysis, more rapid and less subjective analyses of total biomass are often chosen, such as chlorophyll measurements, ATP and carbon determinations. So far, however, there is no rapid method that can replace the counting technique and, at the same time, give information on the proportion of different algae and/or algal groups, information which proves to be so valuable in ecosystem studies.

300

References

ANON. 1955. *Standard methods for the examination of water, sewage, and industrial wastes.* 10th ed. Washington, Am. Publ. Hlth Assoc. 522 p.

ANON. 1966. *Determination of photosynthetic pigments in sea-water,* p. 1–69. Paris, Unesco. (Monogr. oceanogr. Methodol., 1.)

ANON. 1968*a. Instruction manual for oceanographic observations.* 3rd ed., p. D1–D6. Washington, U.S. Hydrogr. Office. (Publ. 607.)

ANON. 1968*b. Zooplankton sampling,* p. 1–174. Paris, Unesco. (Monogr. oceanogr. Methodol., 2.)

ANON. 1969. *Recommended procedures for measuring the productivity of plankton standing stock and related oceanic properties.* Washington, Nat. Acad. Sci. 59 p. (Prepared by the Biological Panel Committee on Oceanography. Div. Earth Sci., Nat. Res. Coun.)

ANON. 1973. Edinburgh Oceanographic Laboratory: Continuous plankton records; a plankton atlas of the North Atlantic and the North Sea. *Bull. mar. Ecol.,* vol. 7, p. 1–174.

ANON. [or Banse, K.]. 1974. *A review of methods used for quantitative phytoplankton studies. Final Report of SCOR Working Group 33,* p. 1–27. (Unesco tech. Pap. mar. Sci., 18.)

AABYE JENSEN, E.; STEEMANN NIELSEN, E. 1953. A water-sampler for biological purposes. *J. Cons. CIEM,* vol. 18, no. 3, p. 296–9.

AHRENS, R. 1971. Untersuchungen zur Verbreitung von Phagen der Gattung *Agrobacterium* in der Ostsee. *Kiel. Meeresforsch.,* vol. 27, no. 1, p. 102–12, pl. 1–9.

AITCHISON, J.; BROWN, J. A. C. 1957. *The lognormal distribution. With special reference to its use in economics.* London, Cambridge Univ. Press. 176 p.

ALLEN, W. E. 1921. A brief study of the range of error in micro-enumeration. *Trans. am. microsc. Soc.,* vol. 40, no. 1, p. 14–25.

——. 1940. Summary of results of twenty years of researches on marine phytoplankton. *Proc. 6th Pacif. Sci. Congr., 1939,* vol. 3, p. 577–83.

ALMEIDA, S. P.; EU, J. K. T. 1976. Water pollution monitoring using matched spatial filters. *Appl. Optics,* vol. 15, no. 2, p. 510–15.

ANRAKU, M. 1956. Some experiments on the variability of horizontal plankton hauls and on the horizontal distribution of plankton in a limited area. *Bull. Fac. Fish. Hokkaido Univ.,* vol. 7, no. 1, p. 1–16.

ARKIN, H.; COLTON, R. R. 1950. *Tables for statisticians.* New York, N.Y., Barnes & Noble. 152 p.

ARON, W. 1958. The use of a large capacity portable pump for plankton sampling with notes on plankton patchiness. *J. mar. Res.,* vol. 16, no. 2, p. 158–73.

——. 1962. Some aspects of sampling the macroplankton. In: J. R. Fraser and J. Corlett (eds.), 'Contributions to Symposium on zooplankton production', *Rapp. P.-v. Réun. CIEM,* vol. 153, p. 29–38.

AVARIA, S. P. 1971. Variaciones mensuales del fitoplancton de la bahía de Valparaiso, entre julio de 1963 y julio de 1966. *Rev. Biol. mar. Estac. Biol. mar. Univ. Chile,* vol. 14, no. 3, p. 15–43.

BAIER, R. E.; GOUPIL, D. W.; PERLMUTTER, S.; KING, R. 1974. Dominant chemical composition of sea-surface films, natural slicks, and foams. *J. Rech. atmos.,* vol. 8, p. 571–600.

BAINBRIDGE, R. 1957. The size, shape and density of marine phytoplankton concentrations. *Biol. Rev. Camb. Philos. Soc.,* vol. 32, no. 1, p. 91–115.

BALECH, E. 1959*a.* Two new genera of Dinoflagellates from California. *Biol. Bull. mar. biol. Lab. Woods Hole, Mass.,* vol. 116, no. 2, p. 195–203.

——. 1959*b*. Tintinnoinea del Mediterráneo. *Trab. Inst. esp. Oceanogr.*, vol. 28, p. 1–88, pl. 1–22.

——. 1962. Tintinnoinea y Dinoflagellata del Pacífico según material de las expediciones Norpac y Downwind del Instituto Scripps de Oceanografía. *Rev. Mus. argent. Cienc. nat. Bernardino Rivadavia (Zool.)*, vol. 7, no. 1, p. 1–253, pl. 1–26.

BALLANTINE, D. 1953. Comparison of the different methods of estimating nanoplankton. *J. mar. biol. Assoc. U.K.*, vol. 32, no. 1, p. 129–47.

BANSE, K. 1976. Rates of growth, respiration and photosynthesis of unicellular algae as related to cell size. A review. *J. Phycol.*, vol. 12, no. 2, p. 135–40.

——. 1977. Determining the carbon-to-chlorophyll ratio of natural phytoplankton. *Mar. Biol.*, vol. 41, no. 3, p. 199–212.

BANSE, K.; FALLS, C. P.; HOBSON, L. A. 1963. A gravimetric method for determining suspended matter in sea water using Millipore[R] filters. *Deep-sea Res. oceanogr. Abstr.*, vol. 10, no. 5, p. 639–42.

BARNES, H. 1951. A statistical study of the variability of catches obtained with two models of the Hardy plankton indicator. *Hull Bull. mar. Ecol.*, vol. 2, no. 16, p. 283–93.

——. 1952. The use of transformations in marine biological statistics. *J. Cons. CIEM*, vol. 18, no. 1, p. 61–71.

——. 1959. *Oceanography and marine biology. A book of techniques.* London, Allen & Unwin. 218 p.

BARNES, H.; HASLE, G. R. 1957. A statistical examination of the distribution of some species of Dinoflagellates in the polluted inner Oslo fjord. *Nytt Mag. Bot.*, vol. 5, p. 113–24.

BARNES, H.; MARSHALL, S. M. 1951. On the variability of replicate plankton samples and some applications of 'contagious' series to the statistical distribution of catches over restricted periods. *J. mar. biol. Assoc. U.K.*, vol. 30, no. 2, p. 233–63.

BARSS, M. S.; WILLIAMS, G. L. 1973. Palynology and nannofossil processing techniques. Ottawa, Geol. Surv. Canada, Dept Energy, Mines and Resources. 25 p. incl. 2 pl. (Pap. 73–26.)

BARTLETT, M. S. 1947. The use of transformations. *Biometrics*, vol. 3, no. 1, p. 39–52.

BARY, B. M. 1953. Sea water discoloration by living organisms. *N.Z. J. Sci. Technol.*, Ser. B, vol. 34, no. 5, p. 393–407.

BAYLOR, E. R.; SUTCLIFFE, W. H. Jr. 1963. Dissolved organic matter in seawater as a source of particulate food. *Limnol. Oceanogr.*, vol. 8, no. 4, p. 369–71.

BAYNE, D. R.; LAWRENCE, J. M. 1972. Separating constituents of natural phytoplankton populations by continuous particle electrophoresis. *Limnol. Oceanogr.*, vol. 17, no. 3, p. 481–9.

BÉ, A. W. H. 1959. A method for rapid sorting of Foraminifera from marine plankton samples. *J. Paleont.*, vol. 33, no. 5, p. 846–8, pl. 118.

——. 1967. Foraminifera. Families: Globigerinidae and Globorotaliidae. In: J. H. Fraser and V. Kr. Hansen (eds.), *Fiches d'identification du zooplancton*. Charlottenlund, Cons. perm. int. Explor. Mer. 9 p. (No. 108.)

BEERS, J. R. 1976. Preservation and laboratory study of Actinopods in plankton samples. In: H. F. Steedman (ed.), *Zooplankton fixation and preservation*, p. 240–9. Paris, Unesco. (Monogr. oceanogr. Methodol., 4.)

BEERS, J. R.; KNOX, C.; STRICKLAND, J. D. H. 1970. A permanent record of plankton samples using holography. *Limnol. Oceanogr.*, vol. 15, no. 6, p. 967–70.

BEERS, J. R.; REEVE, M. R.; GRICE, G. D. 1977. Controlled Ecosystem Pollution Experiment: effect of mercury on enclosed water columns. IV. Zooplankton population dynamics and production. *Mar. Sci. Commun.*, vol. 3, no. 4, p. 355–94.

BEERS, J. R.; REID, F. M. H.; STEWART, G. L. 1975. Microplankton of the North Pacific Central Gyre. Population structure and abundance, June 1973. *Int. Rev. gesamt. Hydrobiol.*, vol. 60, no. 5, p. 607–38.

BEERS, J. R.; STEVENSON, M. R.; EPPLEY, R. W.; BROOKS, E. R. 1971. Plankton populations and upwelling off the coast of Peru, June 1969. *Fish. Bull.*, vol. 69, no. 4, p. 859–76.

BEERS, J. R.; STEWART, G. L. 1969. Micro-zooplankton and its abundance relative to the larger zooplankton and other seston components. *Mar. Biol.*, vol. 4, no. 3, p. 182–9.

——. 1970. The preservation of acantharians in fixed plankton samples. *Limnol. Oceanogr.*, vol. 15, no. 5, p. 825–7.

——. 1971. Micro-zooplankters in the plankton communities of the upper waters of the eastern tropical Pacific. *Deep-sea Res. oceanogr. Abstr.*, vol. 18, no. 9, p. 861–83.

BEERS, J. R.; STEWART, G. L.; STRICKLAND, J. D. H. 1967. A pumping system for sampling small plankton. *J. Fish. Res. Board Can.*, vol. 24, no. 8, p. 1811–18.

BEKLEMISHEV, C. W. 1969. (*Ecology and biogeography of the open ocean.*) Moscow, Nauka. 291 p. (In Russian.)

BELYAEVA, T. V. 1975. (Distribution of diatoms in the area of the Peru current.) *Okeanologiya*, vol. 15, no. 5, p. 874–80. (In Russian.) (Trans.: *Oceanology.*)

BERNHARD, M.; RAMPI, L. 1966. Horizontal microdistribution of marine phytoplankton in the Ligurian sea. *Bot. Gothoburg.*, vol. 3. p. 13–24. (Proc. 5th mar. Biol. Symp., Göteborg 1965.)

BERNHARD, M.; RAMPI, L.; ZATTERA, A. 1967. A phytoplankton component not considered by the Utermöhl method. *Pubbl. Staz. zool. Napoli*, vol. 35, p. 170–214.

——. 1969. La distribuzione del fitoplancton nel mar Ligure. *Pubbl. Staz. zool. Napoli*, vol. 37, suppl. 2, p. 73–114.

BEUM, C. O.; BRUNDAGE, E. G. 1950. A method for analyzing the sociomatrix. *Sociometry*, vol. 13, no. 2, p. 141–5.

BIECHELER, B. 1934. Sur le réseau argentophile et la morphologie de quelques Péridiniens nus. *C.r. Séanc. Soc. Biol. (Paris)*, vol. 115, p. 1039–42.

——. 1952. Recherches sur les Péridiniens. *Bull. biol. Fr. Belg.*, suppl. 36, p. 1–149.

BISALPUTRA, T.; CHENG, J. Y.; TAYLOR, F. J. R.; ANTIA, N. J. 1973. Improved filtration techniques for the concentration and cytological preservation of microalgae for electron microscopy. *Can. J. Bot.*, vol. 51, no. 2, p. 371–7, pl. 1–6.

BLACKBOURN, D. J.; TAYLOR, F. J. R.; BLACKBOURN, J. 1973. Foreign organelle retention by ciliates. *J. Protozool.*, vol. 20, no. 2, p. 286–8.

BLANCHARD, D. C.; PARKER, B. C. 1977. The freshwater to air transfer of microorganisms and organic matter. In: J. Cairns, Jr (ed.), *Aquatic microbial communities*, p. 625–58. New York, N.Y., Garland Publ.

BLANCHARD, D. C.; WOODCOCK, A. H. 1957. Bubble formation and modification in the sea and its meteorological significance. *Tellus*, vol. 9, no. 2, p. 145–58.

BLISS, C. I.; FISHER, R. A. 1953. Fitting the negative binomial distribution to biological data. Note on the efficient fitting of the negative binomial. *Biometrics*, vol. 9, no. 2, p. 176–200.

BOGOYAVLENSKII, A. N. 1962. (On the distribution of heterotrophic micro-organisms in the Indian ocean and in Antarctic waters.) *Okeanologiya*, vol. 2, no. 2, p. 293–7. (In Russian.) (Transl. *Deep-sea Res.*, vol. 11, no. 1, 1964, p. 105–8.)

BONNER, W. A.; HULETT, H. R.; SWEET, R. E.; HERZENBERG, L. A. 1972. Fluorescence activated cell sorting. *Rev, Sci. Instrum.*, vol. 43, no. 3, p. 404–9.

BOURRELLY, P. 1966–70. *Les algues d'eau douce. Initiation à la systématique.* I, *Les algues vertes* (1966), 511 p.; II, *Les algues jaunes et brunes* (1968), 438 p.; III, *Les algues bleues et rouges, les Eugléniens, Péridiniens et Cryptomonadines* (1970), 512 p. Paris, Boubée.

BOWEN, R. A.; ST. ONGE, J. M.; COLTON, J. B. Jr.; PRICE, C. A. 1972. Density-gradient centrifugation as an aid to sorting planktonic organisms, I. Gradient materials. *Mar. Biol.*, vol. 14, no. 3, p. 242–7.

BOYD, C. M.; JOHNSON, G. W. *A plankton counting system for marine biological studies.* (In press.)

BRAARUD, T. 1935. The 'Øst' expedition to the Denmark Strait 1929, II. The phytoplankton and its conditions of growth (including some qualitative data from the Arctic in 1930). *Hvalraad. Skr.*, vol. 10, p. 1–173.

——. 1945. A phytoplankton survey of the polluted waters of inner Oslo fjord. *Hvalraad. Skr.*, vol. 28, p. 1–142.

——. 1961. Cultivation of marine organisms as a means of understanding environmental influences on populations. In: M. Sears (ed.), *Oceanography*, p. 271–98. Washington, Am. Soc. Advanc. Sci. (Publ. 67.)

——. 1962. Species distribution in marine phytoplankton. *J. oceanogr. Soc. Jap.*, p. 628–49. (20th anniv.)

BRAARUD, T.; GAARDER, K. R.; GRØNTVED, J. 1953. The phytoplankton of the North sea and adjacent waters in May 1948. *Rapp. P.-v. Réun. CIEM*, vol. 133, p. 1–87, pl. 1–2.

BRAARUD, T.; GAARDER, K. R.; NORDLI, O. 1958. Seasonal changes in the phytoplankton at various points off the Norwegian west coast. *Fiskeridir. Skr. (Havunders.)*, vol. 12, no. 3, p. 1–77.

BRADLEY, M. V. 1948. A method for making aceto-carmine squashes permanent without removal of the cover slip. *Stain Technol.*, vol. 23, no. 1, p. 41–4.

BRIDGES, C. B. 1937. The vapor method of changing reagents and of dehydration. *Stain Technol.*, vol. 12, no. 2, p. 51–2.

BRILLOUIN, L. 1962. *Science and information theory*. 2nd ed. New York, N.Y. Academic Press. 320 p.

BROENKOW, W. W. 1969. An interface sampler using spring-actuated syringes. *Limnol. Oceanogr.*, vol. 14, no. 2, p. 288–91.

BROOKS, S. H. 1955. The estimation of an optimum subsampling number. *J. am. statist. Assoc.*, vol. 50, no. 270, p. 398–415.

BRUCE, R. H.; AIKEN, J. 1975. The undulating oceanographic recorder. A new instrument system for sampling plankton and recording physical variables in the euphotic zone from a ship underway. *Mar. Biol.*, vol. 32, no. 1, p. 85–97.

BURKOV, V. A. 1972. (*The Pacific ocean, General circulation of the Pacific ocean waters.*) Moscow, Izdat. Nauka. 195 p. (In Russian.)

BURNEY, C. M.; JOHNSON, K. M.; SIEBURTH, J. McN. (Submitted for publication.) Dissolved carbohydrate concentrations and their microbial cycling in the north Atlantic, pt II. Poly-saccharide and total carbohydrate.

BURNEY, C. M.; SIEBURTH, J. McN. 1977. Dissolved carbohydrates in seawater, II. A spectro-photometric procedure for total carbohydrate analysis and polysaccharide estimation. *Mar. Chem.*, vol. 5, no. 1, p. 15–28.

CAIRNS, J. Jr.; DICKSON, K. L.; LANZA, G. R.; ALMEIDA, S. P.; DEL BALZO, D. 1972. Coherent optical spatial filtering of diatoms in water pollution monitoring. *Arch. Mikrobiol.*, vol. 83, p. 141–6.

CAMPBELL, A. S. 1942. The oceanic Tintinnoinea of the plankton gathered during the last cruise of the Carnegie. *Scient. Results Cruise VII Carnegie 1928–29*, Biol., vol. 2, p. 1–163 incl. pl. 1.

CARPENTER, E. J.; GUILLARD, R. R. L. 1971. Intraspecific differences in nitrate half-saturation constants for three species of marine phytoplankton. *Ecology*, vol. 52, no. 1, p. 183–5.

CASSIE, R. M. 1958. Apparatus for investigating spatial distribution of plankton. *N.Z. J. Sci.*, vol. 1, no. 3, p. 436–48.

——. 1959a. An experimental study of factors inducing aggregation in marine plankton. *N.Z. J. Sci.*, vol. 2, no. 3, p. 339–65.

——. 1959b. Micro-distribution of plankton. *N.Z. J. Sci.*, vol. 2, no. 3, p. 398–409.

——. 1959c. Some correlations in replicate plankton samples. *N.Z. J. Sci.*, vol. 2, no. 4, p. 473–84.

——. 1960. Factors influencing the distribution pattern of plankton in the mixing zone between oceanic and harbour waters. *N.Z. J. Sci.*, vol. 3, no. 1, p. 26–50.

——. 1962. Frequency distribution models in the ecology of plankton and other organisms. *J. anim. Ecol.*, vol. 31, no. 1, p. 65–92.

——. 1963. Microdistribution of plankton. *Oceanogr. mar. Biol. ann. Rev.*, vol. 1, p. 223–52.

——. 1968. Sample design. In: Anon. (ed.), *Zooplankton sampling*, p. 105–21. Paris, Unesco. (Monogr. oceanogr. Methodol., 2.)

——. 1971. Sampling and statistics. In: W. T. Edmondson and G. G. Winberg (eds.), *A manual on methods for the assessment of secondary productivity in fresh waters*, p. 174–209. London, International Biological Programme; Oxford, Edinburgh, Blackwell Scient. Publ. (IBP Handbk 17.)

CASSIE, V. 1960. Seasonal changes in diatoms and dinoflagellates off the east coast of New Zealand during 1957 and 1958. *N.Z. J. Sci.*, vol. 3, no. 1, p. 137–72.

CHENG, L. 1975. Marine pleuston—animals at the sea–air interface. *Oceanogr. mar. Biol. ann. Rev.*, vol. 13, p. 181–212, 3 pl.

CHRISTENSEN, T. 1962. Alger. In: T. W. Böcher, M. Lange and T. Sørensen (eds.), *Botanik*, II. *Systematik botanik*, p. 1–178. Copenhagen, Munksgaard.

CLARK, G. (ed.). 1973. *Staining procedures, issued by the Biological Commission*. 3rd ed. Baltimore, Md, Williams & Wilkins. 418 p.

CLARK, W. J.; SIGLER, W. F. 1963. Method of concentrating phytoplankton samples using mem-brane filters. *Limnol. Oceanogr.*, vol. 8, no. 1, p. 127–9.

CLARKE, G. L.; BUMPUS, D. F. 1940. *The plankton sampler. An instrument for quantitative plankton investigations*, p. 1–8. Am. Soc. Limnol. Oceanogr. (Spec. publ. 5.)

CLEVE, P. T. 1897. *A treatise on the phytoplankton of the Atlantic and its tributaries, and on the periodical changes of the plankton of the Skagerak*. Uppsala. 28 p., 4 pl., tabl.

——. 1903. Report on plankton collected by Mr Thorild Wulff during a voyage to and from Bombay. *Ark. Zool.*, vol. 1, p. 329–81, pl. 16–19.

COCHRAN, W. G. 1947. Some consequences when the assumptions for the analysis of variance are not satisfied. *Biometrics*, vol. 3, no. 1, p. 22–38.

——. 1963. *Sampling techniques.* 2nd ed. New York, Wiley & Sons. 413 p.

COHEN, A. L. 1974. Critical point drying. In: M. A. Hayat (ed.), *Principles and techniques of scanning electron microscopy.* Vol. I, p. 44–112. New York, N.Y., Cincinnati, Ohio, Toronto, London, Melbourne, Van Nostrand Reinhold.

COLEBROOK, J. M. 1960. Continuous plankton records: methods of analysis, 1950–59. *Bull. mar. Ecol.,* vol. 5, no. 41, p. 51–64.

——. 1964. Continuous plankton records: a principal component analysis of the geographical distribution of zooplankton. *Bull. mar. Ecol.,* vol. 6, no. 3, p. 78–100.

——. 1966. Continuous plankton records: geographical patterns in the annual fluctuations of abundance of some Copepods in the North sea. In: H. Barnes (ed.), *Some contemporary studies in marine science,* p. 155–61. London, Allen & Unwin.

——. 1969. Variability in the plankton. *Progr. Oceanogr.,* vol. 5, p. 115–25.

——. 1975a. The continuous plankton recorder survey: automatic data processing methods. *Bull. mar. Ecol.,* vol. 8, no. 3, p. 123–42.

——. 1975b. The continuous plankton recorder survey: computer simulation studies of some aspects of the design of the survey. *Bull. mar. Ecol.,* vol. 8, no. 3, p. 143–66.

COLEBROOK, J. M.; ROBINSON, G. A. 1964. Continuous plankton records: annual variations of abundance of plankton, 1948–60. *Bull. mar. Ecol.,* vol. 6, no. 3, p. 52–69, pl. 20–5.

COLEBROOK, J. M.; ROBINSON, G. A. 1965. Continuous plankton records: seasonal cycles of phytoplankton and copepods in the north eastern Atlantic and the North sea. *Bull. mar. Ecol.,* vol. 6, no. 5, p. 123–39, pl. 62–3.

CONN, H. J. (with the assistance of E. H. Stotz, V. M. Emmel, M. A. Darrow, R. D. Lillie and G. G. Glenner). 1961. *Biological stains: a handbook on the nature and uses of the dyes employed in the biological laboratory.* 7th ed. Baltimore, Md, Williams & Wilkins. 355 p.

CONOVER, R. J. 1966. Assimilation of organic matter by zooplankton. *Limnol. Oceanogr.,* vol. 11, no. 3, p. 338–45.

CONOVER, W. J. 1971. *Practical nonparametric statistics.* London, Sydney, Wiley & Sons. 462 p.

CORLISS, J. O. 1974. The changing world of ciliate systematics: historical analysis of past efforts and a newly proposed phylogenetic scheme of classification for the protistan phylum Ciliophora. *Syst. Zool.,* vol. 23, no. 1, p. 91–138.

COULON, C.; ALEXANDER, V. 1972. A sliding-chamber phytoplankton settling technique for making permanent quantitative slides with applications in fluorescent microscopy and autoradiography. *Limnol. Oceanogr.,* vol. 17, no. 1, p. 149–52.

CRIPPEN, R. W.; PERRIER, J. L. 1974. The use of neutral red and Evans blue for live-dead determinations of marine plankton (with comments on the use of rotenone for inhibition of grazing). *Stain Technol.,* vol. 49, no. 2, p. 97–104.

CRISSMAN, H. A.; STEINKAMP, J. A. 1973. Rapid, simultaneous measurement of DNA, protein, and cell volume of single cells from large mammalian cell populations. *J. Cell Biol.,* vol. 59, no. 3, p. 766–71.

CRONBERG, G.; GELIN, C.; LARSSON, K. 1975. Lake Trummen restoration project, II. Bacteria, phytoplankton and phytoplankton productivity. *Verh. int. Ver. theor. angew. Limnol.,* vol. 19, no. 2, p. 1088–96.

CUSHING, D. H.; HUMPHREY, G. F., BANSE, K.; LAEVASTU, T. 1958. Report of the Committee on terms and equivalents. In: Measurements of primary production in the sea. *Rapp. P.-v. Réun. CIEM,* vol. 144, p. 15–16.

CZARNECKI, D. B.; WILLIAMS, H. D. 1972. A new mounting medium for diatoms. *Trans. am. microsc. Soc.,* vol. 91, no. 1, p. 73.

DAGNELIE, P. 1973–75. *Théorie et méthodes statistiques. Applications agronomiques.* 2ème éd. I, 378 p. (1973); II, 463 p. (1975). Gembloux, Presses agron. Gembloux, Duculot.

DAKIN, W. J. 1908. Methods of plankton research. *Proc. Trans. Lpool biol. Soc.,* vol. 22, p. 500–53.

DALE, B. 1976. Cyst formation, sedimentation and preservation: factors affecting dinoflagellate assemblages in recent sediments from Trondheimsfjord, Norway. *Rev. Palaeobot. Palynol.,* vol. 22, p. 39–60.

DALEY, R. J.; HOBBIE, J. E. 1975. Direct counts of aquatic bacteria by a modified epifluorescence technique. *Limnol. Oceanogr.,* vol. 20, no. 5, p. 875–82.

DAUMAS, R. A.; LABORDE, P. L.; MARTY, J. C.; SALIOT, A. 1976. Influence of sampling method on the chemical composition of water surface film. *Limnol. Oceanogr.,* vol. 21, no. 2, p. 319–26.

DAVIS, C. C. 1955. *The marine and freshwater plankton.* Michigan State Univ. Press. 562 p.

305

DAWSON, W. A. 1960. Home-made counting chambers for the inverted microscope. *Limnol. Oceanogr.*, vol. 5, no. 2, p. 235–6.

DEL VECCHIO, V.; D'ARCA SIMONETTI, A. 1959. Il metodo del Most Probable Number (M.P.N.), e la sua importanza nella colimetria delle acque, con particolare riguardo a quelle destinate ad uso potabile. *N. Ann. Ig. Microbiol.*, vol. 10, no. 6, p. 441–81.

DE NOYELLES, F. Jr. 1968. A stained-organism filter technique for concentrating phytoplankton. *Limnol Oceanogr.*, vol. 13, no. 3, p. 562–5.

DERENBACH, J. B. 1970. Zytologische Methoden zur Untersuchung von Phytoplanktonpopulation in ihrem natürlichen Lebensraum. *Kiel. Meeresforsch.*, vol. 26, no. 1, p. 85–104.

DICKMAN, M. 1968. A new method for making permanent phytoplankton mounts. *Hydrobiologia*, vol. 31, no. 2, p. 161–7.

DIXON, W. J.; MASSEY, F. J. Jr. 1957. *Introduction to statistical analysis.* 2nd ed. New York, N.Y., St Louis, Mo, San Francisco, Calif. McGraw-Hill. 488 p.

DODSON, A. N.; THOMAS, W. H. 1964. Concentrating plankton in a gentle fashion. *Limnol. Oceanogr.*, vol. 9, no. 3, p. 455–6.

DOTY, M. S.; OGURI, M. 1958. Selected features of the isotopic carbon primary productivity technique. In: Measurements of primary production in the sea. *Rapp. P.-v. Réun. CIEM*, vol. 144, no. 3, p. 47–55.

DOZIER, B. J.; RICHERSON, P. J. 1975. An improved membrane filter method for the enumeration of phytoplankton. *Verh. int. Ver. theor. angew. Limnol.*, vol. 19, no. 2, p. 1524–9.

DRESSEL, D. M.; HEINLE, D. R.; GROTE, M. C. 1972. Vital staining to sort dead and live copepods. *Chesapeake Sci.*, vol. 13, no. 2, p. 156–9.

DUCE, R. A.; QUINN, J. G.; OLNEY, C. E.; PIOTROWICZ, S. R.; RAY, B. J.; WADE, T. L. 1972. Enrichment of heavy metals and organic compounds in the surface microlayer of Narragansett bay, Rhode Island. *Science* (Wash.), vol. 176, no. 4031, p. 161–3.

DUGDALE, R. C. 1968. Nutrient limitation in the sea: dynamics, identification, and significance. *Limnol. Oceanogr.*, vol. 12 [1967], no. 4, p. 685–95.

DUNSTAN, W. M. 1975. Problems of measuring and predicting influence of effluents on marine phytoplankton. *Environ. Sci. Technol.*, vol. 9, no. 7, p. 635–8.

DURBIN, E. G. 1977. Studies on the autecology of the marine diatom *Thalassiosira nordenskioeldii*, II. The influence of cell size on growth rate, and carbon, nitrogen, chlorophyll *a* and silica content. *J. Phycol.*, vol. 13, no. 2, p. 150–5.

DURBIN, E. G.; KRAWIEC, R. W.; SMAYDA, T. J. 1975. Seasonal studies on the relative importance of different size fractions of phytoplankton in Narragansett bay (USA). *Mar. Biol.*, vol. 32, no. 3, p. 271–87.

DUSSART, B. H. 1965. Les différentes catégories de plancton. *Hydrobiologia*, vol. 26, no. 1–2, p. 72–4 + err.

——. 1966. *Limnologie. L'étude des eaux continentales.* Paris, Gauthier-Villars. 678 p.

DUURSMA, E. K. 1967. A simple, horizontally hauled large volume water-sampling bottle of Van Dorn type. *Deep-sea Res. oceanogr. Abstr.*, vol. 14, no. 1, p. 133–4, 1 pl.

EDDY, J. W.; BANSE, K. 1972. An evaluation of three methods of making permanent marine phytoplankton mounts. Seattle, Dept. Oceanogr. Univ. Wash. 6 p. (Unpubl. rep.)

EKMAN, V. W. 1905. On the use of insulated water-bottles and reversing thermometers. *Publs Circonst., Cons. perm. int. Explor. Mer*, vol. 23, p. 1–28, pl. 1–2.

EPPLEY, R. W. 1972. Temperature and phytoplankton growth in the sea. *Fish. Bull.*, vol. 70, no. 4, p. 1063–85.

EPPLEY, R. W.; REID, F. M. H.; STRICKLAND, J. D. H. 1970. The ecology of the plankton off La Jolla, California, in the period April through September 1967 (ed. J. D. H. Strickland), pt III. Estimates of phytoplankton crop size, growth rate and primary production. *Bull. Scripps Inst. Oceanogr.*, vol. 17, p. 33–42.

EPPLEY, R. W.; ROGERS, J. N.; McCARTHY, J. J. 1969. Half-saturation constants for uptake of nitrate and ammonium by marine phytoplankton. *Limnol. Oceanogr.*, vol. 14, no. 6, p. 912–20.

FAGER, E. W. 1957. Determination and analysis of recurrent groups. *Ecology*, vol. 38, no. 4, p. 586–95.

——. 1963. Communities of organisms. In: M. N. Hill (ed.), *The sea.* Vol. II, p. 415–37. New York, N.Y., London, Interscience–Wiley.

FAGER, E. W.; McGOWAN, J. A. 1963. Zooplankton species groups in the North Pacific. *Science* (Wash.), vol. 140, no. 3566, p. 453–60.

FARINACCI, A. 1971. Round table on calcareous nannoplankton, Roma, September 23–28, 1970. In: A. Farinacci (ed.), *Proc. II Planktonic Conf., Roma 1970,* vol. II, p. 1343–60. Roma, Tecnoscienza.

FAWELL, J. K. 1976. Electronic measuring devices in the sorting of marine zooplankton. In: H. F. Steedman (ed.), *Zooplankton fixation and preservation,* p. 201–6. Paris, Unesco. (Monogr. oceanogr. Methodol., 4.)

FEDOROV, V. D. 1973. (Connection of the phytoplankton species diversity with changes in the mineral nutrient conditions.) *Gidrobiol. Zh.,* vol. 9, no. 3, p. 21–4. (In Russian.) (Transl.: *Hydrobiol. J.)*

FINNEY, D. J. 1948. The Fisher–Yates test of significance in 2×2 contingency tables. *Biometrika,* vol. 35, no. 1–2, p. 145–56.

FINUCANE, J. H.; MAY, B. Z. 1961. Modified Van Dorn water sampler. *Limnol. Oceanogr.,* vol. 6, no. 1, p. 85–7.

FISHER. R. A. 1963. *Statistical methods for research workers.* 13th ed. New York, N.Y., Hafner Publ. 356 p.

——. 1966. *The design of experiments.* 8th ed. Edinburgh, London, Oliver & Boyd. 248 p.

FISHER, R. A.; YATES, F. 1970. *Statistical tables for biological, agricultural and medical research.* 6th ed. Edinburgh, Oliver & Boyd. 146 p.

FJARLIE, R. L. I. 1953. A seawater sampling bottle. *J. mar. Res.,* vol. 12, no. 1, p. 21–30.

FJERDINGSTAD, E. 1965. Taxonomy and saprobic valency of benthic phytomicro-organisms. *Int. Rev. gesamt. Hydrobiol.,* vol. 50, no. 4, p. 475–604.

FLEMING, W. D. 1943. Synthetic mounting medium of high refractive index. *J. R. microsc. Soc.,* ser. 3, vol. 63, no. 1, p. 34–7.

——. 1954. Naphrax: a synthetic mounting medium of high refractive index. New and improved methods of preparation. *J. R. microsc. Soc.,* ser. 3, vol. 74, no. 1, p. 42–4.

FOTT, B. 1971. *Algenkunde.* 2. Aufl. Jena, G. Fischer Verlag. 581 p.

FOURNIER, R. O. 1970. Studies on pigmented microorganisms from aphotic marine environments. *Limnol. Oceanogr.,* vol. 15, no. 5, p. 675–82.

FRACKOWIAK, D.; JANUSZCZYK, L. 1975. Light scattering by suspension of algae. *Bull. Acad. pol. Sci.* (II, sér. Sci. biol.), vol. 22, no. 11, p. 761–8.

FRANCISCO, D. E.; MAH, R. A.; RABIN, A. C. 1973. Acridine orange-epifluorescence technique for counting bacteria in natural waters. *Trans. am. microsc. Soc.,* vol. 92, no. 3, p. 416–21.

FRANKEL, J.; HECKMANN, K. 1968. A simplified Chatton–Lwoff silver impregnation procedure for use in experimental studies with ciliates. *Trans. am. microsc. Soc.,* vol. 87, no. 3, p. 317–21.

FROLANDER, H. F. 1968. Statistical variation in zooplankton numbers from subsampling with a Stempel pipette. *J. Wat. Pollut. Control Fed.,* vol. 40, no. 6, p. R82–R88.

FRONTIER, S. 1969. Sur une méthode d'analyse faunistique rapide du zooplancton. *J. exp. mar. Biol. Ecol.,* vol. 3, no. 1, p. 18–26.

——. 1972. Calcul de l'erreur sur un comptage de zooplancton. *J. exp. mar. Biol. Ecol.,* vol. 8, no. 2, p. 121–32.

——. 1973. Etude statistique de la dispersion du zooplancton. *J. exp. mar. Biol. Ecol.,* vol. 12, no. 3, p. 229–62.

——. 1974. L'analyse factorielle est-elle heuristique en écologie du plancton? *Cah. ORSTOM (Océanogr.),* vol. 12, no. 1, p. 77–81.

——. 1976. Etude de la décroissance des valeurs propres dans une analyse en composantes principales: comparaison avec le modèle du bâton brisé. *J. exp. mar. Biol. Ecol.,* vol. 25, no. 1, p. 67–75.

FRONTIER, S.; IBANEZ, F. 1974. Utilisation d'une cotation d'abondance fondée sur une progression géométrique, pour l'analyse des composantes principales en écologie planctonique. *J. exp. mar. Biol. Ecol.,* vol. 14, no. 3, p. 217–24.

FROST, B. W. 1973. Effects of size and concentration of food particles on the feeding behavior of the planktonic Copepod *Calanus pacificus. Limnol. Oceanogr.,* vol. 17 [1972], no. 6, p. 805–15.

FULWYLER, M. J. 1965. Electronic separation of biological cells by volume. *Science* (Wash.), vol. 150, no. 3698, p. 910–11.

FUNNELL, B. M.; RIEDEL, W. R. (eds.) 1971. *The micropaleontology of oceans.* London, Cambridge Univ. Press. 828 p.

GABE, M. 1968. *Techniques histologiques.* Paris, Masson. 1113 p,

GARRETT, W. D. 1965. Collection of slick forming materials from the sea surface. *Limnol. Oceanogr.*, vol. 10, no. 4, p. 602–5.

GEITLER, L. 1932. Cyanophyceae von Europa unter Berücksichtigung der anderen Kontinente. In: L. Rabenhorst, *Kryptogamen-Flora*, vol. 14, p. 1–1196. Leipzig, Akad. Verlag.

GERARD, R. D.; AMOS, A. F. 1968. A surface-actuated multiple sampler. *Mar. Sci. Instrum.*, vol. 4, p. 682–6.

GERBER, R. P.; MARSHALL, N. 1974. Ingestion of detritus by the lagoon pelagic community at Eniwetok atoll. *Limnol. Oceanogr.*, vol. 19, no. 5, p. 815–24.

GERLACH, D. 1977. *Botanische Mikrotechnik.* 2. Aufl. Stuttgart, G. Thieme. 311 p.

GIBBARD, D. W.; SMITH, D. J.; WELLS, A. 1972. Area sizing and pattern recognition on the Quantimet 720. *Microscope*, vol. 20, no. 1, p. 37–50.

GILBERT, J. Y. 1942. The errors of the Sedgwick–Rafter counting chamber in the enumeration of phytoplankton. *Trans. am. microsc. Soc.*, vol. 61, no. 3, p. 217–26.

GILLBRICHT, M. 1959. Das Phytoplankton im nördlichen nordatlantischen Ozean im Spätwinter und Spätsommer 1950. *Dtsch. hydrogr. Z. (B) Hamb.*, vol. 4, no. 3, p. 90–3.

——. 1962. Über das Auszählen von Planktonschöpfproben. *Helgol. wiss. Meeresunters.*, vol. 8, no. 2, p. 203–18.

GLEASON, H. A. 1922. On the relation between species and area. *Ecology*, vol. 3, no. 2, p. 158–62.

GLOVER, R. S. 1967. The continuous plankton recorder survey of the North Atlantic. In: N. B. Marshall (ed.), *Aspects of marine zoology, Symp. zool. Soc. Lond.*, vol. 19, p. 189–210.

GOLDBERG, E. D.; BAKER, M.; FOX, D. L. 1952. Microfiltration in oceanographic research, I. Marine sampling with the molecular filter. *J. mar. Res.*, vol. 11, no. 2, p. 194–204.

GOLDMAN, J. C.; RYTHER, J. H.; WILLIAMS, L. D. 1975. Mass production of marine algae in outdoor cultures. *Nature* (Lond.), vol. 254, no. 5501, p. 594–5.

GOLTERMAN, H. L.; CLYMO, R. S. (eds.) 1969. *Methods for chemical analysis of fresh waters.* London, International Biological Programme; Oxford, Edinburgh, Blackwell Scient. Publ. 172 p. (IBP Handbk 8.)

GOODALL, D. W. 1966. A new similarity index based on probability. *Biometrics*, vol. 22, no. 4, p. 882–907.

GORHAM, P. R. 1964. Toxic algae. In: D. F. Jackson (ed.), *Algae and man*, p. 307–36. New York, N.Y., Plenum Press.

GOVINDJEE; PAPAGEORGIOU, G.; RABINOWITCH, E. 1967. Chlorophyll fluorescence and photosynthesis. In: G. C. Guilbault (ed.), *Fluorescence. Theory, instrumentation, and practice*, p. 511–64. New York, N.Y., Marcel Dekker.

GRAHAM, H. W. 1942. Studies in the morphology, taxonomy and ecology of the Peridiniales. *Scient. Results Cruise VII Carnegie 1928–1929*, Biol., vol. 3, p. 1–129.

GRAN, H. H. 1902. Das Plankton des norwegischen Nordmeeres von biologischen und hydrographischen Gesichtspunkten behandelt. *Rep. norw. Fish. mar. Invest.*, vol. 2, no. 5, p. 1–222, pl. 1.

——. 1932. Phytoplankton. Methods and problems. *J. Cons. CIEM*, vol. 7, no. 3, p. 343–58.

GRAN, H. H.; BRAARUD, T. 1935. A quantitative study of the phytoplankton in the bay of Fundy and the gulf of Maine (including observations on hydrography, chemistry and turbidity). *J. biol. Board Can.*, vol. 1, no. 5, p. 279–467.

GRANT, B. R.; KERR, J. D. 1970. Phytoplankton numbers and species at Port Hacking station and their relationships to the physical environment. *Aust. J. mar. Freshwat. Res.*, vol. 21, no. 1, p. 34–45.

GRASSÉ, P.-P. (ed.). 1952. *Traité de zoologie.* T. I (1), *Phylogénie Protozoaires: généralités, Flagellés.* Paris, Masson. 1071 p., pl. 1.

GUILLARD, R. R. L. 1973. Division rates. In: J. R. Stein (ed.), *Handbook of phycological methods, Culture methods and growth measurements*, p. 289–311. London, Cambridge Univ. Press.

GUILLARD, R. R. L.; KILHAM, P.; JACKSON, T. A. 1973. Kinetics of silicon-limited growth in the marine diatom *Thalassiosira pseudonana* Hasle and Heimdal (= *Cyclotella nana* Hustedt). *J. Phycol.*, vol. 9, no. 3, p. 233–7.

GUILLARD, R. R. L.; RYTHER, J. H. 1962. Studies of marine planktonic diatoms, I. *Cyclotella nana* Hustedt and *Detonula confervacea* (Cleve) Gran. *Can. J. Microbiol.*, vol. 8, no. 2, p. 229–39.

GUNTER, G.; WILLIAMS, R. H.; DAVIS, C. C.; SMITH, F. G. W. 1948. Catastrophic mass mortality of marine animals and coincident phytoplankton bloom on the west coast of Florida, November 1946 to August 1947. *Ecol. Monogr.*, vol. 18, no. 3, p. 309–24.

HAECKEL, E. 1887. Report of the Radiolaria collected by H.M.S. Challenger during the years 1873–76. *Rep. scient. Results Voyage Challenger*, Zool., vol. 18, 188 + 1803 p., 140 pl., 1 map.

HALLDAL, P.; MARKALI, J.; NAESS, T. 1954. A method for transferring objects from a light microscope to marked areas on electron microscope grids. *Mikroskopie*, vol. 9, no. 5–6, p. 197–200.

HALME, E.; LUKKARINEN, T. 1960. Planktologische Untersuchungen in der Pojo-Bucht und angrenzenden Gewassern. V. Die Ziliaten *Tintinnopsis tubulosa* Levander und *T. brandti* (Nordqv.) Levander. *Ann. zool. Soc. zool.-bot. Fenn.*, vol. 22, no. 2, p. 1–24.

HAMILTON, R. D.; PRESLAN, J. E. 1969. Cultural characteristics of a pelagic marine hymenostome ciliate, *Uronema* sp. *J. exp. mar. Biol. Ecol.*, vol. 4, no. 1, p. 90–9.

HANNA, G. D. 1930. Hyrax, a new mounting medium for diatoms. *J. R. microsc. Soc.*, ser. 3, vol. 50, no. 4, p. 424–6.

——. 1949. A synthetic resin which has unusual properties. *J. R. microsc. Soc.*, ser. 3, vol. 69, no. 1, p. 25–8.

HARDY, A. C. 1939. Ecological investigations with the Continuous Plankton Recorder: object, plan and methods. *Hull Bull. mar. Ecol.*, vol. 1, no. 1, p. 1–57.

HART, T. J. 1966. Some observations on the relative abundance of marine phytoplankton populations in nature. In: H. Barnes (ed.), *Some contemporary studies in marine science*, p. 375–93. London, Allen & Unwin.

HARVEY, G. W. 1966. Microlayer collection from the sea surface: a new method and initial results. *Limnol. Oceanogr.*, vol. 11, no. 4, p. 608–13.

HARVEY, G. W.; BURZELL, L. A. 1972. A simple microlayer method for small samples. *Limnol. Oceanogr.*, vol. 17, no. 1, p. 156–7.

HASLE, G. R. 1950. Phototactic vertical migrations in marine dinoflagellates. *Oikos*, vol. 2, no. 2, p. 162–75.

——. 1954. The reliability of single observations in phytoplankton surveys. *Nytt Mag. Bot.*, vol. 2, p. 121–37.

——. 1959. A quantitative study of phytoplankton from the equatorial Pacific. *Deep-sea Res. oceanogr. Abstr.*, vol. 6, no. 1, p. 38–59.

——. 1969. An analysis of the phytoplankton of the Pacific southern ocean: abundance, composition, and distribution during the Brategg expedition, 1947–48. *Hvalraad. Skr.*, vol. 52, p. 1–168.

——. 1972. The distribution of *Nitzschia seriata* Cleve and allied species. *Nova Hedwigia (Beih.)*, vol. 39, p. 171–90, 1 pl. (First Symp. Recent and Fossil Mar. Diatoms, ed. R. Simonsen.)

——. 1976. The biogeography of some marine planktonic diatoms. *Deep-sea Res. oceanogr. Abstr.*, vol. 23, no. 4, p. 319–38, 6 pl.

——. 1977. The use of electron microscopy in morphological and taxonomical studies. In: D. Werner (ed.), *The biology of diatoms*, p. 18–23. Oxford, London, Edinburgh, Melbourne, Blackwell Scient. Publ. (Botanical Monogrs 13.)

HASLE, G. R.; FRYXELL, G. A. 1970. Diatoms: cleaning and mounting for light and electron microscopy. *Trans. am. microsc. Soc.*, vol. 89, no. 4, p. 469–74.

HATCHER, R. F.; PARKER, B. C. 1974a. Laboratory comparisons of four surface microlayer samplers. *Limnol. Oceanogr.*, vol. 19, no. 1, p. 162–5.

——. 1974b. Microbiological and chemical enrichment of freshwater-surface microlayers relative to the bulk-subsurface water. *Can. J. Microbiol.*, vol. 20, no. 7, p. 1051–7.

HAURY, L. R. 1976a. Small-scale pattern of a California current zooplankton assemblage. *Mar. Biol.*, vol. 37, no. 2, p. 137–57.

——. 1976b. A comparison of zooplankton patterns in the California current and North Pacific Central Gyre. *Mar. Biol.*, vol. 37, no. 2, p. 159–67.

HECKY, R. E.; KILHAM, P. 1974. Environmental control of phytoplankton cell size. *Limnol. Oceanogr.*, vol. 19, no. 2, p. 361–6.

HEIP, C. 1974. A new index measuring evenness. *J. mar. biol. Assoc. U.K.*, vol. 54, no. 3, p. 555–7.

HELLAND-HANSEN, B.; NANSEN, F. 1925. The eastern North Atlantic, *Geofys. Publr*, vol. 4 [1927], p. 1–76, 71 tables/charts.

HENSEN, [V.]. 1887. Ueber die Bestimmung des Planktons oder des im Meere treibenden Materials an Pflanzen und Thieren. *Ber. Komm. wiss. Untersuch. Kiel*, vol. 5, p. 1–108, pl. 1–6, tabl.

HERDMAN, H. F. P. 1963. Water sampling and thermometers. In: M. N. Hill (ed.), *The sea*. Vol. II, p. 124–7. New York, N.Y., London, Interscience.

309

M

HERON, A. C. 1968. Plankton gauze. In: Anon. (ed.), *Zooplankton sampling,* p. 19–25. Paris, Unesco. (Monogr. oceanogr. Methodol., 2.)

HERZENBERG, L. A.; SWEET, R. G.; HERZENBERG, L. A. 1976. Fluorescence-activated cell sorting. *Sci. Am.,* vol. 234, no. 3, p. 108–17.

HIBBERD, D. J. 1977. Observations on the ultrastructure of the cryptomonad endosymbiont of the red-water ciliate *Mesodinium rubrum. J. mar. biol. Assoc. U.K.,* vol. 57, no. 1. p. 45–61, pl. 1–4.

HICKEL, W. 1967. Untersuchungen über die Phytoplanktonblüte in der westlichen Ostsee. *Helgol. wiss. Meeresunters.,* vol. 16, no. 1–2, p. 1–66.

HOBRO, R.; WILLÉN, E. 1975. Phytoplankton countings and volume calculations from the Baltic. A method comparison. *Vatten,* vol. 31, no. 4, p. 317–26.

——. 1977. Phytoplankton countings. Intercalibration results and recommendations for routine work. *Int. Rev. gesamt. Hydrobiol.,* vol. 62, no. 6, p. 805–11.

HOFKER, J. 1930. Besprechung: Kofoid, C. A. und A. S. Campbell, A conspectus of the marine and freshwater Ciliata belonging to the suborder Tintinnoinea . . . . *Naturwiss.,* vol. 18, no. 18, p. 395–6.

HOLLAND, F. A.; CHAPMAN, F. S. 1966. *Pumping of liquids.* New York, N.Y., Reinhold. 406 p.

HOLLANDE, A.; ENJUMET, M. 1960. Cytologie, évolution et systématique des Sphaeroïdés (Radiolaires). *Arch. Mus. natl. Hist. nat. Paris,* sér. 7, vol. 7, p. 1–134, pl. 1–64.

HOLLANDER, M.; WOLFE, D. A. 1973. *Nonparametric statistical methods.* New York, N.Y., Wiley & Sons. 503 p.

HOLLOWAY, J. D.; JARDINE, N. 1968. Two approaches to zoogeography: a study based on the distributions of butterflies, birds and bats in the Indo-Australian area. *Proc. Linn. Soc. Lond.,* vol. 179, no. 2, p. 153–88, 2 pl.

HOLMES, R. W. 1956. The annual cycle of phytoplankton in the Labrador sea, 1950–51. *Bull. Bingham oceanogr. Collect.,* vol. 16, no. 1, p. 1–74.

——. 1962. The preparation of marine phytoplankton for microscopic examination and enumeration on molecular filters. *Spec. scient. Rep. U.S. Fish Wildl. Serv., Fisheries,* vol. 433, p. 1–6.

——. 1966. Light microscope observations on cytological manifestations of nitrate, phosphate, and silicate deficiency in four marine centric diatoms. *J. Phycol.,* vol. 2, no. 4, p. 136–40.

——. 1967. The dissolution of protoplasm in thecate dinoflagellates by ultraviolet induced oxidation. *Stain Technol.,* vol. 42, no. 1, p. 34–5.

HOLMES, R. W.; ANDERSON, G. C. 1963. Size fractionation of C¹⁴-labelled natural phytoplankton communities. In: C. H. Oppenheimer (ed.), *Symposium on marine microbiology,* p. 241–50. Springfield, Thomas.

HOLMES, R. [W.]; NORRIS, R.; SMAYDA, T.; WOOD, E. J. F. 1969. Collection, fixation, identification, and enumeration of phytoplankton standing stock. In: Anon. (ed.), *Recommended procedures for measuring the productivity of plankton standing stock and related oceanic properties,* p. 17–46. Washington, Nat. Acad. Sci. (Biological Panel Committee on Oceanography. Div. Earth Sci., Nat. Res. Council.)

HOLMES, R. W.; REIMANN, B. E. F. 1966. Variation in valve morphology during the life cycle of the marine diatom *Coscinodiscus concinnus. Phycologia,* vol. 5, no. 4, p. 233–44.

HOLMES, R. W.; WIDRIG, T. M. 1956. The enumeration and collection of marine phytoplankton. *J. Cons. CIEM,* vol. 22, no. 1, p. 21–32.

HOLM-HANSEN, O.; PACKARD, T. T.; POMEROY, L. R. 1970. Efficiency of the reverse-flow filter technique for concentration of particulate matter. *Limnol. Oceanogr.,* vol. 15, no. 5, p. 832–5.

HOOK, [R.?]. 1666. An appendix to the directions for seamen, bound for far voyages. *Philos. Trans.,* vol. 1, no. 9, p. 147–9, 1 pl.

HOPPE, H.-G. 1976. Determination and properties of actively metabolizing heterotrophic bacteria in the sea, investigated by means of micro-autoradiography. *Mar. Biol.,* vol. 36, no. 4, p. 291–302.

HOWMILLER, R. P.; SLOEY, W. E. 1969. A horizontal water sampler for investigation of stratified water. *Limnol. Oceanogr.,* vol. 14, no. 2, p. 291–2.

HRBÁČEK, J. 1966. A morphometrical study of some backwaters and fish ponds in relation to the representative plankton samples. (With an appendix by C. O. Junge on depth distribution for quadric surfaces and other configurations.) In: J. Hrbaček (ed.), *Hydrobiological studies.* I, p. 221–65. Prague, Acad. Publ. House.

HUBER-PESTALOZZI, G. 1938–72. Das Phytoplankton des Süsswassers. Systematik und Biologie. In: A. Thienemann (ed.), *Die Binnengewässer*, XVI. Stuttgart, Schweizerbart'sche Verlag. Tl. 1: *Allgemeinen, Blaualgen, Bakterie, Pilze*, 342 p., 66 pl. (1938); Tl. 2: (1) *Chrysophyceen, Farblose Flagellaten, Heterokonten*, 365 p., 107 pl. (1941); (2), *Diatomeen*, 549 p. (1942); Tl. 3: *Cryptophyceen, Chloromonadinen, Peridineen*, 310 p., 69 pl. (1950); Tl. 4: *Euglenophyceen*, 606 p., 114 pl. (1955); Tl. 5: *Chlorophyceen, Volvocales*, 744 p., 158 pl. (1961); Tl. 6 (by B. Fott): *Chlorophyceae, Tetrasporales*, 116 p., 47 pl. (1972).

HULBURT, E. M. 1962. Phytoplankton in the southwestern Sargasso sea and North Equatorial current, February 1961. *Limnol. Oceanogr.*, vol. 7, no. 3, p. 307–15.

——. 1963. The diversity of phytoplanktonic populations in oceanic, coastal and estuarine regions. *J. mar. Res.*, vol. 21, no. 2, p. 81–93.

——. 1964. Succession and diversity in the plankton flora of the western North Atlantic. *Bull. mar. Sci.*, vol. 14, no. 1, p. 33–44.

——. 1966. The distribution of phytoplankton and its relationship to hydrography, between southern New England and Venezuela. *J. mar. Res.*, vol. 24, no. 1, p. 67–81.

HUSTEDT, F. 1927–66. Die Kieselalgen Deutschlands, Österreichs und der Schweiz mit Berücksichtigung der übrigen Länder Europas sowie der angrenzenden Meeresgebiete. In: L. Rabenhorst, *Kryptogamen-Flora*. Vol. 7: Tl. 1, p. 1–920 (1927–30); Tl. 2, p. 1–845 (1931–59); Tl. 3, p. 1–816 (1961–66). Leipzig, Akad. Verlag.

HUTCHESON, K. 1970. A test for comparing diversities based on the Shannon formula. *J. theor. Biol.*, vol. 29, no. 1, p. 151–4.

HUTCHINSON, G. E. 1961. The paradox of plankton. *Am. Nat.*, vol. 95, no. 882, p. 137–45.

——. 1967. *A treatise on limnology*. Vol. II. *Introduction to lake biology and the limnoplankton*. London, Sydney, Wiley & Sons. 1115 p.

IBANEZ, F. 1971. Effet des transformations des données dans l'analyse factorielle en écologie planctonique. *Cah. océanogr.*, vol. 23, no. 6, p. 545–61.

——. 1972. Interprétation de données écologiques par l'analyse des composantes principales: écologie planctonique de la mer du Nord. *J. Cons. CIEM*, vol. 34, no. 3, p. 323–40.

——. 1973a. Méthode d'analyse spatio-temporelle du processus d'échantillonnage en planctologie, son influence dans l'interprétation des données par l'analyse en composantes principales. *Ann. Inst. océanogr., Paris* (nouv. sér.), vol. 49, no. 2, p. 83–111.

——. 1973b. Choix d'une stratégie d'échantillonnage par la théorie des jeux pour l'étude *in situ* d'organismes planctoniques. *Bull. Un. Océanogr. France*, vol. 5, no. 3, p. 18–25.

——. 1974. Une cotation d'abondance réduite à trois classes: justification de son emploi en analyse des composantes principales. Mise en oeuvre et intérêt pratique en planctologie. *Ann. Inst. océanogr., Paris* (nouv. sér.), vol. 50, no. 2, p. 185–98.

——. 1976. Contribution à l'analyse mathématique des événements en écologie planctonique. Optimisations méthodologiques; étude expérimentale en continu à petite échelle de l'hétérogénéité du plancton côtier. *Bull. Inst. océanogr., Monaco*, vol. 72, no. 1431, p. 1–96.

JACCARD, P. 1908. Nouvelles recherches sur la distribution florale. *Bull. Soc. vaud. Sci. nat.*, vol. 44, no. 163, p. 223–70, pl. 10–20.

JACKSON, H. W.; WILLIAMS, L. G. 1962. Calibration and use of certain plankton counting equipment. *Trans. am. microsc. Soc.*, vol. 81, no. 1, p. 96–103.

JANE, F. W. 1942. Methods for the collection and examination of fresh-water algae, with special reference to Flagellates. *J. Queckett microsc. Club*, ser. 4, vol. 1, no. 5, p. 217–29.

JEFFREY, C. 1973. *Biological nomenclature*. London, Edward Arnold. 69 p.

JENSEN, A.; RYSTAD, B.; SKOGLUND, L. 1972. The use of dialysis culture in phytoplankton studies. *J. exp. mar. Biol. Ecol.*, vol. 8, no. 3, p. 241–8.

JENSEN, A.; SAKSHAUG, E. 1970. Producer-consumer relationships in the sea, II. Correlation between *Mytilus* pigmentation and the density and composition of phytoplanktonic populations in inshore waters. *J. exp. mar. Biol. Ecol.*, vol. 5, no. 3, p. 246–53.

JENSEN, W. A. 1962. *Botanical histochemistry. Principles and practice*. San Francisco, Calif., London, W. H. Freeman. 408 p.

JOERIS, L. S. 1964. A horizontal sampler for collection of water samples near the bottom. *Limnol. Oceanogr.*, vol. 9, no. 4, p. 595–8.

JOHANSEN, D. A. 1940. *Plant microtechnique*. New York, N.Y., London, McGraw–Hill. 523 p.

JOHNSON, B. D. 1976. Nonliving organic particle formation from bubble dissolution. *Limnol. Oceanogr.*, vol. 21, no. 3, p. 444–6.

311

JOHNSON, K. M.; BURNEY, C. M.; SIEBURTH, J. McN. (Submitted for publication.) Dissolved carbohydrate concentrations and their microbial cycling in the north Atlantic, pt I. Monosaccharide and dissolved organic carbon.

JOHNSON, K. M.; SIEBURTH, J. McN. 1977. Dissolved carbohydrates in seawater, I. A precise spectrophotometric analysis for monosaccharides. *Mar. Chem.,* vol. 5, no. 1, p. 1–13.

JONES, J. G. 1974. Some observations on direct counts of freshwater bacteria obtained with a fluorescence microscope. *Limnol. Oceanogr.,* vol. 19, no. 3, p. 540–3.

JØRGENSEN, C. B. 1966. *Biology of suspension feeding.* Oxford, Pergamon Press. 357 p.

KAHL, A. 1932. Urtiere oder Protozoa; I, Wimpertiere oder Ciliata (Infusoria); 3, Spirotricha; 2: Unterordnung Oligotricha Bütschli, 1889. In: F. Dahl (ed.), *Die Tierwelt Deutschlands,* vol. 25, p. 487–513. Jena, G. Fischer.

KAMYKOWSKI, D. 1974. Possible interactions between phytoplankton and semi-diurnal internal tides. *J. mar. Res.,* vol. 32, no. 1, p. 67–89.

KARSTEN, G. 1907. Das indische Phytoplankton nach dem Material der Deutschen Tiefsee-Expedition, 1898–1899. *Wiss. Ergebn. Dtsch. Tiefsee-Exped. Valdivia,* II, vol. 2, no. 3, p. 221–548, pl. 35–54.

KAWARADA, Y.; KITOU, M.; FURUHASHI, K.; SANO, A. 1968. Distribution of plankton in the waters neighbouring Japan in 1966 (CSK). *Oceanogr. Mag.,* vol. 20, no. 2, p. 187–212.

KAWARADA, Y.; KITOU, M.; FURUHASHI, K.; SANO, A.; KAROHJI, K.; KURODA, K.; ASAOKA, O.; MATSUZAKI, M.; OHWADA, M.; OGAWA, F. 1966. Distribution of plankton collected on board the research vessels of J.M.A. in 1965 (CSK). *Oceanogr. Mag.,* vol. 18, no. 1–2, p. 91–112.

KERR, S. R. 1974. Theory of size distribution in ecological communities. *J. Fish. Res. Board Can.,* vol. 31, no. 12, p. 1859–62.

KERŽAN, I. 1976. *Katalog pelagičnih alg Jadrana.* 34 p. (Mimeogr.)

KIEFER, D. A. 1973. Fluorescence properties of natural phytoplankton populations. *Mar. Biol.,* vol. 22, no. 3, p. 263–9.

KIERSTEAD, H.; SLOBODKIN, L. B. 1953. The size of water masses containing plankton blooms. *J. mar. Res.,* vol. 12, no. 1, p. 141–7.

KIMBALL, J. F. Jr; WOOD, E. J. F. 1964. A simple centrifuge for phytoplankton studies. *Bull. mar. Sci. Gulf Caribb.,* vol. 14, no. 4, p. 539–44.

KING, J. E.; HIDA, T. S. 1954. Variations in zooplankton abundance in Hawaiian waters, 1950–52. *Spec. scient. Rep. U.S. Fish Wildl. Serv., Fish.,* vol. 118, p. 1–66.

KISSELEV, I. A. 1969. (*Plankton of the seas and continental waters,* I. Introduction and general problem of the planktology.) Leningrad, Izdat. Nauka. 658 p. (In Russian.)

KJENSMO, J. 1967. A water sampler for investigation of 'micro-stratification' in lakes. *Schweiz. Z. Hydrol.,* vol. 29, no. 2, p. 361–5.

KNIGHT-JONES, E. W. 1951. Preliminary studies of nannoplankton and ultraplankton systematics and abundance by a quantitative culture method. *J. Cons. CIEM,* vol. 17, no. 2, p. 140–55.

KNOX, C. 1966. Holographic microscopy as a technique for recording dynamic microscopic subjects. *Science* (Wash.), vol. 153, no. 3739, p. 989–90.

KNUDSEN, M. 1929. A frameless reversing waterbottle. *J. Cons. CIEM,* vol. 4, no. 2, p. 192–3.

KOBLENTZ-MISHKE, O. J.; VOLKOVINSKY, V. V.; KABANOVA, J. G. 1970. Plankton primary production of the world ocean. In: W. S. Wooster (ed.), *Scientific exploration of the South Pacific,* p. 183–93. Washington, Nat. Acad. Sci.

KOFOID, C. A. 1897. Plankton studies, I. Methods and apparatus in use in plankton investigations at the Biological Experiment Station of the University of Illinois. *Bull. Ill. St. Lab. nat. Hist.,* vol. 5, no. 1, p. 1–25, pl. 1–7.

KOFOID, C. A.; CAMPBELL, A. S. 1929. A conspectus of the marine and fresh-water Ciliata belonging to the Suborder Tintinnoinea, with descriptions of new species principally from the Agassiz Expedition to the eastern tropical Pacific, 1904–1905. *Univ. Calif. Publ. Zool.,* vol. 34, p. 1–403.

——. 1939. The Ciliata: the Tintinnoinea. *Bull. Mus. compar. Zool. Harv. Coll.,* vol. 84, p. 1–473, pl. 1–36. (Rep. sci. Results exped. east. trop. Pacif., in charge of Alexander Agassiz . . . , 37.)

KOLKWITZ, R. 1907. Entnahme- und Beobachtungsinstrumente für biologische Wasseruntersuchungen. *Mitt. PrüfAnst. WassVersorg. Abwässerbeseit. Berlin,* vol. '1907', no. 9, p. 111–44.

KOLKWITZ, R.; KRIEGER, H. 1941–44. Zygnemales. In: L. Rabenhorst, *Kryptogamen-Flora,* vol. 13, no. 2, 499 p. Leipzig, Akad. Verlag.

312

KOLKWITZ, R.; MARSSON, M. 1908. Ökologie der pflanzlichen Saprobien. *Ber. Dtsch. bot. Ges.*, vol. 26a, p. 505–19.

KOLLMER, W. E. 1962. The annual cycle of phytoplankton in the waters off Walvis bay 1958. *Mar. Res. Lab. Admin. S.W. Africa, Res. Rep.*, vol. 4, p. 1–44.

KOLTSOVA, T. I.; KONOPLYA, L. A.; MAKSIMOV, V. N.; FEDOROV, V. D. 1971. (Representative sampling in analysis of plankton.) *Gidrobiol. Zh.*, vol. 7, no. 3, p. 109–16. (In Russian.) (Transl.: *Hydrobiol. J.*)

KOVALA, P. E.; LARRANCE, J. D. 1966. *Computation of phytoplankton cell numbers, cell volume, cell surface and plasma volume per liter, from microscopical counts*, p. 1–91. Seattle, Dept. Oceanogr. Univ. Washington. (Spec. rep. 38.)

KOZLOVA, O. G. 1964. (*Diatoms of the Indian and Pacific sectors of the Antarctic.*) Moscow, Izdat. Nauka. 169 p., 6 pl. + map. (In Russian.)

KRIEGER, W. 1933–39. Die Desmidiaceen Europas mit Berücksichtigung der aussereuropäischen Arten. In: L. Rabenhorst, *Kryptogamen-Flora.* Vol. 13 (1), I, p. 1–712, pl. 1–96 (1933–37); II, p. 1–117, pl. 97–142 (1939). Leipzig, Akad. Verlag.

KRUMBEIN, W. C.; PETTIJOHN, F. J. 1938. *Manual of sedimentary petrography.* New York, N.Y., Appleton-Century-Crofts. 549 p.

KRYLOV, V. V. 1968. (On the affinity of species in connection with plankton studies.) *Okeanologiya*, vol. 8, no. 2, p. 301–11. (In Russian.) (Transl.: *Oceanology.*)

KUENZLER, E. J.; PERRAS, J. P. 1965. Phosphatases of marine algae. *Biol. Bull. mar. biol. Lab. Woods Hole, Mass.*, vol. 128, no. 2, p. 271–84.

KUSJMINA, A. I. 1962. (Phytoplankton of the Kuril Straits as an indicator of different water bodies.) In: *Issled. Dalneu. Morei CCCP*, 8 [Explor. Far-East Seas SSSR, 8], p. 6–90, 3 pl. Moscow, Leningrad, Izdat. Nauka. (In Russian.)

KUTKUHN, J. H. 1958. Notes on the precision of numerical and volumetric plankton estimates from small-sample concentrates. *Limnol. Oceanogr.*, vol. 3, no. 1, p. 69–83.

KUWAHARA, A.; ABE, T.; SHIMONO, M.; KASHIWAI, M.; SINODA, M. 1973. A pump-sampler as a method of collecting plankton and its preliminary operation. *J. oceanogr. Soc. Jap.*, vol. 29, no. 3, p. 106–12.

LAURITIS, J. A.; HEMMINGSEN, B. B.; VOLCANI, B. E. 1967. Propagation of *Hantzschia* sp. Grunow daughter cells by *Nitzschia alba* Lewin and Lewin. *J. Phycol.*, vol. 3, no. 4, p. 236–7.

LAVOIE, D. M. 1975. Application of diffusion culture to ecological observations on marine micro-organisms. Kingston, Univ. Rhode Island. 91 p. (MS thesis.)

LAVOIE, D. M.; SIEBURTH, J. McN. (Submitted for publication.) Dissolved carbohydrate concentrations and their microbial cycling in the north Atlantic, pt III. Bacterioplankton and total microbial plankton.

LAWS, E. A. 1975. The importance of respiration losses in controlling the size distribution of marine phytoplankton. *Ecology*, vol. 56, no. 2, p. 419–26.

LEADBEATER, B. S. C. 1972. Identification, by means of electron microscopy, of flagellate nano-plankton from the coast of Norway. *Sarsia*, vol. 49, p. 107–24, pl. 1–4.

——. 1974. Ultrastructural observations on nanoplankton collected from the coast of Yugoslavia and the bay of Algiers. *J. mar. biol. Assoc.U.K.*, vol. 54, no. 1, p. 179–96, pl. 1–7.

LEBOUR, M. V. 1925. *The Dinoflagellates of northern seas.* Plymouth, Marine Biological Association. 250 p., incl. 35 pl.

LEEGAARD, C. 1915. Untersuchungen über einige Planktonciliaten des Meeres. *Nytt. Mag. Naturvid.*, vol. 53, no. 1–2, p. 1–37.

LEGENDRE, L. 1973. Phytoplankton organization in Baie des Chaleurs (Gulf of St Lawrence). *J. Ecol.*, vol. 61, no. 1, p. 135–49.

LEGENDRE, L.; WATT, W. D. 1972. On a rapid technique for plankton enumeration. *Ann. Inst. océanogr. Paris* (nouv. sér.), vol. 48, no. 2, p. 173–7.

LEGENDRE, P.; ROGERS, D. J. 1972. Characters and clustering in taxonomy: a synthesis of two taxi-metric procedures. *Taxon*, vol. 21, no. 5–6, p. 567–606.

LÉGER, G. 1971a. Les populations phytoplanctoniques au point ϕ = 42°47′N, G = 7°29′E Greenwich, bouée-laboratoire du C.O.M.E.X.O./C.N.E.X.O., A. Généralités et premier séjour (21–27 février 1964). *Bull. Inst. océanogr. Monaco*, vol. 69, no. 1412A, p. 1–42, 1 pl.; 1412B, tabl.

——. 1971b. Les populations phytoplanctoniques . . . , B. Deuxième séjour (17–30 juillet 1964). *Bull. Inst. océanogr. Monaco*, vol. 70, no. 1413A, p. 1–41; 1413B, tabl.

Phytoplankton manual

LENZ, J. 1968. Bestandaufnahme A. Plankton. In: C. Schlieper (ed.), *Methoden der meeres-biologischen Forschung*, p. 48–62. Jena, G. Fischer Verlag.
——. 1972. A new type of plankton pump on the vacuum principle. *Deep-sea Res. oceanogr. Abstr.*, vol. 19, no. 6, p. 453–9, 1 pl.
LEVANDOWSKY, M. 1972. An ordination of phytoplankton populations in ponds of varying salinity and temperature. *Ecology*, vol. 53, no. 3, p. 398–407.
LEWIS, W. M. Jr. 1976. Surface/volume ratio: implications for phytoplankton morphology. *Science* (Wash.), vol. 192, no. 4242, p. 885–7.
LIEBMANN, H. 1962. *Handbuch der Frischwasser- und Abwasser-Biologie*, I. München, R. Oldenbourg. 539 p.
LIGHTHART, B. 1969. Planktonic and benthic bacteriovorous Protozoa at eleven stations in Puget Sound and adjacent Pacific ocean. *J. Fish. Res. Board Can.*, vol. 26, no. 2, p. 299–304.
LISITSYN, A. P. 1962. (Using of large-capacity pumps for collecting deep-sea samples.) *Trudy Inst. Okeanol.*, vol. 55, p. 137–68. (In Russian.)
LITTLEFORD, R. A.; NEWCOMBE, C. L.; SHEPHERD, B. B. 1940. An experimental study of certain quantitative plankton methods. *Ecology*, vol. 21, no. 3, p. 309–22.
LLOYD, M.; GHELARDI, R. J. 1964. A table for calculating the 'equitability' component of species diversity. *J. anim. Ecol.*, vol. 33, no. 2, p. 217–25.
LOCICERO, V. R. (ed.) 1975. *Proceedings of the first International Conference on Toxic Dinoflagellate Blooms, November 1974.* Boston, Mass., Wakefield, Mass. Sci. Technol. Fdn. 541 p. + bibliography, by L. A. Loeblich and A. R. Loeblich III, 62 p.
LOEBLICH, A. R. Jr; TAPPAN, H. 1966. Annotated index and bibliography of the calcareous nannoplankton. *Phycologia*, vol. 5, no. 2–3, p. 81–216.
LOHMANN, H. 1908. Untersuchungen zur Feststellung des vollständigen Gehaltes des Meeres an Plankton. *Wiss. Meeresunters. Kiel*, N.F., vol. 10, p. 129–370, pl. A–B, 9–17.
——. 1911. Über das Nannoplankton und die Zentrifugierung kleinster Wasserproben zur Gewinnung desselben in lebendem Zustande. *Int. Rev. gesamt. Hydrobiol. Hydrogr.*, vol. 4, no. 1–2, 1–38, pl. 1–5.
——. 1920. Die Bevölkerung des Ozeans mit Plankton. Nach den Ergebnissen der Zentrifugenfänge während der Ausreise der 'Deutschland' 1911. Zugleich ein Beitrag zur Biologie des Atlantischen Ozeans. Tl. I. *Arch. Biontol., Ges. Naturforsch. Freunde Berlin*, vo.. 4, no. 3, p. 1–470, figs.
LONGHURST, A. R.; WILLIAMS, R. 1976. Improved filtration systems for multiple-serial plankton samplers and their deployment. *Deep-sea Res. oceanogr. Abstr.*, vol. 23, no. 11, p. 1067–73, 4 pl.
LOPEZ BALUJA, L. 1976. (*Phytoplankton of Cuban waters.*) Univ. Moscow. 141 p., 26 tabl. (Doct. thesis.) (In Russian.)
LORENZEN, C. J. 1966. A method for the continuous measurement of *in vivo* chlorophyll concentration. *Deep-sea Res. oceanogr. Abstr.*, vol. 13, no. 2, p. 223–7.
LOVEGROVE, T. 1960. An improved form of sedimentation apparatus for use with an inverted microscope. *J. Cons. CIEM*, vol. 25, no. 3, p. 279–84.
LOVELAND, R. P. 1970. *Photomicrography, a comprehensive treatise.* Vol. II, chap. 20, p. 899–934: Fluorescence microscopy, New York, N.Y., Wiley & Sons.
LUCAS, C. E. 1940. Ecological investigations with the Continuous Plankton Recorder: the phytoplankton in the southern North sea, 1932–1937. *Hull. Bull. mar. Ecol.*, vol. 1, p. 73–170.
——. 1941. Continuous plankton records: phytoplankton in the North sea, 1938–1939; pt I. Diatoms. *Hull Bull. mar. Ecol.*, vol. 2, no. 8, p. 19–46, pl. 6–38.
——. 1942. Continuous plankton records: phytoplankton in the North sea, 1938–39, II. Dinoflagellates, *Phaeocystis* etc. *Hull Bull. mar. Ecol.*, vol. 2, no. 9, p. 47–70, pl. 39–62.
LUMBY, J. R. 1927. The surface sampler, an apparatus for the collection of samples from the sea surface from ships in motion. With a note on surface temperature observations. *J. Cons. CIEM*, vol. 2, no. 3, p. 332–42.
LUND, J. W. G. 1949. Studies on *Asterionella*, I. The origin and nature of the cells producing seasonal maxima. *J. Ecol.*, vol. 37, no. 2, p. 389–419.
——. 1951. A sedimentation technique for counting algae and other organisms. *Hydrobiologia*, vol. 3, no. 4, p. 390–4.
LUND, J. W. G.; KILPLING, C.; LE CREN, E. D. 1958. The inverted microscope method of estimating algal numbers, and the statistical basis of estimation by counting. *Hydrobiologia*, vol. 11, no. 2, p. 143–70.

314

LUND, J. W. G.; TALLING, J. F. 1957. Botanical limnological methods with special reference to the algae. *Bot. Rev.*, vol. 23, no. 8–9, p. 489–583.

MACADAM, R. B. 1971. Preservation of organic-walled microfossil types studied by scanning electron microscopy. *J. Paleont.*, vol. 45, no. 1, p. 140.

MCALICE, B. J. 1970. Observations on the small-scale distribution of estuarine phytoplankton. *Mar. Biol.*, vol. 7, no. 2, p. 100–11.

——. 1971. Phytoplankton sampling with the Sedgwick-Rafter cell. *Limnol. Oceanogr.*, vol. 16, no. 1, p. 19–28.

MCCARTHY, J. J.; TAYLOR, W. R.; LOFTUS, M. E. 1974. Significance of nanoplankton in the Chesapeake bay estuary and problems associated with the measurement of nanoplankton productivity. *Mar. Biol.*, vol. 24, no. 1, p. 7–16.

MCEWEN, G. F.; JOHNSON, M. W.; FOLSOM, T. R. 1954. A statistical analysis of the performance of the Folsom plankton sample splitter, based upon test observations. *Arch. Metorol., Geophys. Bioklimatol.*, ser. A, vol. 7, p. 502–27.

MCGOWAN, J. A. 1971. Oceanic biogeography of the Pacific. In: B. M. Funnell and W. R. Riedel (eds.), *The micropaleontology of oceans*, p. 3–74. London, Cambridge Univ. Press.

MCGOWAN, J. A.; BROWN, D. M. 1966. *A new opening-closing paired zooplankton net.* Scripps Inst. Oceanogr. 56 p. (Unpubl. rep. 66–23.)

MCGOWAN, J. A.; FRAUNDORF, V. J. 1964. A modified heavy fraction zooplankton sorter. *Limnol. Oceanogr.*, vol. 9, no. 1, p. 152–5.

MCINTIRE, C. D.; OVERTON, W. S. 1971. Distributional patterns in assemblages of attached diatoms from Yaquina estuary, Oregon. *Ecology*, vol. 52, no. 5, p. 758–77.

MCINTYRE, A.; BÉ, A. W. H. 1967. Modern Coccolithophoridae of the Atlantic ocean, I. Placoliths and cyrtoliths. *Deep-sea Res. oceanogr. Abstr.*, vol. 14, no. 5, p. 561–97, pl. 1–12.

MCINTYRE, A.; BÉ, A. W. H.; ROCHE, M. B. 1970. Modern Pacific Coccolithophorida: a palaeontological thermometer. *Trans. N.Y. Acad. Sci.*, ser. II, vol. 32, no. 6, p. 720–31.

MCNABB, C. D. 1960. Enumeration of freshwater phytoplankton concentrated on the membrane filter. *Limnol. Oceanogr.*, vol. 5, no. 1, p. 57–61.

MADDUX, W. S.; KANWISHER, J. W. 1965. An in situ particle counter. *Limnol. Oceanogr.*, vol. 10, Suppl., p. R162–8.

MALONE, T. C. 1971. The relative importance of nannoplankton and net-plankton as primary producers in tropical oceanic and neritic phytoplankton communities. *Limnol. Oceanogr.*, vol. 16, no. 4, p. 633–9.

MANTON, I.; LEADBEATER, B. S. C. 1974. Fine-structural observations on six species of *Chrysochromulina* from wild Danish marine nanoplankton, including a description of *C. campanulifera* sp. nov. and a preliminary summary of the nanoplankton as a whole. *Biol. Skr.*, vol. 20, no. 5, p. 1–26, pl. 1–12.

MANTON, I.; PARKE, M. 1960. Further observations on small green Flagellates with special reference to possible relatives of *Chromulina pusilla* Butcher, *J. Mar. biol. Assoc. U.K.*, vol. 39, no. 2, p. 275–98, pl. 1–9.

MARBACH, A.; VARON, M.; SHILO, M. 1976. Properties of marine bdellovibrios. *Microbial Ecol.*, vol. 2, no. 4, p. 284–95.

MARCUSE, S. 1949. Optimum allocation and variance components in nested sampling with an application to chemical analysis. *Biometrics*, vol. 5, no. 3, p. 189–206.

MARGALEF, R. 1956a. Información y diversidad específica en las comunidades de organismos. *Invest. Pesq.*, vol. 3, p. 99–106.

——. 1956b. Estructura y dinámica de la 'purga de mar' en la Ría de Vigo. *Invest. Pesq.*, vol. 5, p. 113–34.

——. 1957a. Variación local e interanual en la secuencia de las poblaciones de fitoplancton de red en las aguas superficiales de la costa mediterránea española. *Invest. Pesq.*, vol. 9, p. 65–95.

——. 1957b. La teoría de la información en ecología. *Mem. R. Acad. Cienc. Art. Barcelona*, vol. 32, no. 13, p. 373–449.

——. 1958. Temporal succession and spatial heterogeneity in phytoplankton. In: A. A. Buzzati-Traverso (ed.), *Perspectives in marine biology*, p. 323–49. Berkeley, Los Angeles, Calif., Univ. Calif. Press; Paris, Union int. Sci. biol.

——. 1961a. Distribución ecológica y geográfica de les especies del fitoplancton marino. *Invest. Pesq.*, vol. 19, p. 81–101.

———. 1961b. Corrélations entre certains caractères synthétiques des populations de phytoplancton. *Hydrobiologia*, vol. 18, no. 1–2, p. 155–64, 1 tabl.

———. 1963. Succession in marine populations. In: R. Vira (ed.), *Advancing frontiers of plant science*, vol. 2, p. 137–88. New Delhi, Inst. Adv. Scient. Cult.

———. 1966. Análisis y valor indicador de las comunidades de fitoplancton mediterráneo. *Invest. Pesq.*, vol. 30, p. 429–82.

———. 1969a. Estudios sobre la distribución a pequeña escala del fitoplancton marino. *Mem. R. Acad. Cienc. Art. Barcelona*, vol. 40, no. 1, p. 1–22, 1 tabl.

———. 1969b. Diversidad de fitoplancton de red en dos áreas del Atlántico. *Invest. Pesq.*, vol. 33, no. 1, p. 275–86.

———. 1969c. Counting. In: R. A. Vollenweider, J. F. Talling and D. F. Westlake (eds.), *A manual on methods for measuring primary production in aquatic environments including a chapter on bacteria*, p. 7–14. London, International Biological Programme; Oxford and Edinburgh, Blackwell Scient. Publ. (IBP Handbk 12.)

———. 1969d. Composición específica del fitoplancton de la costa catalano-levantina (Mediterráneo occidental) en 1962–1967. *Invest. Pesq.*, vol. 33, no. 1, p. 345–80.

———. 1971. Small scale distribution of phytoplankton in western Mediterranean at the end of July. *Pubbl. Staz. zool. Napoli*, vol. 37, Suppl., p. 40–61.

———. 1974. *Ecología*. Barcelona, Omega. 951 p.

MARGALEF, R.; DURÁN, M. 1953. Microplancton de Vigo, de octubre de 1951 a septiembre de 1952. *Publ. Inst. Biol. apl.*, vol. 13, p. 5–78, tabl.

MARGALEF, R.; GONZÁLEZ BERNÁLDEZ, F. 1969. Grupos de especies asociadas en el fitoplancton del mar Caribe (NE de Venezuela). *Invest. Pesq.*, vol. 33, no. 1, p. 287–312.

MARSHALL, K. C. 1976. *Interfaces in microbial ecology*. Cambridge, Mass., London, Harvard Univ. Press. 156 p.

MARSHALL, S. M. 1969. Protozoa. Order: Tintinnida. Family: . . . [etc.]. In: J. H. Fraser and V. Kr. Hansen (eds.), *Fiches d'identification du zooplancton*, no. 117, p. 1–12; no. 118, p. 1–5; no. 119, p. 1–7; no. 120, p. 1–6; no. 121, p. 1–6; no. 122, p. 1–8; no. 123, p. 1–4; no. 124, p. 1–6; no. 125, p. 1–5; no. 126, p. 1–5; no. 127, p. 1–6. Charlottenlund, Cons perm. int. Explor. Mer.

MARTIN, D. F. 1968. *Marine chemistry*, I. *Analytical methods*. New York, N.Y., Marcel Dekker. 280 p.

MEDWIN, H. 1970. *In situ* acoustic measurements of bubble populations in coastal ocean waters. *J. geophys. Res.*, vol. 75, no. 3, p. 599–611.

MEEK, G. A. 1976. *Practical electron microscopy for biologists*. 2nd ed. London, New York, N.Y., Sydney, Toronto, J. Wiley & Sons. 528 p.

MENZEL, D. W. 1977. Summary of experimental results: Controlled Ecosystem Pollution Experiment. *Bull. mar. Sci.*, vol. 27, no. 1, p. 142–5.

MENZEL, D. W.; DUNSTAN, W. M. 1973. Growth measurements by analysis of carbon. In: J. R. Stein (ed.), *Handbook of phycological methods. Culture methods and growth measurements*, p. 313–20. London, Cambridge Univ. Press.

MILLER, C. B. 1970. Some environmental consequences of vertical migration in marine zooplankton. *Limnol. Oceanogr.*, vol. 15, no. 5, p. 727–41.

MILNE, A. 1959. The centric systematic area-sample treated as a random sample. *Biometrics*, vol. 15, no. 2, p. 270–97.

MÖLLER, F.; BERNHARD, M. 1974. A sequential approach to the counting of plankton organisms. *J. exp. mar. Biol. Ecol.*, vol. 15, no. 1, p. 49–68.

MOOD, A. M.; GRAYBILL, F. A.; BOES, D. C. 1974. *Introduction to the theory of statistics*. 3rd ed. New York, N.Y., McGraw-Hill, 564 p.

MOORE, J. K. 1963. Refinement of a method for filtering and preserving marine phytoplankton on a membrane filter. *Limnol. Oceanogr.*, vol. 8, no. 2, p. 304–5.

MOROZOVA-VODIANITSKAYA, N. V. 1954. (Phytoplankton of the Black sea part II.) *Trudy Sevastopol biol. Stat.*, vol. 8, p. 11–99. (In Russian.)

MOSS, B. 1973. Diversity in fresh-water phytoplankton. *Am. Midl. Nat.*, vol. 90, no. 2, p. 341–55.

MOTODA, S. 1959. Devices of simple plankton apparatus. *Mem. Fac. Fish. Hokkaido Univ.*, vol. 7, no. 1–2, p. 73–94.

MULFORD, R. A. 1972. Annual plankton cycle on the Chesapeake bay in the vicinity of Calvert Cliffs, Maryland, June 1969–May 1970. *Proc. Acad. nat. Sci. Philad.*, vol. 124, no. 3, p. 17–40.

MULLIGAN, H. F.; KINGSBURY, J. M. 1968. Application of an electronic particle counter in analysing natural populations of phytoplankton. *Limnol. Oceanogr.*, vol. 13, no. 3, p. 499–506.

MULLIN, M. M.; SLOAN, P. R.; EPPLEY, R. W. 1966. Relationship between carbon content, cell volume, and area in phytoplankton. *Limnol. Oceanogr.*, vol. 11, no. 2, p. 307–11.

MUNRO, A. L. S.; BROCK, T. D. 1968. Distinction between bacterial and algal utilization of soluble substances in the sea. *J. gen. Microbiol.*, vol. 51, no. 1, p. 35–42.

MURPHY, L. S.; GUILLARD, R. R. L. 1976. Biochemical taxonomy of marine phytoplankton by electrophoresis of enzymes, I. The centric diatoms *Thalassiosira pseudonana* and *T. fluviatilis*. *J. Phycol.*, vol. 12, no. 1, p. 9–13.

NAUMANN, E. 1917. Beiträge zur Kenntnis der Teichnannoplanktons, II. Über das Neuston des Süsswassers. *Biol. Zentralbl.*, vol. 37, no. 2, p. 98–106.

NAUWERCK, A. 1963. Die Beziehungen zwischen Zooplankton und Phytoplankton im See Erken. *Symb. bot. Ups.*, vol. 17, no. 5, p. 1–163.

NEWELL, G. E.; NEWELL, R. C. 1963. *Marine plankton: a practical guide.* London, Hutchinson. 221 p.

NEYMAN, J. 1939. On a new class of 'contagious' distributions, applicable in entomology and bacteriology. *Ann. Math. Stat.*, vol. 10, p. 35–57.

NISKIN, S. J. 1962. A water sampler for microbiological studies. *Deep-sea Res. oceanogr. Abstr.*, vol. 9 (Nov./Dec.), p. 501–3.

——. 1964. A reversing-thermometer mechanism for attachment to oceanographic devices. *Limnol. Oceanogr.*, vol. 9, no. 4, p. 591–4.

——. 1968. A deck command multiple water sampler. *Mar. Sci. Instrum.*, vol. 4, p. 19–21.

O'CONNELL, C. P.; LEONG, R. J. H. 1963. A towed pump and shipboard filtering system for sampling small zooplankters. *Spec. scient. Rep. U.S. Fish. Wildl. Serv., Fisheries*, vol. 452, p. 1–19.

ORLÓCI, L. 1975. *Multivariate analysis in vegetation research.* The Hague, Junk. 276 p.

OTTMANN, F. 1965. Une nouvelle bouteille horizontale. *Mar. Geol.*, vol. 3, no. 3, p. 223–6.

PAASCHE, E. 1960. On the relationship between primary production and standing crop of phytoplankton. *J. Cons. CIEM*, vol. 26, no. 1, p. 33–48.

——. 1973a. Silicon and the ecology of marine plankton diatoms, I. *Thalassiosira pseudonana* (*Cyclotella nana*) grown in a chemostat with silicate as limiting nutrient. *Mar. Biol.*, vol. 19, no. 2, p. 117–26.

——. 1973b. The influence of cell size on growth rate, silicon content, and some other properties of four marine diatom species. *Norw. J. Bot.*, vol. 20, no. 2–3, p. 197–204.

PAASCHE, E.; JOHANSSON, S.; EVENSEN, D. L. 1975. An effect of osmotic pressure on the valve morphology of the diatom *Skeletonema subsalsum* (A. Cleve) Bethge. *Phycologia*, vol. 14, no. 4, p. 205–11.

PAERL, H. W. 1973. Detritus in lake Tahoe: structural modification by attached microflora. *Science* (Wash.), vol. 180, no. 4085, p. 496–8.

PALMER, C. M. 1964. Algae in water supplies of the United States. In: D. F. Jackson (ed.), *Algae and man*, p. 239–61. New York, N.Y., Plenum Press.

PALMER, C. M.; MALONEY, T. E. 1954. *A new counting slide for nannoplankton*, p. 1–7. Am. Assoc. Limnol. Oceanogr. (Spec. publ. 21.)

PAQUETTE, R. G.; FROLANDER, H. F. 1957. Improvements in the Clarke-Bumpus plankton sampler. *J. Cons. CIEM*, vol. 22, no. 3, p. 284–8.

PARSONS, T. R. 1969. The use of particle size spectra in determining the structure of a plankton community. *J. oceanogr. Soc. Jap.*, vol. 25, no. 4, p. 172–81.

——. 1974. Controlled Ecosystem Pollution Experiment (CEPEX). *Environ. Conserv.*, vol. 1, p. 224.

PARSONS, T. R.; LeBRASSEUR, R. J.; FULTON, J. D. 1967. Some observations on the dependence of zooplankton grazing on the cell size and concentration of phytoplankton blooms. *J. oceanogr. Soc. Jap.*, vol. 23, no. 1, p. 10–17.

PARSONS, T. R.; SEKI, H. 1969. A short review of some automated techniques for the detection and characterization of particles in sea water. In: *Perspectives in fisheries oceanography. Papers in dedication to Professor M. Uda*, p. 173–7. (*Bull. Jap. Soc. Fish. Oceanogr.*, spec. no.)

PARSONS, T. R.; TAKAHASHI, M. 1973a. Environmental control of phytoplankton cell size. *Limnol. Oceanogr.*, vol. 18, no. 4, p. 511–15.

——. 1973b. *Biological oceanographic processes.* Oxford, New York, N.Y., Toronto, Sydney, Braunschweig, Pergamon Press. 186 p.

317

M*

PASCHER, A. 1913–32. *Die Süsswasser-Flora Mitteleuropas* (= *Die Süsswasser-Flora Deutschlands, Österreichs und der Schweiz*). 1: *Flagellatae I* (A. Pascher und E. Lemmermann), 138 p. (1914); 2: *Flagellatae II* (A. Pascher und E. Lemmermann), 192 p. (1913); 3: *Dinoflagellatae = Peridineae* (A. Schilling), 66 p. (1913); 4: *Chlorophyceae I* (A. Pascher), 506 p. (1927); 5: *Chlorophyceae II* (E. Lemmermann, J. Brunthaler und A. Pascher), 250 p. (1915); 6: *Chlorophyceae III* (W. Heering), 250 p. (1914); 7: *Chlorophyceae IV* (W. Heering), 103 p. (1921); 9: *Zygnemales,* 2. Aufl. (V. Czurda), 232 p. (1932); 10: *Bacillariophyta,* 2 Aufl. (F. Hustedt), 466 p. (1930); 11: *Heterokontae* (A. Pascher, J. Schiller und W. Migula), 250 p. (1925); 12: *Cyanochlorodinae = Chlorobacteriaceae* (L. Geitler und A. Pascher), 481 p. (1925). Jena, G. Fischer Verlag.

——. 1937–39. Heterokonten. In: L. Rabenhorst, *Kryptogamen-Flora.* Vol. 11. Leipzig, Akad. Verlag. 1092 p.

PATTEN, B. C. 1962. Species diversity in net phytoplankton of Raritan bay. *J. mar. Res.,* vol. 20, no. 1, p. 57–75.

PATTEN, B. C.; MULFORD, R. A.; WARINNER, J. E. 1963. An annual phytoplankton cycle in the lower Chesapeake bay. *Chesapeake Sci.,* vol. 4, no. 1, p. 1–20.

PEARSE, A. G. E. 1960. *Histochemistry, theoretical and applied.* Edinburgh, London, Churchill, 998 p.

PIELOU, E. C. 1969. *An introduction to mathematical ecology.* New York, N.Y., London, Sydney, Toronto, Wiley–Interscience. 286 p.

PINON, J.; PIJCK, J. 1975. Water sampling by helicopter. *Rev. int. Océanogr. méd.,* vol. 37–8, p. 153–8.

PLATT, T. 1972. Local phytoplankton abundance and turbulence. *Deep-sea Res. oceanogr. Abstr.,* vol. 19, no. 3, p. 183–7.

——. 1975. The physical environment and spatial structure of phytoplankton populations. *Mém. Soc. R. Sci. Liège,* sér. 6, vol. 7 [1974], p. 9–17.

PLATT, T.; DENMAN, K. L. 1975. A general equation for the mesoscale distribution of phytoplankton in the sea. *Mém. Soc. R. Sci. Liège,* sér. 6, vol. 7 [1974], p. 31–42.

——. 1978. The structure of pelagic marine ecosystems. *Rapp. P.-v. Réun. CIEM,* vol. 173, p. 60–5.

PLATT, T.; DICKIE, L. M.; TRITES, R. W. 1970. Spatial heterogeneity of phytoplankton in a nearshore environment. *J. Fish. Res. Board Can.,* vol. 27, no. 8, p. 1453–73.

PLATT, T.; FILION, C. 1973. Spatial variability of the productivity: biomass ratio for phytoplankton in a small marine basin. *Limnol. Oceanogr.,* vol. 18, no. 5, p. 743–9.

PLATT, T.; SUBBA RAO, D. V. 1970. Energy flow and species diversity in a marine phytoplankton bloom. *Nature* (Lond.), vol. 227, no. 5262, p. 1059–60.

POMEROY, L. R.; JOHANNES, R. E. 1968. Occurrence and respiration of ultraplankton in the upper 500 metres of the ocean. *Deep-sea Res. oceanogr. Abstr.,* vol. 15, no. 3, p. 381–91, 1 tabl.

POSTUMA, J. A. 1971. *Manual of planktonic Foraminifera.* Amsterdam, Elsevier. 420 p., 162 pl.

POUCHET, G. 1883. Contribution à l'histoire des Cilio-Flagellés. *J. Anat. Physiol.,* vol. 19, p. 399–455, pl. 18–21.

POWELL, T. M.; RICHERSON, P. J.; DILLON, T. M.; AGEE, T. A.; DOZIER, B. J.; GODDEN, A.; MYRUP, L. O. 1975. Spatial scales of current speed and phytoplankton biomass fluctuations in lake Tahoe. *Science* (Wash.), vol. 189, no. 4208, p. 1088–90.

PRAKASH, A.; SKOGLUND, L.; RYSTAD, B.; JENSEN, A. 1973. Growth and cell-size distribution of marine planktonic algae in batch and dialysis cultures. *J. Fish. Res. Board Can.,* vol. 30, no. 2, p. 143–55.

PRATT, D. M. 1959. The phytoplankton of Narragansett bay. *Limnol. Oceanogr.,* vol. 4, no. 4, p. 425–40.

PRICE, C. A.; MENDIOLA-MORGENTHALER, L. R.; GOLDSTEIN, M.; BREDEN, E. N.; GUILLARD, R. R. L. 1974. Harvest of planktonic marine algae by centrifugation into gradients of silica in the CF-6 continuous-flow zonal rotor. *Biol. Bull. mar. biol. Lab. Woods Hole, Mass.,* vol. 147, no. 1, p. 136–45.

PRINGSHEIM, E. G. 1959. Phagotrophie. In: W. Ruhland (ed.), *Handbuch der Pflanzenphysiologie,* vol. 11, p. 179–197. Berlin, Göttingen, Heidelberg.

PROSHKINA-LAVRENKO, A. I. 1955 (*Plankton diatoms of the Black sea.*) Moscow, Leningrad, Izdat. Nauka. 222 p., 8 pl. (In Russian.)

RAE, K. M. 1952. Continuous plankton records: explanation and methods, 1946–1949. *Hull. Bull. mar. Ecol.,* vol. 3, no. 21, p. 135–55, pl. 9–16.

RAJ, H. D. 1977. *Microcyclus* and related ring-forming bacteria. *Critical reviews in microbiology,* vol. 5, no. 3. Cleveland, CRC Press, p. 243–69.

RASSOULZADEGAN, F.; GOSTAN, J. 1976. Répartition des Ciliés pélagiques dans les eaux de Ville-franche-sur-Mer. Remarques sur la dispersion du microzooplancton en mer et à l'intérieur des échantillons dénombrés par la méthode d'Utermöhl. *Ann. Inst. océanogr., Paris* (nouv. sér.), vol. 52, no. 2, p. 175–87.

RATCLIFFE, J. F. 1968. The effect on the *t* distribution of non-normality in the sampled population. *Appl. Stat.,* vol. 17, no. 1, p. 42–8.

REEVES, R. G. (ed.). 1975. *Manual of remote sensing.* Falls Church, Va, Am. Soc. Photogrammetry. 2144 p.

REID, F. M. H.; FUGLISTER, E.; JORDAN, J. B. 1970. The ecology of the plankton off La Jolla, California, in the period April through September, 1967 (ed. J. D. H. Strickland), part V. Phytoplankton. *Bull. Scripps Inst. Oceanogr.,* vol. 17, p. 51–66.

REID, P. C. 1975. Large scale changes in North sea phytoplankton. *Nature* (Lond.), vol. 257, no. 5523, p. 217–19.

REIMANN, B. E. F.; LEWIN, J. C. 1964. The diatom genus *Cylindrotheca* Rabenhorst (with a reconsideration of *Nitzschia closterium*). *J. R. microsc. Soc.,* ser. 3, vol. 83, no. 3, p. 283–96.

REIMANN, B. E. F.; LEWIN, J.; VOLCANI, B. E. 1965. Studies on the biochemistry and fine structure of silica shell formation in diatoms, I. The structure of the cell wall of *Cylindrotheca fusiformis* Reimann and Lewin. *J. Cell Biol.,* vol. 24, no. 1, p. 39–55.

RENZ, G. W. 1976. The distribution and ecology of Radiolaria in the central Pacific; plankton and surface sediments. *Bull. Scripps Inst. Oceanogr.,* vol. 22, p. 1–267.

REYSSAC, J.; ROUX, M. 1972. Communautés phytoplanctoniques dans les eaux de Côte d'Ivoire. Groupes d'espèces associées. *Mar. Biol.,* vol. 13, no. 1, p. 14–33.

RICKER, W. E. 1937. Statistical treatment of sampling processes useful in the enumeration of plankton organisms. *Arch. Hydrobiol.,* vol. 31, no. 1, p. 68–84.

RIEDEL, W. R. 1971. Systematic classification of polycystine Radiolaria. In: B. M. Funnell and W. R. Riedel (eds.), *The micropaleontology of oceans,* p. 649–61. London, Cambridge Univ. Press.

RIEDEL, W. R.; FOREMAN, H. P. 1961. Type specimens of North American paleozoic Radiolaria. *J. Paleont.,* vol. 35, no. 3, p. 628–32.

RILEY, G. A. 1957. Phytoplankton of the north central Sargasso sea. *Limnol. Oceanogr.,* vol. 2, no. 3, p. 252–70.

——. 1976. A model of plankton patchiness. *Limnol. Oceanogr.,* vol. 21, no. 6, p. 873–80.

RILEY, J. P. 1965. Analytical chemistry of seawater. In: J. P. Riley and G. Skirrow (eds.), *Chemical oceanography,* vol. II, p. 295–424. London, New York, N.Y., Acad. Press.

ROBINSON, G. A. 1961. Contribution towards a plankton atlas of the north-eastern Atlantic and the North sea, pt I. Phytoplankton. *Bull. mar. Ecol.,* vol. 5, no. 42, p. 81–9, pl. 15–20.

——. 1965. Continuous plankton records: contribution towards a plankton atlas of the North Atlantic and the North Sea, pt IX. Seasonal cycles of phytoplankton. *Bull. mar. Ecol.,* vol. 6, no. 4, p. 104–22, pl. 26–61.

——. 1970. Continuous plankton records: variation in the seasonal cycle of phytoplankton in the North Atlantic. *Bull. mar. Ecol.,* vol. 6, no. 9, p. 333–45.

RODHE, W.; VOLLENWEIDER, R. A.; NAUWERCK, A. 1958. The primary production and standing crop of phytoplankton. In: A. A. Buzzati-Traverso (ed.), *Perspectives in marine biology,* p. 299–322. Berkeley, Los Angeles, Calif., Univ. Calif. Press; Paris, Union int. Sci. biol.

RODINA, A. G. 1961. Microbiological methods in application to hydrobiology. *Verh. int. Ver. theor. angew. Limnol.,* vol. 14, no. 2, p. 831–7, 3 pl.

——. 1967. Variety and destruction of lake detritus. *Verh. int. Ver. theor. angew. Limnol.,* vol. 16 [1966], no. 3, p. 1513–17, 1 pl.

ROGERS, D. J.; TANIMOTO, T. T. 1960. A computer program for classifying plants. *Science* (Wash.), vol. 132, no. 3434, p. 1115–18.

ROMEIS, B. 1968. *Mikroskopische Technik.* München, Oldenburg. 757 p.

ROUND, F. E. 1973. The problem of reduction of cell size during diatom cell division. *Nova Hedwigia,* vol. 23 [1972], no. 2–3, p. 291–303.

RUTTNER, F. 1963. *Fundamentals of limnology.* 3rd ed. Toronto, Toronto Univ. Press. 295 p.

RYTHER, J. H. 1969. Photosynthesis and fish production in the sea. *Science* (Wash.), vol. 166, no. 3901, p. 72–6.

RYTHER, J. H.; DUNSTAN, W. M.; TENORE, K. R.; HUGUENIN, J. E. 1972. Controlled eutrophication. Increasing food production from the sea by recycling human wastes. *BioScience*, vol. 22, no. 3, p. 144–52.

RYTHER, J. H.; GUILLARD, R. R. L. 1962. Studies on marine planktonic diatoms, III. Some effects of temperature on respiration of five species. *Can. J. Microbiol.*, vol. 8, p. 447–53.

SACHS, K. N. Jr. 1965. Removal of ash from plankton samples concentrated by ignition. *Deep-sea Res. oceanogr. Abstr.*, vol. 12, no. 5, p. 697.

SACHS, K. N. Jr; CIFELLI, R.; BOWEN, V. T. 1964. Ignition to concentrate shelled organisms in plankton samples. *Deep-sea Res. oceanogr. Abstr.*, vol. 11, no. 4, p. 621–2.

ST. ONGE, J. M.; PRICE, C. A. 1975. Automatic sorting of ichthyoplankton; factors controlling plankton density in gradients of silica. *Mar. Biol.*, vol. 29, no. 2, p. 187–94.

SAKSHAUG, E. 1970. Quantitative phytoplankton investigations in near-shore water masses. *Skr. K. nor. Vidensk. Selsk.*, vol. '1970', no. 3, p. 1–8.

SALONEN, K. 1974. Effectiveness of cellulose ester and perforated polycarbonate membrane filters in separating bacteria and phytoplankton. *Ann. bot. Fenn.*, vol. 11, no. 2, p. 133–5.

SANFORD, G. R.; SANDS, A.; GOLDMAN, C. R. 1969. A settle-freeze method for concentrating phytoplankton in quantitative studies. *Limnol. Oceanogr.*, vol. 14, no. 5, p. 790–4.

SARACENI, C.; RUGGIU, D. 1969. Techniques for sampling water and phytoplankton. In: R. A. Vollenweider, J. F. Talling and D. F. Westlake (eds.), *A manual on methods for measuring primary production in aquatic environments including a chapter on bacteria*, p. 5–7. London, International Biological Programme; Oxford, Edinburgh, Blackwell Scient. Publ. (IBP Handbk 12.)

SARJEANT, W. A. S. 1974. *Fossil and living dinoflagellates.* London, New York, N.Y., Acad. Press. 182 p.

SCHEWIAKOFF, W. 1926. Die Acantharia des Golfes von Neapel. *Fauna Flora Golfo Napoli*, vol. 37, 755 p., 46 pl., 22 tabl.

SCHILLER, J. 1931–37. Dinoflagellatae (Peridineae) in monographischer Behandlung. In: L. Rabenhorst, *Kryptogamen-Flora*, vol. 10 (3): Tl. 1, p. 1–617 (1931–33); Tl. 2, p. 1–590 (1935–37). Leipzig, Akad. Verlag.

SCHINK, D. R.; ANDERSON, M. C. 1969. Bag samplers for collecting thirty tons of deep-ocean water. *Mar. technol. Soc. J.*, vol. 3, no. 5, p. 49–58.

SCHULTZ, J. S.; GERHARDT, P. 1969. Dialysis culture of microorganisms: design, theory, and results. *Bacteriol. Rev.*, vol. 33, no. 1, p. 1–47.

SCHWOERBEL, J. 1966. *Methoden der Hydrobiologie.* Stuttgart, Franckh'sche Verlag. 207 p. (Transl.: *Methods of hydrobiology.* Pergamon, 1970. 200 p.)

SELBY, S. M. (ed.). 1968. *Standard mathematical tables.* 16th ed. Cleveland, Ohio, CRC Press. 692 p.

SEMINA, H. J. 1962. (Phytoplankton from the central Pacific collected along the meridian 174°W, pt I. Methods and taxonomy.) *Trudy Inst. Okeanol.*, vol. 58, p. 3–26. (In Russian.)

——. 1968. Movements of water and the size-groups of phytoplankton. *Sarsia*, vol. 34, p. 267–72.

——. 1969. (The size of phytoplankton cells along 174°W in the Pacific ocean.) *Okeanologiya*, vol. 9, no. 3, p. 479–87. (In Russian.) (Transl.: *Oceanology*.)

——. 1971. Oceanographic conditions affecting the cell size of phytoplankton. In: B. M. Funnell and W. R. Riedel (eds.), *The micropaleontology of oceans*, p. 89. London, Cambridge Univ. Press.

——. 1972. The size of phytoplankton cells in the Pacific Ocean. *Int. Rev. gesamt. Hydrobiol.*, vol. 57, no. 2, p. 177–205.

——. 1974a. The spatial pattern of cell size of phytoplankton differs from the patterns of its species and abundance. *Marine plankton and sediments*, p. 66. Kiel. (Abstr. vol.)

——. 1974b. (*Pacific phytoplankton.*) Moscow, Izdat. Nauka. 239 p. (In Russian.)

——. 1976. (Spatial pattern of life forms and size groups of phytoplankton in the ocean.) In: G. P. Andrusaitis (ed.), (*Congress of All-Union Hydrobiological Society, Riga 1974*), p. 120–1. (In Russian.)

SEMINA, H. J.; TARKHOVA, I. A. 1972. Ecology of phytoplankton in the North Pacific ocean. In: A. Y. Takenouti (ed.), *Biological oceanography of the northern North Pacific ocean*, p. 117–24. Tokyo, Idemitsu Shoten.

SEMINA, H. J.; TARKHOVA, I. A.; TRUONG NGOC AN. 1976. Patterns of phytoplankton distribution, cell size, species composition and abundance. *Mar. Biol.*, vol. 37, no. 4, p. 389–95.

320

SERFLING, R. E. 1949. Quantitative estimation of plankton from small samples of Sedgwick–Rafter cell mounts of concentrate samples. *Trans. am. microsc. Soc.*, vol. 68, no. 3, p. 185–99.

SHANNON, C. E.; WEAVER, W. 1949. *The mathematical theory of communication.* Urbana, Chicago, Ill., London, Univ. Illinois Press. 125 p.

SHELDON, R. W. 1972. Size separation of marine seston by membrane and glass–fiber filters. *Limnol. Oceanogr.*, vol. 17, no. 3, p. 494–8.

SHELDON, R. W.; PARSONS, T. R. 1967a. *A practical manual on the use of the Coulter counter in marine science.* Coulter Electr. Sales Comp. 66 p.

——. 1967b. A continuous size spectrum for particulate matter in the sea. *J. Fish. Res. Board Can.*, vol. 24, no. 5, p. 909–15.

SHELDON, R. W.; PRAKASH, A.; SUTCLIFFE, W. H. Jr. 1972. The size distribution of particles in the ocean. *Limnol. Oceanogr.*, vol. 17, no. 3, p. 327–40.

SHELDON, R. W.; SUTCLIFFE, W. H. Jr. 1969. Retention of marine particles by screens and filters. *Limnol. Oceanogr.*, vol. 14, no. 3, p. 441–4.

SHELDON, R. W.; SUTCLIFFE, W. H. Jr.; PRAKASH, A. 1973. The production of particles in the surface waters of the ocean with particulate reference to the Sargasso sea. *Limnol. Oceanogr.*, vol. 18, no. 5, p. 719–33.

SHOLKOVITZ, E. R. 1970. A free vehicle bottom-water sampler. *Limnol. Oceanogr.*, vol. 15, no. 4, p. 641–4.

SIEBURTH, J. McN. 1963. A simple form of the ZoBell bacteriological sampler for shallow water. *Limnol. Oceanogr.*, vol. 8, no. 4, p. 489–92.

——. 1965. Bacteriological samplers for air–water and water–sediment interfaces. In: *Ocean science and ocean engineering 1965*, vol. 2, p. 1064–8. Mar. Technol. Soc./Am. Soc. Oceanogr. Limnol. Washington.

——. 1971. An instance of bacterial inhibition in oceanic surface water. *Mar. Biol.*, vol. 11, no. 1, p. 98–100.

——. 1976. Bacterial substrates and productivity in marine ecosystems. *A. Rev. Ecol. Syst.*, vol. 7, p. 259–85.

——. 1977. Convener's report on the informal session on biomass and productivity of micro-organisms in planktonic ecosystems. *Helgol. wiss. Meeresunters.*, vol. 30, no. 1–4, p. 697–704. (Int. Helgol. Symp. Ecosyst. Res.)

——. *Sea microbes,* chap. 4: Sampling. New York, N.Y., London, Oxford University Press. (In press.)

SIEBURTH, J. McN.; FREY, J. A.; CONOVER, J. T. 1963. Microbiological sampling with a piggy-back device during routine Nansen bottle casts. *Deep-sea Res. oceanogr. Abstr.*, vol. 10, no. 6, p. 757–8, 1 pl.

SIEBURTH, J. McN.; JOHNSON, K. M.; BURNEY, C. M.; LAVOIE, D. M. 1977. Estimation of in situ rates of heterotrophy using diurnal changes in dissolved organic matter and growth rates of microplankton in diffusion culture. *Helgol. wiss. Meeresunters.*, vol. 30, no. 1–4, p. 565–74. (Int. Helgol. Symp. Ecosyst. Res.)

——. Dissolved carbohydrate concentrations and their microbial cycling in the north Atlantic, pt IV. Influence of the microbial plankton. (Submitted for publication.)

SIEBURTH, J. McN.; SMETACEK, V.; LENZ, J. Pelagic ecosystem structure: heterotrophic compart-ments of the plankton and their relationship to plankton size-fractions. (In press.)

SIEBURTH, J. McN.; WILLIS, P.-J.; JOHNSON, K. M.; BURNEY, C. M.; LAVOIE, D. M.; HINGA, K. R.; CARON, D. A.; FRENCH, F. W. III; JOHNSON, P. W.; DAVIS, P. G. 1976. Dissolved organic matter and heterotrophic microneuston in the surface microlayers of the North Atlantic. *Science* (Wash.). vol. 194. no. 4272, p. 1415–18.

SIMONSEN, R. 1974. The diatom plankton of the Indian Ocean Expedition of R.V. 'Meteor' 1964–65. *Meteor Forschungsergeb.* (D. Biol.), vol. 19, p. 1–66, pl. 1–41.

SIMPSON, E. H. 1949. Measurement of diversity. *Nature* (Lond.), vol. 163, no. 4148, p. 688.

SKULBERG, O. M. 1975. Observation and monitoring of water quality by use of experimental biological methods. *Verh. int. Ver. theor. angew. Limnol.*, vol. 19, no. 3, p. 2053–63.

SMAYDA, T. J. 1958. Biogeographical studies of marine phytoplankton. *Oikos,* vol. 9, no. 2, p. 158–91.

——. 1963. A quantitative analysis of the phytoplankton of the Gulf of Panama, I. Results of the regional phytoplankton surveys during July and November 1957 and March 1958. *Bull. inter-am. trop. Tuna Comm.*, vol. 7, no. 3, p. 191–253.

321

——. 1965. A quantitative analysis . . . , II. On the relationship between C14 assimilation and the diatom standing crop. *Bull. inter-am. trop. Tuna Comm.*, vol. 9, no. 7, p. 465–531.

——. 1966. A quantitative analysis . . . , III. General ecological conditions, and the phytoplankton dynamics at 8°45′N, 79°23′W from November 1954 to May 1957. *Bull. inter-am. trop. Tuna Comm.*, vol. 11, no. 5, p. 355–612.

——. 1970. The suspension and sinking of phytoplankton in the sea. *Oceanogr. mar. Biol. Ann. Rev.*, vol. 8, p. 353–414, 1 pl.

——. 1973. The growth of *Skeletonema costatum* during a winter-spring bloom in Narragansett bay, Rhode Island. *Norw. J. Bot.*, vol. 20, no. 2–3, p. 219–47.

SMAYDA, T. J.; BOLEYN, B. J. 1965. Experimental observations on the flotation of marine diatoms, I. *Thalassiosira* cf. *nana*, *Thalassiosira rotula*, and *Nitzschia seriata*. *Limnol. Oceanogr.*, vol. 10, no. 4, p. 499–509.

——. 1966. Experimental observations . . . , II. *Skeletonema costatum* and *Rhizosolenia setigera*. *Limnol. Oceanogr.*, vol. 11, no. 1, p. 18–34.

SMITH, R. K. 1967. Ignition and filter methods of concentrating shelled organisms. *J. Paleont.*, vol. 41, no. 5, p. 1288–91.

SNEATH, P. H. A.; SOKAL, R. R. 1973. *Numerical taxonomy. The principles and practice of numerical classification.* San Francisco, Calif., W. H. Freeman, 573 p.

SNEDECOR, G. W. 1956. *Statistical methods.* 5th ed. Ames, Iowa, Iowa Univ. Press. 534 p.

SOKAL, R. R.; ROHLF, F. J. 1969. *Biometry. The principles and practice of statistics in biological research.* San Francisco, Calif., W. H. Freeman. 776 p.

SOROKIN, YU. I. 1971a. On the role of bacteria in the productivity of tropical oceanic waters. *Int. Rev. gesamt. Hydrobiol.*, vol. 56, no. 1, p. 1–48.

——. 1971b. (On bacterial numbers and production in the water column of the central Pacific.) *Okeanologiya*, vol. 11, no. 1, p. 105–16. (In Russian.) (Transl.: *Oceanology*.)

SOROKIN, YU. I.; SUKHANOVA, I. N.; KONOVALOVA, G. V.; PAVELYEVA, E. B. 1975. (Primary production and phytoplankton in the area of equatorial divergence in the eastern part of the Pacific ocean.) *Trudy Inst. Okeanol.*, vol. 102, p. 108–22. (In Russian.)

SOURNIA, A. 1968. Variations saisonnières et nycthémérales du phytoplancton marin et de la production primaire dans une baie tropicale, à Nosy-Bé (Madagascar). *Int. Rev. gesamt. Hydrobiol.*, vol. 53, no. 1, p. 1–76.

——. 1969. Cycle annuel du phytoplancton et de la production primaire dans les mers tropicales. *Mar. Biol.*, vol. 3, no. 4, p. 287–303.

——. 1973. La production primaire planctonique en Méditerranée. Essai de mise à jour. *Bull. Etud. comm. Méditerr.*, p. 1–128, 1 tabl. (No. spéc. 5.)

——. 1974. Circadian periodicities in natural populations of marine phytoplankton. *Adv. mar. Biol.*, vol. 12, p. 325–89.

SOURNIA, A.; CACHON, J.; CACHON, M. 1975. Catalogue des espèces et taxons infraspécifiques de Dinoflagellés marins actuels publiés depuis la révision de J. Schiller, II. Dinoflagellés parasites ou symbiotiques. *Arch. Protistenk.*, vol. 117, no. 1–2, p. 1–19.

SPILHAUS, A. F.; MILLER, A. R. 1948. The sea sampler. *J. mar. Res.*, vol. 7, no. 3, p. 370–85.

SPURR, A. R. 1969. A low-viscosity epoxy resin embedding medium for electron microscopy. *J. ultrastruct. Res.*, vol. 26, no. 1–2, p. 31–43.

STAFLEU, F. A. (ed.). 1972. *International code of botanical nomenclature adopted by the eleventh International botanical congress, Seattle, August 1969 (Regnum vegetabile 82).* Utrecht, Int. Assoc. Plant Taxonomy. 426 p. (In English, French and German.)

STARMACH, K. 1963–74. *Flora slodkowodna polski.* 1: (K. Starmach), 271 p. (1963); 2: *Cyanophyta, Glaucophyta* (K. Starmach), 807 p. (1966); 4: *Cryptophyceae, Dinophyceae, Raphidophyceae* (K. Starmach), 520 p. (1974); 5: *Chrysophyta I* (K. Starmach), 598 p. (1968); 6: *Chrysophyta II* (K. Siemińska), 610 p. (1964); 7: *Xanthophyceae* (K. Starmach), 393 p. (1968); 10: *Chlorophyta III* (K. Starmach and J. Siemińska), 751 p. (1972); 11: *Chlorophyta IV* (T. Mrozińska-Webb), 659 p. (1969); 12A: *Chlorophyta V* (K. Starmach and J. Siemińska), 431 p. (1972). Warsaw, Polska Akad. Nauka.

STEARN, W. T. 1973. *Botanical Latin. History, grammar, syntax, terminology and vocabulary.* 2nd ed. Newton Abbot, David & Charles. 566 p.

STEEDMAN, H. F. (ed.). 1976. *Zooplankton fixation and preservation.* Paris, Unesco. 350 p. (Monogr. oceanogr. Methodol., 4.)

STEELE, J. 1973. Patchiness. *CUEA Newsl.*, vol. 2, no. 4, p. 3–7.

——. 1976. Patchiness. In: D. H. Cushing and J. J. Walsh (eds.), *The ecology of the seas*, p. 98–115. Oxford, London, Edinburgh, Melbourne, Blackwell Scient. Publ.

STEEMANN NIELSEN, E. 1933. Über quantitative Untersuchung von marinem Plankton mit Utermöhls umgekehrten Mikroskop. *J. Cons. CIEM*, vol. 8, no. 2, p. 201–10.

STEIN, J. R. (ed.). 1973. *Handbook of phycological methods. Culture methods and growth measurements*. London, Cambridge Univ. Press. 448 p., 10 pl.

STEINKAMP, J. A.; FULWYLER, M. J.; COULTER, J. R.; HIEBERT, R. D.; HORNEY, J. L.; MULLANEY, P. F. 1973. A new multiparameter separator for microscopic particles and biological cells. *Rev. sci. Instrum.*, vol. 44, no. 9, p. 1301–10.

STEPHENS, K. 1962. Improved tripping mechanism for plastic water samplers. *Limnol. Oceanogr.*, vol. 7, no. 4, p. 484.

STOSCH, H. A. VON. 1969. Dinoflagellaten aus der Nordsee, I. Über *Cachonina niei* Loeblich (1968), *Gonyaulax grindleyi* Reinecke (1967) und eine Methode zur Darstellung von Peridineenpanzern. *Helgol. wiss. Meeresunters.*, vol. 19, no. 4, p. 558–68.

——. 1974. Pleurax, seine Synthese und seine Verwendung zur Einbettung und Darstellung der Zellwände von Diatomeen, Peridineen und anderen Algen, sowie für eine neue Methode zur Elektivfärbung von Dinoflagellaten-Panzern. *Arch. Protistenk.*, vol. 116, no. 1–2, p. 132–41.

STRATHMANN, R. R. 1967. Estimating the organic carbon content of phytoplankton from cell volume or plasma volume. *Limnol. Oceanogr.*, vol. 12, no. 3, p. 411–18.

STRICKLAND, J. D. H. 1960. Measuring the production of marine phytoplankton. *Bull. Fish. Res. Board Can.*, vol. 122, p. 1–172.

——. 1971. Microbial activity in aquatic environments. *Symp. Soc. gen. Microbiol.*, vol. 21, p. 231–53.

STRICKLAND, J. D. H.; PARSONS, T. R. 1972. A practical handbook of seawater analysis. 2nd ed. *Bull. Fish. Res. Board Can.*, vol. 167, 311 p.

STRICKLAND, J. D. H.; SOLÓRZANO, L.; EPPLEY, R. W. 1970. The ecology of the plankton off La Jolla, California, in the period April through September 1967 (ed. J. D. H. Strickland), I. General introduction, hydrography and chemistry. *Bull. Scripps Inst. Oceanogr.*, vol. 17, p. 1–22.

STUDENT. 1907. On the error of counting with a haemacytometer. *Biometrika*, vol. 5, no. 3, p. 351–60.

SUKHANOVA, I. N. 1973. (Vertical structure of phytocoenosis in the NE region of the Indian ocean.) In: V. V. Shuleikin (ed.), *(Tropical zones of the world ocean)*, p. 238–43. Moscow, Izdat. Nauka. (In Russian.) (Transl. *Mar. Sci. Commun.*, vol. 2, no. 6, p. 375–86.)

——. 1976. (The qualitative composition and quantitative distribution of the phytoplankton in the northeastern Indian ocean.) *Trudy Inst. Okeanol.*, vol. 105, p. 55–82. (In Russian.)

SUTCLIFFE, W. H. Jr; BAYLOR, E. R.; MENZEL, D. W. 1963. Sea surface chemistry and Langmuir circulation. *Deep-sea Res. oceanogr. Abstr.*, vol. 10, no. 3, p. 233–43, 2 pl.

SUTCLIFFE, W. H. Jr; SHELDON, R. W.; PRAKASH, A.; GORDON, D. C. Jr. 1971. Relations between wind speed, Langmuir circulation and particle concentration in the ocean. *Deep-sea Res. oceanogr. Abstr.*, vol. 18, no. 6, p. 639–43.

SVERDRUP, H. U. 1953. On conditions for the vernal blooming of phytoplankton. *J. Cons. CIEM*, vol. 18, no. 3, p. 287–95.

SVERDRUP, H. U.; JOHNSON, M. W.; FLEMING, R. H. 1942. *The oceans. Their physics, chemistry and general biology*. New York, N.Y., Prentice-Hall. 1087 p.

SWAROOP, S. 1956. Estimation of bacterial density of water samples. Methods of attaining international comparability. *Bull. Wld Hlth Org.*, vol. 14, p. 1089–1107.

SWIFT, E. 5th. 1967. Cleaning diatom frustules with ultraviolet radiation and peroxide. *Phycologia*, vol. 6, no. 2–3, p. 161–3.

SYVERTSEN, E. H. 1977. *Thalassiosira gravida* and *T. rotula*: ecology and morphology. *Nova Hedwigia (Beih.)*, vol. 54, p. 99–112, incl. pl. 1–4. (Fourth Symp. Recent and Fossil Mar. Diatoms, ed. R. Simonsen.)

TAGUCHI, S. 1976. Relationships between photosynthesis and cell size of marine diatoms. *J. Phycol.*, vol. 12, no. 2, p. 185–9.

TANGEN, K. 1976. A device for safe sedimentation using the Utermöhl technique. *J. Cons. CIEM*, vol. 36, no. 3, p. 282–4.

TATE, M. W.; CLELLAND, R. C. 1957. *Nonparametric and shortcut statistics*. Danville, Ill., Interstate Printers Publ. 171 p.

TAYLOR, F. J. R. 1971. Scanning electron microscopy of thecae of the dinoflagellate genus *Ornithocercus*. *J. Phycol.*, vol. 7, no. 3, p. 249–58.

——. 1973. Applications of the scanning electron microscope to the study of tropical microplankton. *J. mar. biol. Assoc. India*, vol. 14 [1972], no. 1, p. 55–60, pl. 1–2.

——. 1976a. Appendix: Shipboard and curating techniques. Appendix: Treatment of micro- and nanoplankton samples. In: H. F. Steedman (ed.), *Zooplankton fixation and preservation*, p. 32–3. Paris, Unesco. (Monogr. oceanogr. Methodol., 4.)

——. 1976b. Flagellates. In: H. F. Steedman (ed.), *Zooplankton fixation and preservation*, p. 259–64. Paris, Unesco. (Monogr. oceanogr. Methodol., 4.)

——. 1976c. Flagellates. Appendix: Preparation for examination by scanning electron microscopy (SEM). In: H. F. Steedman (ed.), *Zooplankton fixation and preservation*, p. 265–7. Paris, Unesco. (Monogr. oceanogr. Methodol., 4.)

——. 1976d. Dinoflagellates from the International Indian Ocean Expedition. A report on material collected by the R.V. 'Anton Bruun' 1963–1964. *Bibliotheca bot.*, vol. 132, p. 1–234, pl. 1–46.

TAYLOR, F. J. R.; BLACKBOURN, D. J.; BLACKBOURN, J. 1971. The red-water Ciliate *Mesodinium rubrum* and its 'incomplete symbionts': a review including new and ultrastructural observations. *J. Fish. Res. Board Can.*, vol. 28, no. 3, p. 391–407, 8 pl.

TAYLOR, L. R. 1961. Aggregation, variance and the mean. *Nature* (Lond.), vol. 189, no. 4766, p. 732–5.

TAYLOR, V. I.; BAUMANN, P.; REICHELT, J. L.; ALLEN, R. D. 1974. Isolation, enumeration, and host range of marine bdellovibrios. *Arch. Mikrobiol.*, vol. 98, no. 2, p. 101–14.

TCHAN, Y. T. 1953. Study of soil algae, I. Fluorescence microscopy for the study of algae. *Proc. Linn. Soc. N.S.W.*, vol. 77 [1952], no. 5–6, p. 265–9.

THOMAS, M. 1949. A generalization of Poisson's binomial limit for use in ecology. *Biometrika*, vol. 36, no. 1–2, p. 18–25.

THOMAS, W. H.; SEIBERT, D. L. R.; TAKAHASHI, M. 1977. Controlled Ecosystem Pollution Experiment: effect of mercury on enclosed water columns, III. Phytoplankton population dynamics and production. *Mar. Sci. Commun.*, vol. 3, no. 4, p. 331–54.

THORRINGTON-SMITH, M. 1971. West Indian ocean phytoplankton: a numerical investigation of phytohydrographic regions and their characteristic phytoplankton associations. *Mar. Biol.*, vol. 9, no. 2, p. 115–37.

THRONDSEN, J. 1969a. A simple micropipette for use with the Wild M40 and the Zeiss plankton microscopes. *J. Cons. CIEM*, vol. 32, no. 3, p. 430–2.

——. 1969b. Flagellates of Norwegian coastal waters. *Nytt. Mag. Bot.*, vol. 16, no. 3–4, p. 161–216.

——. 1970a. A small sedimentation chamber for use on the Wild M40 and the Zeiss plankton microscope. *J. Cons. CIEM*, vol. 33, no. 2, p. 297–8.

——. 1970b. A non-toxic water sampler for shallow waters (0–50 m). *J. Cons. CIEM*, vol. 33, no. 2, p. 298–300.

TRANTER, D. J.; SMITH, P. E. 1968. Filtration performance. In: Anon. (ed.), *Zooplankton sampling*, p. 27–56. (Monogr. oceanogr. Methodol., 2.)

TRAVERS, A.; TRAVERS, M. 1971. Utilisation à bord d'un navire de la méthode d'Utermöhl pour la numération du microplancton. *Rapp. P.-v. Réun. CIESMM*, vol. 20, no. 3, p. 295–6.

——. 1975. Catalogue du microplancton du golfe de Marseille. *Int. Rev. gesamt. Hydrobiol.*, vol. 60, no. 2, p. 251–76.

TRAVERS, M. 1971. Diversité du microplancton du golfe de Marseille en 1964. *Mar. Biol.*, vol. 8, no. 4, p. 308–43.

TRÉGOUBOFF, G. 1953. Classe des Radiolaires. In: P.-P. Grassé (ed.), *Traité de zoologie*, vol. I (2), p. 321–436. Paris, Masson.

TRÉGOUBOFF, G.; ROSE, M. 1957. *Manuel de planctonologie méditerranéenne*. I (*Texte*), 587 p.; II (*Planches*), 207 pl. Paris, Centre nat. Rech. scient.

TUNGATE, D. S. 1967. A new sedimentation chamber. *J. Cons. CIEM*, vol. 31, no. 2, p. 284–5.

UEHLINGER, V. 1964. Etude statistique des méthodes de dénombrement planctonique. *Arch. Sci. Genève*, vol. 17, no. 2, p. 121–223, tabl.

URSIN, E. 1973. On the prey size preferences of cod and dab. *Medd. Danm. Fisk. Havunders.*, n.s., vol. 7, p. 85–98.

USACHEV, P. I. 1961. (Quantitative method of sampling and examination of phytoplankton.) *Trudy vses. Gidrobiol. Obshch.*, vol. 11, p. 411–15. (In Russian.)

UTERMÖHL, H. 1931. Neue Wege in der quantitativen Erfassung des Planktons (mit besonderer Berücksichtigung des Ultraplanktons). *Verh. int. Ver. theor. angew. Limnol.,* vol. 5, no. 2, p. 567–96.

——. 1958. Zur Vervollkommnung der quantitativen Phytoplankton-Methodik. *Mitt. int. Ver. theor. angew. Limnol.,* vol. 9, p. 1–38, pl. 1.

VAN DORN, W. G. 1957. Large-volume water sampler. *Trans. am. geophys. Un.,* vol. 37, no. 6, p. 682–4.

VARGO, G. 1968. Studies on phytoplankton ecology in tropical and subtropical environments of the Atlantic ocean. Pt 2. Quantitative studies of phytoplankton distribution in the straits of Florida and its relation to physical factors. *Bull. mar. Sci.,* vol. 18, no. 1, p. 5–60.

VENRICK, E. L. 1969. *Thr distribution and ecology of oceanic diatoms in the North Pacific.* San Diego, Calif., Univ. Calif. 655 p. (Abstract: *Diss. Abstr.,* B, vol. 30, no. 5, p. 2330.) (Ph.D. thesis.)

——. 1971. Recurrent groups of diatoms in the North Pacific. *Ecology,* vol. 52, no. 4, p. 614–25.

——. 1972a. The statistics of subsampling. *Limnol. Oceanogr.,* vol. 16 [1971], no. 5, p. 811–18.

——. 1972b. Small-scale distribution of oceanic diatoms. *Fish. Bull.,* vol. 70, no. 2, p. 363–72.

VENRICK, E. L.; BEERS, J. R.; HEINBOKEL, J. F. 1977. Possible consequences of containing microplankton for physiological rate measurements. *J. exp. mar. Biol. Ecol.,* vol. 26, no. 1, p. 55–76.

VENRICK, E. L.; McGOWAN, J. A.; MANTYLA, A. W. 1973. Deep maxima of photosynthetic chlorophyll in the Pacific ocean. *Fish. Bull.,* vol. 71, no. 1, p. 41–52.

VERDUIN, J. 1951. A comparison of phytoplankton data obtained by a mobile sampling method with those obtained from a single station. *Am. J. Bot.,* vol. 38, no. 1, p. 5–11.

VINOGRADOVA, L. A. 1962. (Qualitative and quantitative distribution of phytoplankton in different water masses of the Norwegian sea in October 1958.) *Okeanol. Issled.,* vol. 5, p. 140–53. (In Russian.)

——. 1973. (Seasonal development of phytoplankton and the feed base of herbivorous copepods in the tropical Atlantic.) In: V. V. Shuleikin (ed.), (*Tropical zones of the world ocean*), p. 243–50. Moscow, Izdat. Nauka. (In Russian.)

VOLK, R. 1901. Über die bei der hamburgischen Elbe-Untersuchung angewandten Methoden zur quantitativen Ermittelung des Planktons. *Mitt. naturhist. Mus. Hamb.,* vol. 18 [1900], no. 2, p. 135–82, pl. 1–3.

——. 1906. Hamburgische Elbe-Untersuchung, VIII. Studien über die Einwirkung der Trockenperiode im Sommer 1904 auf die biologischen Verhältnisse der Elbe bei Hamburg. Mit einem Nachtrag über chemische und planktologische Methoden. *Mitt. naturhist. Mus. Hamb.,* vol. 23 [1905], no. 2, p. 1–101, 3 pl., 1 tabl., 1 map.

VOLLENWEIDER, R. A. 1956. Beitrag zur Theorie der Schwarmbildung. Mathematisch-ökologische Analyse. *Mem. Ist. ital. Idrobiol.,* vol. 9, p. 265–72.

VOLLENWEIDER, R. A.; TALLING, J. F.; WESTLAKE, D. F. (eds.). 1969. *A manual on methods for measuring primary production in aquatic environments, including a chapter on bacteria.* London, International Biological Programme; Oxford, Edinburgh, Blackwell Scient. Publ. 213 p. (IBP handbk 12.)

WAILES, G. H. 1937. *Canadian Pacific fauna, 1. Protozoa (1a, Lobosa; 1b, Reticulosa; 1c, Heliozoa; 1d, Radiolaria).* Toronto, Biol. Board Can./Univ. Toronto Press. 14 p.

——. 1939. *Canadian Pacific fauna, 1. Protozoa (1e, Mastigophora).* Toronto, Fish. Res. Board Can./Univ. Toronto Press. 45 p.

——. 1943. *Canadian Pacific fauna, 1. Protozoa (1f, Ciliata; 1g, Suctoria).* Toronto, Fish. Res. Board Can./Univ. Toronto Press. 46 p.

WALL, D.; DALE, B. 1968. Modern Dinoflagellate cysts and evolution of the Peridiniales. *Micropaleontology,* vol. 14, no. 3, p. 265–304.

WALLIS, W. A.; ROBERTS, H. V. 1956. *Statistics: a new approach.* Glencoe, Ill., Free Press. 646 p.

WALSH, J. J. 1971. Relative importance of habitat variables in predicting the distribution of phytoplankton at the ecotone of the Antarctic upwelling ecosystem. *Ecol. Monogr.,* vol. 41, no. 4, p. 291–309.

——. 1972. Implications of a systems approach to oceanography. *Science* (Wash.), vol. 176, no. 4038, p. 969–75.

WALSH, J. J.; DUGDALE, R. C. 1971. A simulation model of the nitrogen flow in the Peruvian upwelling system. *Invest. Pesq.,* vol. 35, no. 1, p. 309–30.

WATSON, S. W.; NOVITSKY, T. J.; QUINBY, H. L.; VALOIS, F. W. 1977. Determination of bacterial number and biomass in the marine environment. *Appl. Environ. Microbiol.,* vol. 33, no. 4, p. 940–6.

WEILER, C. S.; CHISHOLM, S. W. 1976. Phased cell division in natural populations of marine dino-flagellates from shipboard cultures. *J. exp. mar. Biol. Ecol.,* vol. 25, no. 3, p. 239–47.

WEISS, R. F. 1971. Flushing characteristics of oceanographic sampling bottles. *Deep-sea Res. oceanogr. Abstr.,* vol. 18, no. 6, p. 653–6.

WELCH, P. S. 1952. *Limnology.* 2nd ed. New York, N.Y., Toronto, London, McGraw-Hill. 538 p.

WEST, W.; WEST, G. S. 1904–23. *A monograph of the British Desmidiaceae.* I, p. 1–224, pl. 1–32 (1904); II, p. 1–206, pl. 33–64 (1905); III, p. 1–274, pl. 65–95 (1908); IV, p. 1–191, pl. 96–128 (1912); V, p. 1–300, pl. 129–66 (1923). London, Ray Soc.

WETZEL, R. G. 1975. *Limnology.* Philadelphia, Pa, London, Toronto, W. B. Saunders. 743 p.

WHITTAKER, R. H.; GAUCH, H. G. Jr. 1973. Evaluation of ordination techniques. In: R. H. Whittaker (ed.), *Ordination and classification of communities,* p. 287–321. The Hague, Junk. (Handbook of vegetation, pt V.)

WIBORG, K. F. 1948. Experiments with the Clarke-Bumpus plankton sampler and with a plankton pump in the Lofoten area in northern Norway. *Fisk. Direkt. Skr., Ser. Havunders.,* vol. 9, no. 2, p. 1–22.

WIEBE, P. H.; HOLLAND, W. R. 1968. Plankton patchiness: effects on repeated net tows. *Limnol. Oceanogr.,* vol. 13, no. 2, p. 315–21.

WIEBE, W. J.; BANCROFT, K. 1975. Use of the adenylate energy charge ratio to measure growth state of microbial communities. *Proc. nat. Acad. Sci. U.S.,* vol. 72, no. 6, p. 2112–15.

WIEBE, W. J.; POMEROY, L. R. 1972. Microorganisms and their association with aggregates and detritus in the sea: a microscopic study. In: U. Melchiorri-Santolini and J. W. Hopton (eds.), *Detritus and its role in aquatic ecosystems. Mem. Ist. ital. Idrobiol.,* vol. 29, Suppl., p. 325–52.

WIED, G. L.; BAHR, G. F.; BARTELS, P. H. 1970. Automatic analysis of cell images by TICAS. In: G. L. Wied and G. F. Bahr (eds.), *Automated cell identification and cell sorting,* p. 195–360. New York, N.Y., Acad. Press.

WILKINSON, M. C.; ELLIS, R.; CALLAWAY, S. 1974. Evaluation of the use of the Quantimet 720 in the determination of the size distributions of monodisperse latices. *Microscope,* vol. 22, p. 229–45.

WILLÉN, E. 1976. A simplified method of phytoplankton counting. *Brit. phycol. J.,* vol. 11, no. 3, p. 265–78.

WILLÉN, T. 1975. Biological long-term investigations of Swedish lakes. *Verh. int. Ver. theor. angew. Limnol.,* vol. 19, no. 2, p. 1117–24.

WILLIAMS, C. B. 1964. *Patterns in the balance of nature and related problems in quantitative ecology.* London, New York, N.Y., Acad. Press. 324 p.

WILLIAMSON, M. H. 1961. An ecological survey of a scottish herring fishery, IV. Changes in the plankton during the period 1949 to 1959. With appendix: a method for studying the relation of plankton variations to hydrography. *Bull. mar. Ecol.,* vol. 5, no. 48, p. 207–29.

——. 1963. The relation of plankton to some parameters of the herring population of the north-western North sea. *Rapp. P.-v. Réun. CIEM,* vol. 154, p. 179–85.

WILLINGHAM, C. A.; BUCK, J. D. 1965. A preliminary comparative study of fungal contamination in non-sterile water samplers. *Deep-sea Res. oceanogr. Abstr.,* vol. 12, no. 5, p. 693–5, 2 pl.

WIMPENNY, R. S. 1936. The size of diatoms, I. The diameter variation of *Rhizosolenia styliformis* Brightw. and *R. alata* Brightw. in particular and of pelagic marine diatoms in general. *J. mar. biol. Assoc. U.K.,* vol. 21, no. 1, p. 29–60, pl. 1.

——. 1946. The size of diatoms. II. Further observations on *Rhizosolenia styliformis* (Brightwell). *J. mar. biol. Assoc. U.K.,* vol. 26, no. 3, p. 271–84, pl. 5.

——. 1956. The size of diatoms, III. The cell width of *Biddulphia sinensis* Greville from the southern North sea. *J. mar. biol. Assoc. U.K.,* vol. 35, no. 2, p. 375–86.

——. 1966a. The size of diatoms, IV. The cell diameters in *Rhizosolenia styliformis* var. *oceanica. J. mar. biol. Assoc. U.K.,* vol. 46, no. 3, p. 541–6.

——. 1966b. *The plankton of the sea.* London, Faber & Faber. 426 p., 17 pl.

WINER, B. J. 1962. *Statistical principles in experimental design.* New York, N.Y., McGraw-Hill. 907 p.

WINSOR, C. P.; CLARKE, G. L. 1940. A statistical study of variation in the catch of plankton nets. *J. mar. Res.,* vol. 3, no. 1, p. 1–34.

WOELKERLING, W. J.; KOWAL, R. R.; GOUGH, S. B. 1976. Sedgwick–Rafter cell counts: a procedural analysis. *Hydrobiologia,* vol. 48, no. 2, p. 95–107.

WOOD, E. J. F. 1955. Fluorescent microscopy in marine microbiology. *J. Cons. CIEM,* vol. 21, no. 1, p. 6–7.

——. 1962. A method for phytoplankton study. *Limnol. Oceanogr.,* vol. 7, no. 1, p. 32–5.

——. 1965. *Microbial ecology.* London, Chapman & Hall; New York, N.Y., Reinhold. Publ. 243 p., 14 pl.

——. 1968. Studies of phytoplankton ecology in tropical and subtropical environments of the Atlantic ocean. Pt 3. Phytoplankton communities in the Providence channels and the Tongue of the ocean. *Bull. mar. Sci.,* vol. 18, no. 2, p. 481–543.

WOOD, E. J. F.; OPPENHEIMER, C. H. 1962. Note on fluorescence microscopy in marine microbiology. *Z. allgem. Mikrobiol.,* vol. 2, no. 2, p. 164–5.

WOODWARD, R. L. 1957. How probable is the Most Probable Number? *J. am. Wat. Wks Ass.,* vol. 49, p. 1060–8.

WROBLEWSKI, J. S. 1977. Vertically migrating herbivorous plankton. Their possible role in the creation of small scale phytoplankton patchiness in the ocean. In: N. R. Anderson and B. J. Zahuranec (eds.), *Oceanic sound scattering prediction,* p. 817–47. New York, N.Y., Plenum Press. (Marine science, 5.)

WROBLEWSKI, J. S.; O'BRIEN, J. J. 1976. A spatial model of phytoplankton patchiness. *Mar. Biol.,* vol. 35, no. 2, p. 161–75.

WROBLEWSKI, J. S., O'BRIEN, J. J.; PLATT, T. 1975. On the physical and biological scales of phytoplankton patchiness in the ocean. *Mém. Soc. R. Sci. Liège,* ser. 6, vol. 7 [1974], p. 43–57.

WÜST, G. 1932. Programm, Ausrüstung, Methoden der Serienmessungen. *Wiss. Ergebn. Dtsch. Atlant. Exped. Meteor 1925–27,* vol. 4, no. 1, p. 1–59, pl. 1–3.

ZACHARY, A. 1974. Isolation of bacteriophages of the marine bacterium *Beneckea natriegens* from coastal salt marshes. *Appl. Microbiol.,* vol. 27, no. 5, p. 980–2.

ZAIKA, V. E.; ANDRYUSCHENKO, A. A. 1969. (Taxonomic diversity of phyto- and zooplankton in the Black sea.) *Gidrobiol. Zh.,* vol. 5, no. 3, p. 12–19. (In Russian.) (Transl.: *Hydrobiological J.)*

ZAITSEV, YU. P. 1971. *Marine neustonology* (transl. from Russian). Jerusalem, Israel Progr. Scient. Transl. 207 p. (Ref. 5976.)

ZEITZSCHEL, B. 1970. The quantity, composition and the distribution of suspended particulate matter in the gulf of California. *Mar. Biol.,* vol. 7, no. 4, p. 305–18.

——. 1978. Oceanographic factors influencing the distribution of plankton in space and time. *Micropaleontology,* vol. 24, no. 1, p. 139–59.

ZIMMERMANN, R.; MEYER-REIL, L.-A. 1974. A new method for fluorescence staining of bacterial populations on membrane filters. *Kiel. Meeresforsch.,* vol. 30, no. 1, p. 24–7, pl. 1.

ZOBELL, C. E. 1941. Apparatus for collecting water samples from different depths for bacteriological analysis. *J. mar. Res.,* vol. 4, no. 3, p. 173–88.

——. 1946. *Marine microbiology: a monograph on hydrobacteriology.* Waltham, Mass., Chronica Botanica. 240 p.

Addresses of manufacturers cited

The following list includes those manufacturers whose products are mentioned in the manual. In no case can this be considered as a complete list, nor as a qualitative selection of manufacturers whose products are relevant to phytoplankton studies. Mention of a given product and its manufacturers does not imply that it should be considered as preferable to others available at that time or developed since.

For the convenience of the reader, publishers specialized in the reprinting of older taxonomical literature are included here.

American Optical Corp.
Scientific Instrument Division, Sugar & Eggert Roads, Buffalo, New York 14215 (United States)

Antiquariaat Junk
Dr R. Schierenberg & Sons B.V., Valderstraat 10, Postbox 5 (Netherlands)

A. Asher & Co. B.V.
Keizergracht 526, Amsterdam 1002 (Netherlands)

Bausch & Lomb Inc.
Scientific Optical Products Div., 77476 Bausch Street, Rochester, New York 14602 (United States)
Analytical Systems Div., 820 Linden Avenue, Rochester, New York 14625 (United States)

Beckman Instruments Inc.
2500 Harbor Blvd, Fullerton, California 92634 (United States)
Spinco Div., 1117 California Avenue, Palo Alto, California 94304 (United States)

Becton Dickinson Electronics Laboratory
506 Clyde Avenue, Mountain View, California 94043 (United States)

Bergen Nautik
Strangt 18, P.O. Box 1231, 5001 Bergen (Norway)

Bio/Physics Systems Inc.
Baldwin Place Road, Mahopac, New York 10541 (United States)

Rudolf Brand
Postfach 310, 6980 Wertheim (Federal Republic of Germany)

Cambridge Instruments Company Inc.
40 Robert Pitt Drive, Monsey, New York 10952 (United States)

Clay Adams
Division of Beckton, Dickinson & Co., 199 Webro Road, Parsippany, New Jersey 07054 (United States)

Coulter Electronics Inc.
590 West 20 Street, Hialeah, Florida 33010 (United States)

J. Cramer
Postfach 48, 3306 Lehre (Federal Republic of Germany)

Custom Research and Development Inc.
8500 Mt Vernon Road, Auburn, California 95603 (United States)

DuPont Instruments
DuPont de Nemours Co., Wilmington, Delaware 19898 (United States)

Eastman Kodak Co.
343 State Street, Rochester, New York 14650 (United States)

Fisher Scientific Co.
711 Forbes Avenue, Pittsburgh, Pennsylvania 15219 (United States)
(Fisher Scientific GmbH, Imhofstrasse 3, 8 Munich 40, Federal Republic of Germany)

Flatters & Garnett Ltd
309 Oxford Road, Manchester M13 9PQ (United Kingdom)
Fuyo Sangyo Co. Ltd
Kyodo Bldg, Higashikonya-cho, Kanda, Chiyoda-ku, Tokyo (Japan)
Gelman Instrument Co.
600 S. Wagner Road, Ann Arbor, Michigan 48106 (United States)
General Oceanics Inc.
5535 Northwest Seventh Avenue, Miami, Florida 33127 (United States)
G. T. Gurr Ltd/Baird and Tatlock Ltd
Freshwater Road, Chadwell Heath, Romford, Essex RM1 1HA (United Kingdom)
Hausser Scientific
Blue Bell, Pennsylvania 19422 (United States)
Hydro-Bios Apparatebau GmbH
2300 Kiel-Holtenau, Am Jägersberg 5–7 (Federal Republic of Germany)
Hydro Products Co.
P.O. Box 2528, San Diego, California 92112 (United States)
Imanco Ltd
Image Analysing Computers Ltd, Melbourn, Royston, Hertfordshire SG8 6EJ (United Kingdom)
Institute of Oceanographic Sciences
Natural Environment Research Council, Brock Road, Wormley, Godalming, Surrey GU8 5UB (United Kingdom)
InterOcean Systems Inc.
3510 Kurtz Street, San Diego, California 92110 (United States)
Johnson Reprint Corporation
111 Fifth Avenue, New York, New York 10003 (United States)
Kahl Scientific Instrument Corp.
P.O. Box 1166, El Cajon, California 92022 (United States)
Kimble Products Division
Owens-Illinois Inc., P.O. Box 1035, Toledo, Ohio 43666 (United States)
Otto Koeltz Science Publishers
Postfach 1380, Herrnwaldstrasse 6, 624 Koenigstein/Taunus (Federal Republic of Germany)
Kressilk Products Inc.
Monterey Park, California 91754 (United States)
Laboratoire Océanographique
Skovkrogen 8, 2920 Charlottenlund (Denmark)
E. Leitz GmbH
Postfach 2020, 6330 Wetzlar (Federal Republic of Germany)

(E. Leitz Inc., Rockleigh, New Jersey 07647, United States)
A.B. Lars Ljungberg & Co.
Svetsarvägen 4, 17183 Solna (Sweden)
Machinator AB
Storgaten 30/2, 75331 Uppsala (Sweden)
Mécabolier S.A.
57 Avenue de la République, 94290 Ville-neuve-le-Roi (France)
E. Merck (Reagenzien, Diagnostica, Chemikalien)
Postfach 4119, 61 Darmstadt 1 (Federal Republic of Germany)
(Merck & Co. Inc., Rahway, New Jersey 07065, United States)
Millipore Corp.
Ashby Road, Bedford, Massachusetts 01370 (United States)
Monsanto Co.
800 N. Lindbergh Blvd, St Louis, Missouri 63166 (United States)
(Monsanto Ltd, Monsanto House, 10 Victoria Street, London SW1H OE2, United Kingdom)
Nereïdes
Office d'Instrumentation Hydrographique, 66 Blvd de Mondétour, 91400 Orsay, (France)
Newark Wire Cloth Company
351 Verona Avenue, Newark, New Jersey 07104 (United States)
Nippon-Nakano Bolting Cloth Co. Ltd
Yofuka-Kaikau Building, Tokyo (Japan)
Northern Biological Supplies
31 Cheltenham Avenue, Ipswich, Suffolk IP1 4LN (United Kingdom)
Nuclepore Corp.
7035 Commerce Circle, Pleasanton, California 94566 (United States)
Pacific Scientific Co., HIAC Instruments Division
P.O. Box 3007, 4719 West Brooks Street, Montclair, California 91763 (United States)
Particle Data Inc.
P.O. Box 265, Elmhurst, Illinois 60126 (United States)
Plastok
Rosemount, Oxton, Birkenhead L43 5SN (United Kingdom)
Prolabo
12 Rue Pelée, BP. N.200, 75526 Paris 11 (France)
Rigosha & Co. Ltd
4-10-1-chome, Kaji-cho, Chiyoda-ku, Tokyo (Japan)
Rohm & Haas Co.
Independence Mall West, Philadelphia, Pennsylvania 19105 (United States)

Sartorius-Membranfilter GmbH
Postfach 142, 3400 Göttingen (Federal Republic of Germany)
(Sartorius Filters Inc., 803 Grandview Drive, San Francisco, California 94080, United States)

Schott-Ruhrglas GmbH
Rheinallee 109, 65 Mainz (Federal Republic of Germany)
(Schott Optical Glass Inc., 400 York Avenue, Duyea, Pennsylvania 18642, United States)

W. Schreck (?)
In der Witz 19, 6238 Hofheim (Taunus) (Federal Republic of Germany)

Schweiz. Seidengazefabrik AG/Swiss Silk Bolting Cloth Mfg Co. Ltd
9425 Thal SG (Switzerland)

Selas Flotronics
1957 Pioneer Road, Huningdon Valley, Pennsylvania 19006 (United States)

Henry Simon Ltd
Special Products Division, P.O. Box 31, Stockport, Cheshire SK3 0RT (United Kingdom)

Telefunken AEG
Neue Strasse 113/115, Postfach 1145, 79 Ulm (Federal Republic of Germany)

Arthur H. Thomas Co.
Third & Vine Street, Box 779, Philadelphia, Pennsylvania 19105 (United States)

Toa Electric Co.
Kobe (Japan)

Tobler, Ernst & Traber Inc.
71 Murray Street, New York, New York 10007 (United States)

Tripette & Renaud S.A.
39 Rue J.-J.-Rousseau, 75038 Paris 01 (France)

Tsurumi-Seiki Co. Ltd
1506 Tsurumi-cho, Tsurumi-ku, Yokohama (Japan)

Turtox-Cambosco, Macmillan Science Co. Inc.
8200 S. Hoyne Avenue, Chicago, Illinois 60620 (United States)

Union Générale des Gazes à Bluter S.A. (UGB)
42360 Panissières (France)

Unitron Instruments Inc.
101 Crossways Park West, Woodbury, New York 11797 (United States)

VEB Transformratoren und Rontegemwerk
48 Overbeckstrasse, 8030 Dresden (German Democratic Republic)

Vereinigte Seidenwebereien AG/United Silk Mills GmbH
Abt. Technische Gewebe, 4152 Kempen, 4 St Hubert, Speefeld 7 (Federal Republic of Germany)

Whatman Inc.
9 Bridewell Place, Clifton, New Jersey 07014 (United States)

Wild Heerbrugg AG
9435 Heerbrugg (Switzerland)
(Wild Heerbrugg Instruments Inc., 465 Smith Street, Farmingdale, New York 11735, United States)

Wildlife Supply Company
301 Case Street, Saginaw, Michigan 48602 (United States)

Carl Zeiss
Postfach 1369–1380, 7082 Oberkochen, Württ. (Federal Republic of Germany)
(Carl Zeiss Inc., 444 Fifth Avenue, New York, New York 10018, United States)

Züricher Beuteltuchfabrik AG/Zürich Bolting Cloth Mfg Co. Ltd
8803 Rüschlikon, Switzerland

Contributors

J. R. Beers, Institute of Marine Resources A-018, University of California at San Diego, La Jolla, California 92093 (United States)

C. M. Boyd, Department of Oceanography, Dalhousie University, Halifax, Nova Scotia B3H 4J1 (Canada)

J. M. Colebrook, Institute for Marine Environmental Research, Prospect Place, The Hoe, Plymouth PL1 3DH (United Kingdom)

A. N. Dodson, Institute of Marine Resources A-003, Scripps Institution of Oceanography, University of California at San Diego, La Jolla, California 92093 (United States)

R. O. Fournier, Department of Oceanography, Dalhousie University, Halifax, Nova Scotia B3H 4J1 (Canada)

R. R. L. Guillard, Woods Hole Oceanographic Institution, Woods Hole, Massachusetts 02543 (United States)

G. R. Hasle, Department of Marine Biology and Limnology, Section of Marine Botany, University of Oslo, P.O. Box 1069, Blindern, Oslo 3 (Norway)

B. R. Heimdal, Biological Station, Espegrend, Blomsterdalen 5065 (Norway)

A. R. Hiby, Institute for Marine Environmental Research, Prospect Place, The Hoe, Plymouth PL1 3DH (United Kingdom)

B. S. C. Leadbeater, Department of Plant Biology, University of Birmingham, P.O. Box 363, Birmingham B15 2TT (United Kingdom)

L. Legendre, Groupe Interuniversitaire de Recherches Océanographiques du Québec, Département de Biologie, Université Laval, Québec G1K 7P4, Québec (Canada)

P. Legendre, Centre de Recherches en Sciences de l'Environnement, Université du Québec à Montréal, C.P. 8888, Montréal H3C 3P8, Québec (Canada)

R. Margalef, Universidad de Barcelona, Facultad de Biología, Cátedra de Ecología, Av. José Antonio 585, Barcelona 7 (Spain)

B. C. Parker, Department of Biology, Virginia Polytechnic Institute and State University, Blacksburg, Virginia 24061 (United States)

F. M. H. Reid, Institute of Marine Resources A-018, University of California at San Diego, La Jolla, California 92093 (United States)

G. A. Robinson, Institute for Marine Environmental Research, Prospect Place, The Hoe, Plymouth PL1 3DH (United Kingdom)

H. J. Semina, Institute of Oceanology, Academy of Sciences, 23 Krasikova, Moscow 117218 (U.S.S.R.)

R. W. SHELDON, Marine Ecology Laboratory, Bedford Institute of Oceanography, Dartmouth, Nova Scotia B2Y 4A2 (Canada)

J. McN. SIEBURTH, Graduate School of Oceanography and Department of Microbiology, University of Rhode Island, Narrangansett and Kingston Village, Rhode Island 02881 (United States)

T. J. SMAYDA, Graduate School of Oceanography, University of Rhode Island, Kingston, Rhode Island 02881 (United States)

A. SOURNIA, Laboratoire d'Ichtyologie Générale et Appliquée, Muséum National d'Histoire Naturelle, 57 Rue Cuvier, 75231 Paris 05 (France)

I. N. SUKHANOVA, Institute of Oceanology, Academy of Sciences, 23 Krasikova, Moscow 117218 (U.S.S.R.)

K. TANGEN, Department of Marine Biology and Limnology, Section of Marine Botany, University of Oslo, P.O. Box 1069, Blindern, Oslo 3 (Norway)

F. J. R. TAYLOR, Institute of Oceanography, Department of Botany, University of British Columbia, Vancouver, British Columbia V6T 1W5 (Canada)

W. H. THOMAS, Institute of Marine Resources A-003, Scripps Institution of Oceanography, University of California at San Diego, La Jolla, California 92093 (United States)

J. THRONDSEN, Department of Marine Biology and Limnology, Section of Marine Botany, University of Oslo, P.O. Box 1069, Blindern, Oslo 3 (Norway)

G. A. VARGO, Marine Ecosystems Research Laboratory, Graduate School of Oceanography, University of Rhode Island, Kingston, Rhode Island 02881 (United States)

E. L. VENRICK, Scripps Institution of Oceanography, P.O. Box 109, La Jolla, California 92093 (United States)

E. Willén, Limnologiska Institutionen, Uppsala Universitet, Box 557, 751 22 Uppsala 1 (Sweden)

T. WILLÉN, Limnologiska Institutionen, Uppsala Universitet, Box 557, 751 22 Uppsala 1 (Sweden)

B. ZEITZSCHEL, Institut für Meereskunde an der Universität Kiel, Düsternbrooker Weg 20, 23 Kiel 1 (Federal Republic of Germany)

Subject index

The user is urged to read 'How to Use the Manual' (page xiv)
Italic numbers refer to illustrations